Lectures on Numerical Methods

I. P. MYSOVSKIH

Leningrad State University

Lectures on
Numerical Methods

Translated by L. B. Rall
Mathematics Research Center,
U. S. Army
University of Wisconsin

WOLTERS-NOORDHOFF PUBLISHING   GRONINGEN 1969
THE NETHERLANDS

ISBN 978-94-011-7485-5   ISBN 978-94-011-7483-1 (eBook)
DOI 10.1007/978-94-011-7483-1

CONTENTS

PREFACE

The course of lectures on numerical methods (part I) given by the author to students in the numerical third of the course of the mathematics-mechanics department of Leningrad State University is set down in this volume. Only the topics which, in the opinion of the author, are of the greatest value for numerical methods are considered in this book. This permits making the book comparatively small in size, and, the author hopes, accessible to a sufficiently wide circle of readers. The book may be used not only by students in daily classes, but also by students taking correspondence courses and persons connected with practical computation who desire to improve their theoretical background.

The author is deeply grateful to V. I. Krylov, the organizer of the course on numerical methods (part I) at Leningrad State University, for his considerable assistance and constant interest in the work on this book, and also for his attentive review of the manuscript.

The author is very grateful to G. P. Akilov and I. K. Daugavet for a series of valuable suggestions and observations.

The Author

Chapter I

NUMERICAL SOLUTION OF EQUATIONS

In this chapter, methods for the numerical solution of equations of the form

$$P(x) = 0,$$

will be considered, where $P(x)$ is in general a complex-valued function. Although it is true that an "explicit" solution of this equation may be given in certain very rare cases, even so, the formulas which are obtained for this are ordinarily very involved and consequently difficult to use. Because of this, it is of considerable importance to have methods for finding approximate solutions of the equation considered. The following numerical methods for the solution of equations will be presented below: The secant (chord) method, the method of iterations, Newton's method, Lobačevskiĭ's method, and factorization methods, including Lin's method. The method of iterations and Newton's method will also be given for systems of equations.

1. Finding an initial approximation

Strictly speaking, the terminology "approximate solution" is meaningless if one does not define the accuracy of approximation, in so far as any two numbers are approximately equal. Nevertheless, all of the methods considered in this chapter, with the exception of Lobačevskiĭ's method, may be successful only if a more or less close approximation to the actual solution is known. It is just such an approximation that we mean when we speak of an initial approximation to a solution of the equation

$$P(x) = 0. \tag{1.1}$$

Graphical methods are frequently applied to finding an initial approximation to the solution. Suppose, for example, that $P(x)$ is a real function, and the determination of a real solution of equation (1.1) is being discussed. Graphical methods are based on the fact that real solutions of equation (1.1) are represented by the points of intersection of the graph of the function $y = P(x)$ and the x-axis. By constructing the graph of the function $y = P(x)$, we may determine approximate values of the abscissas of the points of intersection of the graph with the x-axis. These approximate values for the abscissas may thus be taken as initial approximations to the solutions of equation (1.1).

Sometimes, it is expedient to write equation (1.1) in the form

$$P_1(x) = P_2(x).$$

The real solutions of such an equation are the abscissas of the points of intersection of the graphs of the functions $y = P_1(x)$ and $y = P_2(x)$.

Example 1. We seek an initial approximation to the smallest positive solution of the equation

$$x = \tan x.$$

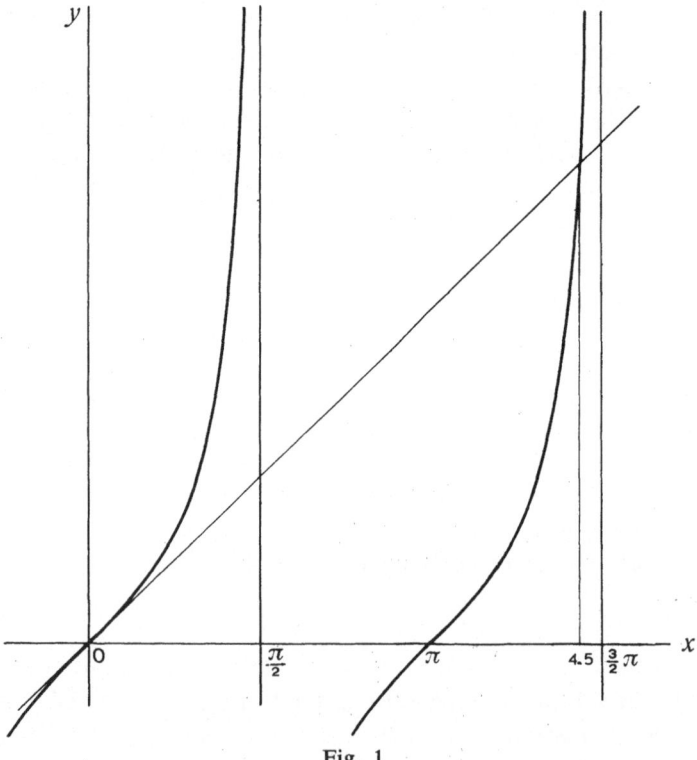

Fig. 1

We take $P_1(x) = x$ and $P_2(x) = \tan x$. The graphs of the functions $y = x$ and $y = \tan x$ are depicted in fig. 1. From the figure, it is seen that $x_0 = 4.5$ may be taken as the initial approximation.
One may also construct a table of the values of the function $y = P(x)$ to determine an initial approximation.

Example 2. One seeks an initial approximation to a solution of the equation

$$P(x) \equiv x^3 + x - 1 = 0.$$

We construct a table of the values of $P(x)$:

x	$-\infty$	-1	0	1	$+\infty$
$P(x)$	$-\infty$	-3	-1	1	$+\infty$

It is seen from the table that 0 or 1 may be taken as the initial approximation x_0.
Graphical methods may also be applied to finding initial approximations to complex solutions. It is necessary to separate the real and imaginary parts of $P(x)$:

$$P(x) = X(u, v) + i Y(u, v), \qquad x = u + iv,$$

and construct the graphs of the curves defined by the equations

$$X(u, v) = 0, \qquad Y(u, v) = 0,$$

in the (u, v) coordinate system. The points of intersection of these curves, considered as points in the complex plane of the variable $x = u + iv$, are solutions of equation (1.1). In this way, initial approximations to the complex solutions of (1.1) may be determined on the basis of a graphical method.

Example 3. We seek an initial approximation to a complex solution of the equation

$$P(x) \equiv x^4 - 2.7x^3 + 4x^2 - 3.3x + 1 = 0. \tag{1.2}$$

To separate the real and imaginary parts, we set $x = u + iv$ and write $P(u + iv)$ in powers of iv in the neighborhood of u by Taylor's formula:

$$P(x) = P(u) + ivP'(u) - \frac{v^2}{2} P''(u) - \frac{iv^3}{6} P'''(u) + \frac{v^4}{24} P^{(IV)}(u).$$

We obtain

$$P(x) = X(u, v) + iY(u, v),$$

where

$$X(u, v) = P(u) - \frac{P''(u)}{2} v^2 + v^4,$$

$$Y(u, v) = P'(u)v - \frac{P'''(u)}{6} v^3.$$

The graphs of the curves defined by the equations

$$X(u, v) = 0, \qquad Y(u, v) = 0, \qquad\qquad (1.3)$$

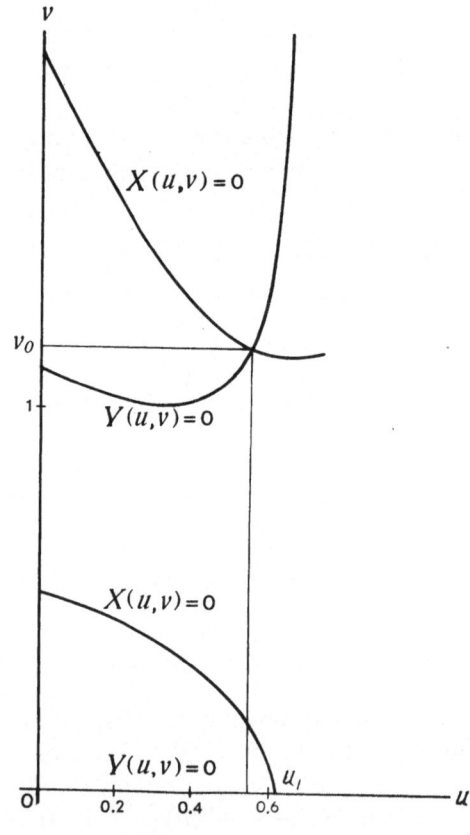

Fig. 2

are depicted in fig. 2 for u in the interval $0 \leq u \leq 0.6$, and $v \geq 0$. It is seen from the figure that

$$x_0 = u_0 + iv_0 = 0.52 + 1.16i$$

may be taken as an initial approximation to a complex solution of equation (1.2), and $u_1 = 0.66$ may be taken as an approximation to a real solution.

The graphs of the curves defined by equations (1.3) may be constructed in the following manner. For given u, the first equation of (1.3) is a quadratic equation in v^2, and the second equation may be factored into two equations: $v = 0$ and a linear equation in v^2. Except for signs, the coefficients of these equations are the numbers

$$P(u), P'(u), \frac{P''(u)}{2}, \frac{P'''(u)}{6}.$$

These numbers are calculated to two decimal places for $u = (0.2)k$, $k = 0, 1, 2, 3$, after which the non-negative solutions of the equations

$$X((0.2)k; v) = 0, \qquad Y = ((0.2)k; v) = 0,$$
$$k = 0, 1, 2, 3,$$

are found.

It is easy to see that the method applied in this example is general. It may be applied to finding an initial approximation to a complex solution of an algebraic equation of any degree n. If n (or $n-1$) is even, then the problem reduces to finding the real solutions of two equations, where the first equation is of degree $n/2$ (or $(n-1)/2$) in v^2, and the second equation factors into the equation $v = 0$ and an equation of degree $(n-2)/2$ (or $(n-1)/2$) in v^2.

This method requires considerable calculation, so before applying it to the equation $P_n(x) = 0$, where $P_n(x)$ is a polynomial of degree n with real coefficients, one should find all of the real solutions $x_1, x_2, \ldots, x_k$ of the equation, form the polynomial

$$Q_k(x) = (x - x_1)(x - x_2) \ldots (x - x_k),$$

and perform the division of $P_n(x)$ by $Q_k(x)$:

$$P_n(x) = Q_k(x)R_{n-k}(x).$$

We may consider $n - k$ to be even and ≥ 2. If $n - k = 2$, then $R_{n-k}(x)$

is a quadratic trinomial, and its roots are easy to find. If $n-k \geq 4$, then the method discussed may be applied to find initial approximations to the roots of the polynomial $R_{n-k}(x)$ (and, hence, to the roots of the polynomial $P_n(x)$).

Sometimes it is possible to select an equation which is close to (1.1) with solutions which are easy to find. Then it is natural to take its solutions as initial approximations to the solutions of equation (1.1).

Example 4. An initial approximation is sought for the solution of the equation

$$\sum_{k=0}^{\infty} \left(\frac{x}{2}\right)^{2k} - 1 = 0.$$

Replacing the series by the partial sum of its first two terms, we obtain the quadratic equation

$$\frac{x}{2} + \frac{x^2}{4} - 1 = 0.$$

We take the positive solution $-1+\sqrt{5} \simeq 1.2$ of this equation as an initial approximation to the solution of the original equation. The negative solution falls outside of the limits of the interval of convergence of the series which enters into the equation.

Graphical methods may be used to find initial approximations to real solutions of systems of two equations in two unknowns

$$P(\xi, \eta) = 0, \qquad Q(\xi, \eta) = 0,$$

where $P(\xi, \eta)$ and $Q(\xi, \eta)$ are real functions.

Example 5. We seek initial approximations to the solutions of the system

$$P(\xi, \eta) \equiv \xi^3 + \eta^3 - 4 = 0,$$
$$Q(\xi, \eta) \equiv \xi^4 + \eta^2 - 3 = 0.$$

The graphs of the functions defined by the equations

$$\xi^3 + \eta^3 - 4 = 0, \qquad \xi^4 + \eta^2 - 3 = 0,$$

are shown in fig. 3. The coordinates of the points of intersection of the graphs give the solutions of the system. It is seen from the figure that

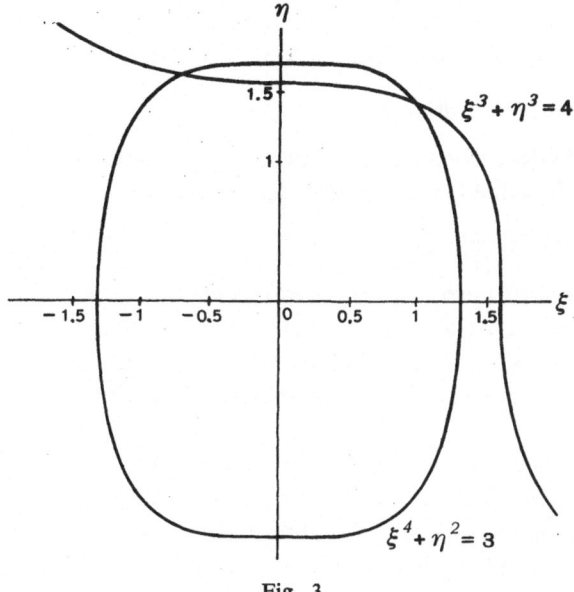

Fig. 3

the system has two solutions. As initial approximations to these solutions, one may take

$$\xi_0^{(1)} = 1, \qquad \eta_0^{(1)} = 1.4;$$
$$\xi_0^{(2)} = -0.7, \qquad \eta_0^{(2)} = 1.6.$$

The problem of determining initial approximations to the solutions of systems of equations in three or more unknowns presents considerable difficulties.

In the case of a system of three equations

$$P(\xi, \eta, \zeta) = 0,$$
$$Q(\xi, \eta, \zeta) = 0, \qquad (1.4)$$
$$R(\xi, \eta, \zeta) = 0,$$

where P, Q, R are real functions, initial approximations to its real solutions may be found in the following way. We consider the set of numbers $Z = \{\zeta\}$, where ζ is a real number such that the system of equations consisting of the first two equations of (1.4),

$$P(\xi, \eta, \zeta) = 0,$$
$$Q(\xi, \eta, \zeta) = 0, \qquad (1.5)$$

is solvable for ξ and η. The set Z is nonempty if system (1.4) has at least one real solution. We denote the solutions of system (1.5) for $\zeta \in Z$ by

$$\xi = \xi(\zeta), \qquad \eta = \eta(\zeta). \tag{1.6}$$

If system (1.5) does not have a unique solution, then the functions (1.6) will be multiple-valued. We consider the function of ζ,

$$\phi(\zeta) = R(\xi(\zeta), \eta(\zeta), \zeta),$$

where $\xi(\zeta)$ and $\eta(\zeta)$ are defined by formulas (1.6). The function $\phi(\zeta)$ is defined on the set Z, and may be multiple-valued. If the system (1.4) has a real solution, then the function $\phi(\zeta)$ has a real root. Conversely, to each real root of $\phi(\zeta)$, there corresponds a real solution of the system (1.4).

Thus, the problem of finding an initial approximation to a solution of system (1.4) is reduced to graphical or tabular determination of a root of the function $\phi(\zeta)$. From it, approximate values of the functions (1.6), which are approximate values of solution of the systems (1.5), may be obtained graphically.

For the sake of definiteness, we eliminated the first two unknowns by using the first two equations of system (1.4). In actual computations, we would, of course, eliminate the two unknowns by using the pair of equations of the system (1.4) for which the system analogous to (1.5) would be the simplest.

2. The secant method

The secant (chord) method may be used for finding real solutions of the equation

$$P(x) = 0, \tag{2.1}$$

where it is assumed that the function $P(x)$ is real and continuous. Suppose that an interval $[a, b]$ is known for which the condition

$$P(a)P(b) < 0$$

is satisfied. Then, equation (2.1) has a solution interior to this interval. We shall assume that equation (2.1) has only one solution in $[a, b]$. Uniqueness is known to hold if $P(x)$ is strictly monotone in $[a, b]$, in particular, if it is differentiable and $P'(x) > 0$ (or $P'(x) < 0$) in $[a, b]$.

We replace the graph of the function $y = P(x)$ on the interval $[a, b]$ by the straight line passing through the points $(a, P(a))$ and $(b, P(b))$ (see fig. 4). We denote the point of intersection of this line with the x-axis by x_1. We take the point x_1 as the first approximation to the solution. If $P(x_1) \neq 0$, then either $P(x_1)P(a) < 0$, or $P(x_1)P(b) < 0$. Suppose, for example, that $P(x_1)P(b) < 0$. Working with the interval $[a_1, b_1] \equiv [x_1, b]$, we obtain a further approximation x_2 to the solution, etc.

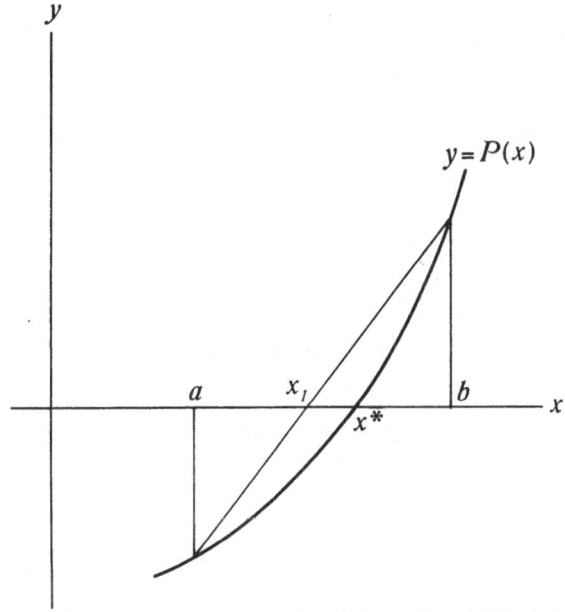

Fig. 4

The equation of the line passing through the points $(a, P(a))$ and $(b, P(b))$ may be written as:

$$y = P(a) + \frac{P(b) - P(a)}{b - a}(x - a).$$

The point of its intersection with the x-axis is

$$x_1 = a - \frac{P(a)}{P(a, b)}, \tag{2.2}$$

where

$$P(a, b) = \frac{P(b) - P(a)}{b - a}.$$

This is the computational formula for applying the secant method.

It can be proved that $x_n \to x^*$ as $n \to \infty$, where x^* is the solution sought. The calculation of solutions by the secant method may be laid out in the two following diagrams:

Diagram 1

x	$\ldots$	$P(x)$
a	$\ldots$	$P(a)$
b	$\ldots$	$P(b)$
x_1	$\ldots$	$P(x_1)$

Diagram 2

n	a_n	b_n	$b_n - a_n$	$P(a_n)$	$P(b_n)$	$P(b_n) - P(a_n)$	$P(a_n, b_n)$	$\Delta_n = \\ = -\dfrac{P(a_n)}{P(a_n, b_n)}$	$x_{n+1} = \\ = x_n + \Delta_n$
0	a	b	$b - a$	$P(a)$	$P(b)$	$P(b) - P(a)$	$P(a, b)$	Δ_0	x_1
1	a_1	b_1							

First, we calculate the values of $P(a)$ and $P(b)$. All entries for these calculations are made on the first two lines in diagram 1. Then, we fill out the line corresponding to $n = 0$ in diagram 2. We obtain x_1. Now, we can complete the third line in diagram 1 and obtain $P(x_1)$. Depending on the sign of $P(x_1)$, we determine a_1 and b_1, and fill out the line $n = 1$ of diagram 2. We obtain x_2, etc.

Example. We calculate the real solution of the equation

$$P(x) \equiv x^3 + x - 1 = 0.$$

We choose $a = 0$ and $b = 1$. The results of the calculations are shown in tables 1 and 2. After we find that $P(0.6364) = -0.1059$, we see that is useless to take

$$[a_2, b_2] = [0.6364; 1],$$

since 1, the right-hand end of this interval, is far from the solution ($P(1) = 1$). Therefore, we choose

$$[a_2, b_2] = [0.6364, 0.7],$$

after first verifying that $P(0.7) > 0$.

Table 1

x	x^2	x^3	$P(x)$
0.0000	0.0000	0.0000	−1.0000
1.0000	1.0000	1.0000	1.0000
0.5000	0.2500	0.1250	−0.3750
0.6364	0.4050	0.2577	−0.1059
0.7000	0.4900	0.3430	0.0430
0.6816	0.4646	0.3167	−0.0017
0.6823	0.4655	0.3176	−0.0001

Table 2

n	a_n	b_n	$b_n - a_n$	$P(a_n)$	$P(b_n)$	$P(b_n) - -P(a_n)$	$P(a_n, b_n)$	Δ_n	$x_{n+1} = =x_n + \Delta_n$
0	0.0000	1.0	1.0000	−1.0000	1.000	2.0000	2.000	0.5000	0.5000
1	0.5000	1.0	0.5000	−0.3750	1.000	1.3750	2.750	0.1364	0.6364
2	0.6364	0.7	0.0636	−0.1059	0.043	0.1489	2.341	0.0452	0.6816
3	0.6816	0.7	0.0184	−0.0017	0.043	0.0447	2.429	0.0007	0.6823
4	0.6823	0.7	0.0177	−0.0001	0.043	0.0431	2.435	0.0000	0.6823

Since x_4 and x_5 agree to four decimal places, we terminate the calculation. In order to obtain a more accurate value of the solution, it would be necessary to perform further calculations with numbers with more figures.

3. The method of iterations

The method of iterations may be applied to finding both real and complex solutions of the equation

$$P(x) = 0.$$

It is assumed that one has an initial approximation x_0 to the solution of the equation. In order to apply the method of iterations, it is necessary that the equation be equivalent to an equation of the form

$$x = \phi(x). \tag{3.1}$$

Such a transformation may be performed in many ways, for example:

$$x = x + CP(x), \qquad C \neq 0.$$

The method of iterations consists of the following. We determine the sequence of numbers $\{x_n\}$:

$$x_{n+1} = \phi(x_n), \qquad n = 0, 1, 2, \ldots. \tag{3.2}$$

If the sequence $\{x_n\}$ has a limit x^* and $\phi(x)$ is a continuous function, then x^* is a solution of equation (3.1). In order to verify this, it is sufficient to pass to the limit as $n \to \infty$ in the relationship (3.2). The numbers x_n may be taken as approximate values of the solution.

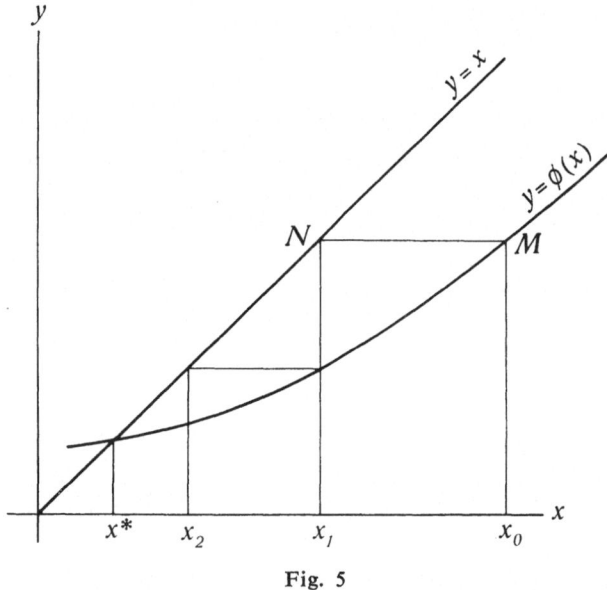

Fig. 5

It should be remarked at once that the sequence $\{x_n\}$ does not always converge, by far. Some theorems will be proved below which give sufficient conditions for the convergence of the method of iterations.

We can give a geometric interpretation of the method of iterations in the case that the function $\phi(x)$ on the right-hand side of equation (3.1) is real and a real solution is being sought. The graphs of the functions

$y = x$ and $y = \phi(x)$ are depicted in fig. 5. The abscissa x^* of the point of their intersection is the desired solution, and x_0 is the initial approximation. In order to determine x_1, the first approximation to the solution, we pass a line parallel to the y-axis from x_0 to its intersection with the curve $y = \phi(x)$. The point of intersection is denoted by M. From the point M, we draw a line parallel to the x-axis to its intersection with the line $y = x$ at the point N. Obviously, x_1 is the abscissa of the point N. We now prove a theorem on the convergence of the method of iterations.

Theorem 1. Suppose that for the equation (3.1) $x = \phi(x)$ and the initial approximation x_0, the following conditions are satisfied:

1) for any x', x'' in the disk

$$|x - x_0| \leqq \delta, \tag{3.3}$$

the function $\phi(x)$ satisfies the inequality

$$|\phi(x') - \phi(x'')| \leqq q|x' - x''|, \tag{3.4}$$

where $0 < q < 1$:

2) the inequality

$$\frac{m}{1-q} \leqq \delta, \tag{3.5}$$

is satisfied, where $m = |x_0 - \phi(x_0)|$.

Then, equation (3.1) has a unique solution x^* in the disk (3.3), to which the successive approximations x_n converge, the rate of convergence being determined by the inequality

$$|x_n - x^*| \leqq \frac{m}{1-q} q^n. \tag{3.6}$$

Proof. We shall establish that the inequality

$$|x_k - x_{k-1}| \leqq mq^{k-1} \tag{3.7}$$

holds for $k = 1, 2, 3, \ldots$.

Note that the satisfaction of inequality (3.7) for $k = 1, 2, \ldots, N$ implies that $x_1, x_2, \ldots, x_N$ belong to the disk (3.3). For $1 \leqq p \leqq N$, we have

$$|x_p - x_0| \leqq |x_p - x_{p-1}| + |x_{p-1} - x_{p-2}| + \ldots + |x_1 - x_0| \leqq$$

$$\leqq mq^{p-1} + mq^{p-2} + \ldots + m < \frac{m}{1-q} \leqq \delta,$$

from which

$$|x_p - x_0| \leqq \frac{m}{1-q} \leqq \delta, \qquad (3.8)$$

and this means that x_p belongs to the disk (3.3).

Inequality (3.7) will be proved by the method of mathematical induction. For $k = 1$, inequality (3.7) holds:

$$|x_1 - x_0| = |\phi(x_0) - x_0| = m$$

(it becomes an equation). We assume that inequality (3.7) holds for $k = 1, 2, \ldots, n$, and show that it holds for $k = n+1$.

It has been noted that it follows from the inductive hypothesis that x_1, $x_2, \ldots, x_n$ lie in the disk (3.3). By virtue of the definition of the sequence (3.2), we have

$$x_{n+1} - x_n = \phi(x_n) - \phi(x_{n-1}).$$

Since x_n and x_{n-1} lie in the disk (3.3), the right-hand side of this equation may be estimated by the use of condition (3.4):

$$|x_{n+1} - x_n| = |\phi(x_n) - \phi(x_{n-1})| \leqq q|x_n - x_{n-1}|.$$

By the inductive hypothesis, inequality (3.7) holds for $k = n$, and we obtain

$$|x_{n+1} - x_n| \leqq qmq^{n-1} = mq^n.$$

This proves the assertion. Inequality (3.7) holds for all integers $k = 1, 2, 3, \ldots$, and, consequently, all of the x_k lie in the disk (3.3). On the basis of (3.7), we have

$$|x_{n+p} - x_n| \leqq |x_{n+p} - x_{n+p-1}| + \ldots + |x_{n+1} - x_n| \leqq$$

$$\leqq mq^{n+p-1} + \ldots + mq^n < \frac{mq^n}{1-q},$$

or

$$|x_{n+p} - x_n| < \frac{mq^n}{1-q}, \qquad (3.9)$$

for any $p = 1, 2, 3, \ldots$.

Since $0 < q < 1$, the sequence $\{x_n\}$ is fundamental (i.e., a Cauchy sequence), and thus has a limit x^*. As the sequence $\{x_n\}$ lies inside of the disk (3.3), x^* is also contained in it. By virtue of the continuity of $\phi(x)$, which follows from condition (3.4), x^* is a solution of equation (3.1). We shall prove that x^* is unique in the disk (3.3). If it is assumed that $\tilde{x}$ is any solution of equation (3.1) in the disk (3.3), then it will be shown that $x^* = \tilde{x}$. We obtain from condition (3.4) that

$$|\tilde{x}-x^*| = |\phi(\tilde{x})-\phi(x^*)| \leqq q|\tilde{x}-x^*|,$$

which is only possible for $\tilde{x} = x^*$, since $0 < q < 1$.

In order to establish inequality (3.6), which characterizes the rate of convergence, it is sufficient to pass to the limit as $p \to \infty$ in inequality (3.9). The theorem is proved.

A solution x^* of equation (3.1) is sometimes called a *fixed point* of the function $\phi(x)$.

If $\phi(x)$ is a function of a complex variable which is differentiable in the disk (3.3), and for which the inequality

$$|\phi'(x)| \leqq q < 1, \tag{3.10}$$

holds at all points of the disk, then the function $\phi(x)$ satisfies condition 1) of theorem 1.

Actually, suppose that x' and x'' belong to the disk (3.3). From the formula

$$\phi(x'')-\phi(x') = \int_{x'}^{x''} \phi'(x)\mathrm{d}x,$$

where the integral is taken along the line segment joining the points x' and x'', we obtain

$$|\phi(x')-\phi(x'')| \leqq \max |\phi'(x)||x'-x''| \leqq q|x'-x''|.$$

If the function $\phi(x)$ is real and a real solution is sought, then it is natural to take a real initial approximation x_0. In this case, theorem 1 remains valid if the disk (3.3) is replaced by the interval

$$|x-x_0| \leqq \delta$$

on the real axis.

We note that the condition

$$\max |\phi'(x)| \leqq q < 1$$

is essential for the convergence of the method of iterations. An example of a real function $\phi(x)$ with a real fixed point is illustrated in fig. 6. From the figure, $\phi'(x) > 1$ in the vicinity of the fixed point, so $\{x_n\}$ does not converge to x^*. Furthermore, it is evident that $\{x_n\}$ will not converge to x^* no matter how close that x_0 is chosen to x^*, so long as $x_0 \neq x^*$.

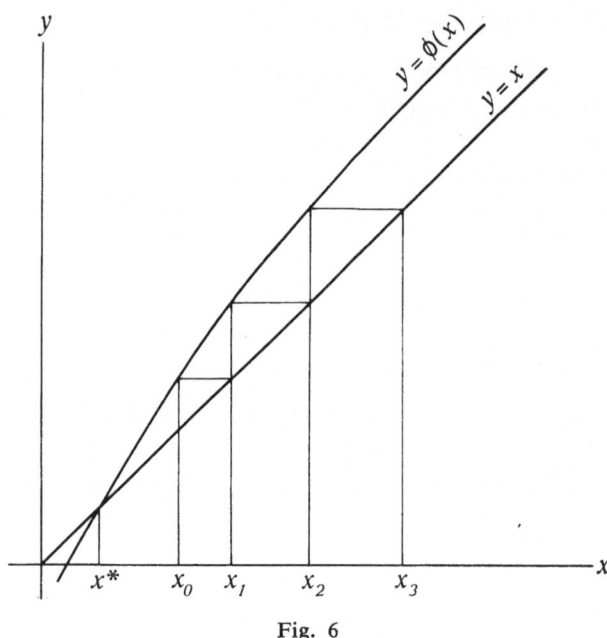

Fig. 6

The results of the calculations which are necessary to perform may be written down according to the diagram below:

n	x_n	...	$\phi(x_n)$
0	x_0	...	$\phi(x_0)$
1	x_1	...	$\phi(x_1)$

It is necessary to continue the computation until two successive approximations x_n and x_{n+1} agree to the required accuracy. This rule may be used with certainty if $|\phi'(x)| \leq q < 1$ in the neighborhood of a solution [in the disk (3.3)], and q is not close to one. If q is close to one, then it is possible that the successive values x_n and x_{n+1} are close to each other, but far from the solution. We consider the equation

$$x = 0.999x.$$

Here, $\phi(x) = 0.999x$ and $q = 0.999$. We shall calculate the solution of this equation by the method of iterations, taking $x_0 = 1$ as the initial approximation. Then, $x_1 = 0.999$ and $x_0 - x_1 = 0.001$, while at the same time, the solution of the equation is $x^* = 0$, which differs from x_1 by 0.999.

In computational practice, the application of theorem 1 on the convergence of the method of iterations is subject to the following difficulties. In actual computation of a solution of (3.1), using the method of iterations, the arithmetic is performed with a given number of decimal places (or significant figures); hence, already at the first step of the method of iterations, we do not obtain $x_1 = \phi(x_0)$, as a rule, but some value $\tilde{x}_1$ close to x_1. The difference $\tilde{x}_1 - x_1 = \Gamma_0$ we call the *round-off* (or *rounding*) *error*. At the second step of the method of iterations, the determination of $x_2 = \phi(x_1)$ is required. However, since x_1 is unknown, the error in the calculation of x_2 arises from two sources:
1) The argument $\tilde{x}_1$ has been substituted for x_1 into the function $\phi(x)$;
2) $\phi(\tilde{x}_1)$ is calculated with a round-off error, since only approximate equality to $\phi(\tilde{x}_1)$ is obtained for a given value of $\tilde{x}_1$:

$$\tilde{x}_2 - \phi(\tilde{x}_1) = \Gamma_1,$$

where Γ_1 is the round-off error.

Therefore, in actual computation, we obtain the sequence $\{\tilde{x}_n\}$ satisfying the equations

$$\tilde{x}_{n+1} = \phi(\tilde{x}_n) + \Gamma_n, \qquad n = 0, 1, 2, \ldots,$$

where the Γ_n are the round-off errors. The conclusion of theorem 1 refers to the sequence $\{x_n\}$. Without additional assumptions, this conclusion does not hold for the sequence $\{\tilde{x}_n\}$. This sequence, generally speaking, does not even converge to the solution. This will always be true, for example, when the solution is an irrational number.

We now prove a theorem on the behavior of the sequence $\{\tilde{x}_n\}$. For definiteness, we assume that x_0 and the function $\phi(x)$ are real.

Theorem 2. For equation (3.1), suppose that the initial approximation x_0 and the sequence $\{\tilde{x}_n\}$ defined by the equations

$$\tilde{x}_{n+1} = \phi(\tilde{x}_n) + \Gamma_n, \quad \tilde{x}_0 = x_0, \quad n = 0, 1, 2, \ldots, \tag{3.11}$$

satisfy the following conditions:

1) For arbitrary x' and x'' in the interval

$$|x-x_0| \leq \delta, \tag{3.12}$$

$\phi(x)$ satisfies the inequality

$$|\phi(x')-\phi(x'')| \leq q|x'-x''|, \qquad 0 < q < 1; \tag{3.13}$$

2) the numbers Γ_n which enter into (3.11) satisfy the inequalities

$$|\Gamma_n| \leq \Gamma q_1^n, \qquad n = 0, 1, 2, \ldots, \tag{3.14}$$

where $0 < q_1 \leq 1$;

3) the inequality

$$\frac{1}{1-q}(m+\Gamma) \leq \delta, \tag{3.15}$$

is satisfied, where $m = |x_0 - \phi(x_0)|$.

Then, equation (3.1) has a unique solution x^* in the interval (3.12). If $0 < q_1 < 1$, then the sequence $\{\tilde{x}_n\}$ converges to x^*. If $q_1 = 1$, then $\tilde{x}_n$, considered as an approximation to x^*, satisfies the inequality

$$|\tilde{x}_n - x^*| \leq \frac{1}{1-q}(\Gamma + mq^n). \tag{3.16}$$

Proof. The existence and uniqueness of the solution x^* in the interval (3.12) follows at once from theorem 1, as its hypotheses are obviously satisfied.

We shall prove that the inequality

$$|\tilde{x}_k - x_k| \leq \Gamma \sum_{i=1}^{k} q^{k-i} q_1^{i-1} \tag{3.17}$$

holds for $k = 1, 2, 3, \ldots$. Since $0 < q_1 \leq 1$, then

$$\sum_{i=1}^{k} q^{k-i} q_1^{i-1} \leq \sum_{i=1}^{k} q^{k-i} < \frac{1}{1-q}, \tag{3.18}$$

and thus the inequality

$$|\tilde{x}_k - x_k| < \frac{\Gamma}{1-q} \tag{3.19}$$

follows from (3.17).

We note that if $\tilde{x}_k$ satisfies inequality (3.17), then $\tilde{x}_k$ lies in the interval (3.12). Indeed,

$$|\tilde{x}_k - x_0| \leq |\tilde{x}_k - x_k| + |x_k - x_0| < \frac{\Gamma}{1-q} + \frac{m}{1-q} \leq \delta.$$

Here, inequality (3.19), inequality (3.8) for $p = k$, and condition (3.15) have been used.

Inequality (3.17) will be proved by mathematical induction. For $n = 0$, we obtain from (3.11) and (3.14) that

$$|\tilde{x}_1 - x_1| = |\Gamma_0| \leq \Gamma,$$

so that (3.17) holds for $k = 1$. We shall assume that inequality (3.17) is satisfied for $k = n$, and prove that it is valid for $k = n+1$.

Subtracting equation (3.2) term by term from equation (3.11), we obtain

$$\tilde{x}_{n+1} - x_{n+1} = \phi(\tilde{x}_n) - \phi(x_n) + \Gamma_n.$$

Since $\tilde{x}_n$ and x_n lie in the interval (3.12), we obtain, on the basis of (3.13) and (3.14), that

$$|\tilde{x}_{n+1} - x_{n+1}| \leq |\phi(\tilde{x}_n) - \phi(x_n)| + |\Gamma_n| \leq q|\tilde{x}_n - x_n| + \Gamma q_1^n.$$

Using inequality (3.17) for $k = n$:

$$|\tilde{x}_{n+1} - x_{n+1}| \leq q\Gamma \sum_{i=1}^{n} q^{n-i}q_1^{i-1} + \Gamma q_1^n = \Gamma \sum_{i=1}^{n+1} q^{n+1-i}q_1^{i-1}.$$

This proves inequality (3.17) for $k = n+1$.

We obtain, with the aid of inequalities (3.17) and (3.6), that

$$|\tilde{x}_n - x^*| \leq |\tilde{x}_n - x_n| + |x_n - x^*| \leq \Gamma \sum_{i=1}^{n} q^{n-i}q_1^{i-1} + \frac{m}{1-q} q^n. \qquad (3.20)$$

If $0 < q_1 < 1$, then it follows from inequality (3.20) that

$$|\tilde{x}_n - x^*| \to 0 \quad \text{as} \quad n \to \infty,$$

since as $n \to \infty$,

$$\sum_{i=1}^{n} q^{n-i}q_1^{i-1} \leq n[\max(q, q_1)]^{n-1} \to 0.$$

If $q_1 = 1$, then inequality (3.16) is obtained from (3.20). The theorem is proved.

The number Γ_n may be considered to be the round-off error arising at the $(n+1)$st step of the method of iterations. Theorem 2 gives an estimate for the error in $\tilde{x}_n$ as an approximation to x^* in the case that the round-off errors Γ_n do not exceed a given number Γ in absolute value. The theorem also allows one to assert that $\tilde{x}_n \to x^*$ as $n \to \infty$ if the round-off error decreases with the speed of a geometric progression.

For $\Gamma = 0$, one obtains the "exact method of iterations". In this case, theorem 2 coincides with theorem 1.

It follows that all theorems on error estimation and the convergence of numerical methods may be formulated in the same way as theorem 2. Nevertheless, we shall not do this in what follows, but shall limit ourselves instead to the study of exact methods. The questions of error estimation and the calculation of round-off error will be considered in Chapter IV for certain methods for numerical integration of ordinary differential equations.

Example. We seek the solution of the equation

$$P(x) \equiv x^3 - 4x^2 + 10x - 10 = 0$$

which is located in the interval [1, 2]. Since $P(1) = -3$ and $P(2) = 2$, we take $x_0 = 1.5$ as the initial approximation. We write the equation in the form

$$x = \tfrac{1}{10}(-x^3 + 4x^2 + 10),$$

so that

$$\phi(x) = -0.1x^3 + 0.4x^2 + 1.$$

Theorem 1 on the convergence of the method of iterations will be applied. In choosing the quantity δ, the following must be kept in mind: If δ is chosen very small, then it might result that inequality (3.5) is not satisfied; if δ is taken to be large, then it might happen that the condition that $|\phi'(x)| \leq q < 1$ is not satisfied. In our case,

$$m = |1.5 - \phi(1.5)| = 0.0625.$$

The value of the derivative

$$\phi'(x) = -0.3x^2 + 0.8x$$

is calculated for $x = 1.5$, and $\phi'(1.5) = q_0 = 0.525$ is obtained. If $\phi'(x)$ were constant in the vicinity of x_0, one could take

$$\frac{m}{1-q} = \frac{0.0625}{1-0.525} \cong 0.13$$

for δ. Since the derivative is actually not constant, we shall try choosing $\delta = 0.2$. It is easy to verify that max $|\phi'(x)|$ on the interval

$$|x-1.5| \leq 0.2 \tag{3.21}$$

is attained for $x = \frac{4}{3}$, and $\phi'(\frac{4}{3}) = \frac{8}{15}$. Consequently, one takes $q = \frac{8}{15}$. We have

$$\frac{m}{1-q} = \frac{0.0625}{1-\frac{8}{15}} = \frac{15}{112} = 0.1339 \ldots < 0.2.$$

For $\delta = 0.2$ and $q = \frac{8}{15}$, the hypotheses of theorem 1 are satisfied, and thus the sequence $\{x_n\}$ converges, with the error estimate

$$|x_n - x^*| \leq \frac{15}{112}(\frac{8}{15})^n.$$

We shall show how to apply the estimate (3.16) of theorem 2 to this example. In order to indicate the value of the upper bound Γ of the round-off errors Γ_n, we duly give an exact description of the numerical process. The computation will be performed in the following manner. We choose $x_0 = 1.5$ as the initial approximation. The computed approximations $\tilde{x}_n (n \geq 0)$ will be given as decimal fractions with four decimal places. The approximation $\tilde{x}_{n+1}$ is calculated accumulated to four decimal places by the formula

$$\tilde{x}_{n+1} = \{1 + 0.4(\tilde{x}_n^2) - [0.1(\tilde{x}_n^2)]\tilde{x}_n\}. \tag{3.22}$$

The results of the calculations are given in table 3.

Table 3

n	$\tilde{x}_n$	$(\tilde{x}_n^2)$	$[0.1(\tilde{x}_n^2)]$	$\phi(\tilde{x}_n)+\Gamma_n$
0	1.5000	2.2500	0.2250	1.5625
1	1.5625	2.4414	0.2441	1.5952
2	1.5952	2.5447	0.2545	1.6119
3	1.6119	2.5982	0.2598	1.6205
4	1.6205	2.6260	0.2626	1.6249
5	1.6249	2.6403	0.2640	1.6271
6	1.6271	2.6475	0.2648	1.6281
7	1.6281	2.6507	0.2651	1.6287
8	1.6287	2.6527	0.2653	1.6290
9	1.6290	2.6536	0.2654	1.6291
10	1.6291	2.6540	0.2654	1.6292
11	1.6292	2.6543	0.2654	1.6293
12	1.6293	2.6546	0.2655	1.6293
13	1.6293			

We proceed to explain formula (3.22). First, $\tilde{x}_n^2$ is calculated and rounded to four decimal places according to the usual rule (if the fifth decimal place is 5, and all of the places following it are zero, then the rounding is done to the nearest number ending in an even digit). The result of rounding $\tilde{x}_n^2$ is denoted in parentheses by $(\tilde{x}_n^2)$, so that

$$(\tilde{x}_n^2) = \tilde{x}_n^2 + \varepsilon_n, \tag{3.23}$$

where $|\varepsilon_n| \leqq \varepsilon$, $\varepsilon = 0.5 \times 10^{-4}$.

Then, $(\tilde{x}_n^2)$ is multiplied by 0.1, and the result of rounding is denoted by square brackets; consequently,

$$[0.1(\tilde{x}_n^2)] = 0.1(\tilde{x}_n^2) + \delta_n, \tag{3.24}$$

where $|\delta_n| \leqq \varepsilon$. Finally, we calculate the sum

$$1 + 0.4(\tilde{x}_n^2) - [0.1(\tilde{x}_n^2)]\tilde{x}_n$$

and round it to four decimal places. The result of the rounding is denoted by braces:

$$\{1 + 0.4(\tilde{x}_n^2) - [0.1(\tilde{x}_n^2)]\tilde{x}_n\} = 1 + 0.4(\tilde{x}_n^2) - [0.1(\tilde{x}_n^2)]\tilde{x}_n + \gamma_n, \tag{3.25}$$

where $|\gamma_n| \leqq \varepsilon$.

Taking equations (3.23), (3.24), and (3.25) into account, equation (3.22) may be written in the form

$$\tilde{x}_{n+1} = 1 + 0.4(\tilde{x}_n^2 + \varepsilon_n) - [0.1(\tilde{x}_n^2 + \varepsilon_n) + \delta_n]\tilde{x}_n + \gamma_n,$$

(here the parentheses and brackets have their ordinary meaning), or

$$\tilde{x}_{n+1} = \phi(\tilde{x}_n) + 0.4\varepsilon_n - (0.1\varepsilon_n + \delta_n)\tilde{x}_n + \gamma_n.$$

It is seen from this equation that for our method of calculation,

$$\Gamma_n = 0.4\varepsilon_n - (0.1\varepsilon_n + \delta_n)\tilde{x}_n + \gamma_n, \tag{3.26}$$

where the quantities ε_n, δ_n, γ_n do not exceed $\varepsilon = 0.5 \times 10^{-4}$ in absolute value.

With the aid of (3.26), we obtain the inequality

$$|\Gamma_n| \leqq (1.4 + 1.1|\tilde{x}_n|)\varepsilon.$$

Assuming that $\tilde{x}_n$ lies in the interval (3.21), then $|\tilde{x}_n| \leqq 1.7$, and, consequently,

$$|\Gamma_n| \leqq (1.4 + 1.1 \times 1.7)\varepsilon < 1.7 \times 10^{-4}.$$

Thus, one may take $\Gamma = 1.7 \times 10^{-4}$. The hypotheses of theorem 2 are satisfied for this Γ (and the values of δ and q indicated earlier), since inequality (3.15) holds:

$$\frac{1}{1-q}(m+\Gamma) = \tfrac{15}{7}(0.0625+1.7\times10^{-4}) < 0.2.$$

In particular, we obtain from the theorem that the $\tilde{x}_n$ lie in the interval (3.21), so that our assumption about the $\tilde{x}_n$ is justified.
The estimate (3.16) may be written as:

$$|\tilde{x}_n-x^*| \leq \tfrac{15}{7}[1.7\times10^{-4}+0.0625(\tfrac{8}{15})^n]. \tag{3.27}$$

This is an *a priori bound*, in as much as it is obtained before one proceeds with the calculations. Inequality (3.27) permits one to assert that $|\tilde{x}_{13}-x^*| \leq 5\times10^{-4}$. As we shall see, $|\tilde{x}_{13}-x^*| < 0.7\times10^{-4}$, and thus the estimate (3.27) is pessimistically large. This is because the quantities Γ_n are significantly less than $\Gamma = 1.7\times10^{-4}$, as is shown in table 4.

Table 4

n	0	1	2	3	4	5	6	7	8	9	10	11	12
$10^4\Gamma_n$	0	1.07	−0.41	0.19	0.38	0.02	−1.14	−0.22	−0.26	−0.78	−0.29	0.20	−0.30

The so-called *a posteriori* error estimates, which are obtained after an approximate solution of a problem has already been found, are more precise. We give an example of an a posteriori estimate. Suppose that $\tilde{x}$ is a numerical approximation to a solution of the equation $P(x) = 0$. We shall assume that $\tilde{x}$ and $P(x)$ are real. We shall also suppose that $P'(x)$ is continuous and does not vanish in some interval $[a, b]$ which contains $\tilde{x}$ in its interior. For definiteness, we shall assume that $P(\tilde{x}) < 0$ and $P'(x) > 0$ on $[a, b]$. We set

$$p = \min_{[a, b]} P'(x).$$

Then, the curve $y = P(x)$ is located above the line

$$y = P(\tilde{x})+p(x-\tilde{x}).$$

If the point of intersection X of this line with the x-axis lies in the interval $[a, b]$:

$$X = \tilde{x}-\frac{P(\tilde{x})}{p} \leq b, \tag{3.28}$$

then there is a unique solution x^* of the equation $P(x) = 0$ in the interval $[\tilde{x}, X]$, and we obtain

$$0 < x^* - \tilde{x} \leq \frac{-P(\tilde{x})}{p} \leq b, \qquad (3.29)$$

Inequality (3.29) is an a posteriori bound for $x^* - \tilde{x}$.

We carry out the error estimation for the approximate value $\tilde{x} = \tilde{x}_{13} = = 1.6293$ of the solution which was obtained in our example. We set $P(x) = x - \phi(x)$ and employ the bound (3.29). One has

$$0 < -P(\tilde{x}_{13}) < 0.31 \times 10^{-4}.$$

We choose the interval (3.21) for $[a, b]$. On it,

$$P'(x) = 1 - \phi'(x) \geq 1 - \tfrac{8}{15} = \tfrac{7}{15},$$

so that $p \geq \tfrac{7}{15}$. With the aid of (3.29), we obtain

$$0 < x^* - \tilde{x}_{13} < 0.7 \times 10^{-4}.$$

Inequality (3.28) is obviously satisfied.

4. The method of iterations for systems of equations

We consider a system of k equations in k unknowns $\xi_1, \xi_2, \ldots, \xi_k$,

$$\left.\begin{array}{l} P_1(\xi_1, \xi_2, \ldots, \xi_k) = 0, \\ P_2(\xi_1, \xi_2, \ldots, \xi_k) = 0, \\ \cdots\cdots\cdots\cdots\cdots \\ P_k(\xi_1, \xi_2, \ldots, \xi_k) = 0, \end{array}\right\} \qquad (4.1)$$

where, generally speaking, the equations are nonlinear. In order to apply the method of iterations, the system (4.1) must be transformed into an equivalent system of the form

$$\left.\begin{array}{l} \xi_1 = \phi_1(\xi_1, \xi_2, \ldots, \xi_k), \\ \xi_2 = \phi_2(\xi_1, \xi_2, \ldots, \xi_k), \\ \cdots\cdots\cdots\cdots\cdots \\ \xi_k = \phi_k(\xi_1, \xi_2, \ldots, \xi_k). \end{array}\right\} \qquad (4.2)$$

The functions $\phi_j(\xi_1, \xi_2, \ldots, \xi_k)$, $j = 1, 2, \ldots, k$, will be assumed to be real, and we shall seek real solutions of the system (4.2). We shall suppose that the initial approximations

$$\xi_1^{(0)}, \xi_2^{(0)}, \ldots, \xi_k^{(0)}, \qquad (4.3)$$

to the solutions of the system (4.2) are known, where the $\zeta_j^{(0)}$
$(j = 1, 2, \ldots, k)$ are real numbers.

The method of iterations for the system (4.2) consists of the following. From the initial approximations (4.3), one obtains the first approximations

$$\begin{aligned}
\xi_1^{(1)} &= \phi_1(\xi_1^{(0)}, \xi_2^{(0)}, \ldots, \zeta_k^{(0)}), \\
\xi_2^{(1)} &= \phi_2(\xi_1^{(0)}, \xi_2^{(0)}, \ldots, \zeta_k^{(0)}), \\
&\cdots\cdots\cdots\cdots\cdots \\
\xi_k^{(1)} &= \phi_k(\xi_1^{(0)}, \xi_2^{(0)}, \ldots, \zeta_k^{(0)}).
\end{aligned}$$

In the same way, one obtains the $(n+1)$st approximations from the nth approximations by

$$\left.\begin{aligned}
\xi_1^{(n+1)} &= \phi_1(\xi_1^{(n)}, \xi_2^{(n)}, \ldots, \zeta_k^{(n)}), \\
\xi_2^{(n+1)} &= \phi_2(\xi_1^{(n)}, \xi_2^{(n)}, \ldots, \zeta_k^{(n)}), \\
&\cdots\cdots\cdots\cdots\cdots \\
\xi_k^{(n+1)} &= \phi_k(\xi_1^{(n)}, \xi_2^{(n)}, \ldots, \zeta_k^{(n)}), \\
& n = 0, 1, 2, 3, \ldots
\end{aligned}\right\} \tag{4.4}$$

We shall establish sufficient conditions for the convergence of the method of iterations for the system (4.2). For this purpose, it is necessary for us to generalize theorem 1, which was proved in § 3.

A set X of elements of arbitrary nature is called a *metric space* if to each pair of elements x', x'' of the set, there corresponds a real number $\rho(x', x'')$, called the *distance* between the elements x' and x'', which satisfies the following conditions:

1) $\rho(x', x'') \geqq 0$, and $\rho(x', x'') = 0$ if and only if $x' = x''$;
2) $\rho(x', x'') = \rho(x'', x')$;
3) $\rho(x', x'') \leqq \rho(x', x''') + \rho)x''', x'')$.

A sequence $\{x_n\}$ of elements of the metric space X is said to *converge* if there exists an element $x^* \in X$ such that

$$\lim_{n \to \infty} \rho(x_n, x^*) = 0.$$

A sequence $\{x_n\}$ of elements of X is said to be *fundamental* (or a *Cauchy sequence*) if

$$\lim_{n \to \infty} \rho(x_{n+p}, x_n) = 0,$$

for all $p = 1, 2, 3, \ldots$.

It is easy to see that any convergent sequence is fundamental. The converse is not true: There exist metric spaces in which not all fundamental sequence are convergent. Because of this, the following definition is made: A metric space X is said to be *complete* if every fundamental sequence of elements of X converges.

The set of elements x of the metric space X which satisfy the condition

$$\rho(x, x_0) \leq \delta,$$

is called the (closed) *sphere* with center x_0 and radius δ.

The simplest example of a complete metric space is the set of real numbers, where the distance beween the numbers x' and x'' is taken to be the absolute value of the difference between these two numbers,

$$\rho(x', x'') = |x' - x''|.$$

Another example of a complete metric space is the set of complex numbers, where the distance between two complex numbers is taken to be the modulus of the difference of the complex numbers.

We mention one additional example of a complete metric space. Let X be the set of vectors x with k components,

$$x = (\xi_1, \xi_2, \ldots, \xi_k),$$

where $\xi_1, \xi_2, \ldots, \xi_k$ are real numbers. Let

$$x' = (\xi'_1, \xi'_2, \ldots, \xi'_k) \text{ and } x'' = (\xi''_1, \xi''_2, \ldots, \xi''_k).$$

We set

$$\rho(x', x'') = \max_{1 \leq j \leq k} |\xi'_j - \xi''_j|. \tag{4.5}$$

It is not difficult to verify that $\rho(x', x'')$ satisfies the conditions enumerated earlier, which must be satisfied by a distance. The resulting k-dimensional vector space in which the distance is defined by formula (4.5) will be denoted by m_k. It is easy to prove that the metric space m_k is complete.

Suppose that X is a complete metric space. We consider the equation

$$x = \phi(x), \tag{4.6}$$

where $\phi(x)$ is an operator which maps each element x of the space X into an element y in the same space X. The following theorem holds.

Theorem. For the equation (4.6) and the element x_0, taken as an initial approximation, suppose that the following conditions are satisfied:

1) For any x', x'' in the sphere

$$\rho(x, x_0) \leqq \delta, \tag{4.7}$$

the inequality

$$\rho(\phi(x'), \phi(x'')) \leqq q\rho(x', x''), \tag{4.8}$$

holds, where $0 < q < 1$;

2) the inequality

$$\frac{m}{1-q} \leqq \delta, \tag{4.9}$$

is satisfied, where $m = \rho(x_0, \phi(x_0))$.

Then, equation (4.6) has a unique solution x^* in the sphere (4.7), to which the sequence of approximations x_n defined by the formula

$$x_{n+1} = \phi(x_n)$$

converges, and for which the error estimate

$$\rho(x_n, x^*) \leqq \frac{mq^n}{1-q}$$

holds.

The proof of this theorem is exactly the same as the proof of the theorem in § 3. The only difference is that it is necessary to write $\rho(x, y)$ in place of the modulus $|x - y|$.

The system (4.2) may be considered to be the equation

$$x = \phi(x) \tag{4.10}$$

in the space m_k. The operator $\phi(x)$ maps each vector

$$x = (\xi_1, \xi_2, \ldots, \xi_k),$$

into a vector

$$y = (\eta_1, \eta_2, \ldots, \eta_k),$$

the components of which are defined by the formulas

$$\eta_1 = \phi_1(\xi_1, \xi_2, \ldots, \xi_k),$$
$$\eta_2 = \phi_2(\xi_1, \xi_2, \ldots, \xi_k),$$
$$\cdot \cdot \cdot \cdot \cdot \cdot \cdot \cdot \cdot \cdot \cdot \cdot \cdot$$
$$\eta_k = \phi_k(\xi_1, \xi_2, \ldots, \xi_k).$$

We now apply the general theorem on the convergence of the method of iterations to equation (4.10), or, what is the same, to the system (4.2). In the space m_k, the sphere (4.7), $\rho(x, x_0) \leq \delta$, is represented by the k-dimensional cube

$$\max_{1 \leq j \leq k} |\xi_j - \xi_j^{(0)}| \leq \delta, \tag{4.11}$$

with center at the point $x_0 = (\xi_1^{(0)}, \xi_2^{(0)}, \ldots, \xi_k^{(0)})$. We shall assume that the functions $\phi_i(\xi_1, \xi_2, \ldots, \xi_k)$ have continuous partial derivatives $\partial\phi_i/\partial\xi_j$ in the cube (4.11).

We give a sufficient condition which guarantees the satisfaction of the inequality (4.8). Suppose that the points

$$x' = (\xi_1', \xi_2', \ldots, \xi_k') \quad \text{and} \quad x'' = (\xi_1'', \xi_2'', \ldots, \xi_k'')$$

lie in the cube (4.11). Using the mean-value theorem,

$$\phi_i(\xi_1', \xi_2', \ldots, \xi_k') - \phi_i(\xi_1'', \xi_2'', \ldots, \xi_k'') =$$

$$= \sum_{j=1}^{k} \frac{\partial\phi_i(\xi_1' + \theta(\xi_1'' - \xi_1'), \ldots, \xi_k' + \theta(\xi_k'' - \xi_k'))}{\partial\xi_j} (\xi_j' - \xi_j'').$$

Evidently, one has that

$$|\phi_i(\xi_1', \xi_2', \ldots, \xi_k') - \phi_i(\xi_1'', \xi_2'', \ldots, \xi_k'')| \leq$$

$$\leq \rho(x', x'') \sum_{j=1}^{k} \left| \frac{\partial\phi_i(\xi_1' + \theta(\xi_1'' - \xi_1'), \ldots, \xi_k' + \theta(\xi_k'' - \xi_k'))}{\partial\xi_j} \right| \leq$$

$$\leq \rho(x', x'') \max_{\rho(x, x_0) \leq \delta} \sum_{j=1}^{k} \left| \frac{\partial\phi_i(\xi_1, \xi_2, \ldots, \xi_k)}{\partial\xi_j} \right|,$$

from which we obtain the inequality

$$\rho(\phi(x'), \phi(x'')) \leq \rho(x', x'') \max_{1 \leq i \leq k} \left\{ \max_{\rho(x, x_0) \leq \delta} \sum_{j=1}^{k} \left| \frac{\partial\phi_i(\xi_1, \xi_2, \ldots, \xi_k)}{\delta\xi_j} \right| \right\}.$$

Thus, we may take the coefficient of $\rho(x', x'')$ on the right-hand side of the last inequality as q in inequality (4.8) (if it is less than one), and we obtain the following theorem on the convergence of the method of iterations for system (4.2).

Theorem. Suppose that these conditions are satisfied:

1) $$q = \max_{1 \leq i \leq k} \left\{ \max_{\rho(x, x_0) \leq \delta} \sum_{j=1}^{k} \left| \frac{\partial\phi_i(\xi_1, \xi_2, \ldots, \xi_k)}{\partial\xi_j} \right| \right\} < 1,$$

where the inside maximum is taken over all points of the cube (4.11);

2) $\quad \dfrac{1}{1-q} \max_{1 \leq i \leq k} |\xi_i^{(0)} - \phi_i(\xi_1^{(0)}, \xi_2^{(0)}, \ldots, \xi_k^{(0)})| \leq \delta.$

Then, the system (4.2) has a unique solution $x^* = (\xi_1^*, \xi_2^*, \ldots, \xi_k^*)$ in the cube (4.11), to which the sequence of approximations (4.4) converges, with the error estimate given by

$$\max_{1 \leq i \leq k} |\xi_i^{(n)} - \xi_i^*| \leq \dfrac{1}{1-q} \max_{1 \leq i \leq k} |\xi_i^{(0)} - \phi_i(\xi_1^{(0)}, \xi_2^{(0)}, \ldots, \xi_k^{(0)})| q^n.$$

5. Numerical evaluation of polynomials and their derivatives

Suppose that the polynomial

$$P_n(x) = a_0 x^n + a_1 x^{n-1} + \ldots + a_n, \tag{5.1}$$

is given, the coefficients $a_0, a_1, \ldots, a_n$ of which are real numbers. The evaluation of this polynomial is required for $x = x_0$. We first consider the case that x_0 is a real number.

We carry out the division of $P_n(x)$ by $x - x_0$, and obtain

$$a_0 x^n + a_1 x^{n-1} + \ldots + a_n =$$
$$= (s_0 x^{n-1} + s_1 x^{n-2} + \ldots + s_{n-1})(x - x_0) + s_n. \tag{5.2}$$

Clearly,

$$s_n = P_n(x_0),$$

and, consequently, in order to evaluate $P_n(x_0)$, it is sufficient to calculate s_n.

We equate the coefficients of like powers of x on the left and right sides of the relationship (5.2):

$$a_0 = s_0,$$
$$a_1 = s_1 - s_0 x_0,$$
$$a_2 = s_2 - s_1 x_0,$$
$$\cdots \cdots \cdots \cdots$$
$$a_n = s_n - s_{n-1} x_0.$$

Solving these equations for $s_0, s_1, \ldots, s_n$, we obtain

$$\left.\begin{aligned}
s_0 &= a_0, \\
s_1 &= a_1 + s_0 x_0, \\
s_2 &= a_2 + s_1 x_0, \\
&\cdot \cdot \cdot \cdot \cdot \cdot \cdot \cdot \cdot \\
s_n &= a_n + s_{n-1} x_0.
\end{aligned}\right\} \tag{5.3}$$

The recurrence relations (5.3) provide a sequential determination of the numbers $s_0, s_1, \ldots, s_n$. The numbers $s_0, s_1, \ldots, s_{n-1}$ are the coefficients of the quotient in the division of $P_n(x)$ by $x - x_0$. This method for evaluation of $P_n(x_0)$ is commonly referred to as Horner's procedure.

Equation (5.2) may be used to check the validity of the calculation. In fact, setting $x = 1$ in it, we obtain

$$a_0 + a_1 + \ldots + a_n = (s_0 + s_1 + s_{n-1})(1 - x_0) + s_n.$$

In order to determine whether or not this equation is satisfied, it is necessary to find the sum $P_n(1)$ of the coefficients of the polynomial $P_n(x)$, the sum

$$S = s_0 + s_1 + \ldots + s_{n-1},$$

and to calculate the quantity

$$P_n(1) - S(1 - x_0).$$

For the calculation to be valid, the equation

$$P_n(1) - S(1 - x_0) = s_n$$

must hold.

The equation

$$a_n - a_{n-1} + a_{n-2} - \ldots + (-1)^n a_0 =$$
$$= (s_{n-1} - s_{n-2} + \ldots + (-1)^{n-1} s_0)(-1 - x_0) + s_n,$$

which is obtained by setting $x = -1$ in (5.2), may also be used for checking.

The evaluation of $P_n(x_0)$ by means of the relationships (5.3) requires that n multiplications be performed. If $P_n(x_0)$ is evaluated in the ordinary way instead, then $2n - 1$ multiplications will be necessary: $n - 1$ multiplications for the evaluation of the powers

$$x_0^2, x_0^3, \ldots, x_0^n,$$

and n multiplications for the evaluation of the products

$$a_0 x_0^n, a_1 x_0^{n-1}, \ldots, a_{n-1} x_0.$$

Example 1. We shall evaluate the polynomial

$$P_4(x) = x^4 + 3.8491 x^3 - 4.9126 x^2 + 1.8312 x - 2.3844$$

for $x = 1.2134$. We have:

1	3.8491	−4.9126	1.8312	−2.3844
	1.2134	6.1428	1.4927	4.0332
1	5.0625	1.2302	3.3239	1.6488

We obtain that $P_4(1.2134) = 1.6488$.
To check the validity of the calculation, we find that

$$P_4(1) = -0.6167, \ S = s_0 + s_1 + s_2 + s_3 = 10.6166,$$

$$P_4(1) - S(1 - x_0) = -0.6167 + 10.6166 \times 0.2134 = 1.6489.$$

The number $P_4(1) - S(1 - x_0) = 1.6489$ differs from s_4 by only one unit in the fourth decimal place, so the evaluation is accepted as being obtained accurately.
We now go to the question of the evaluation of the derivatives of the polynomial $P_n(x)$ for $x = x_0$ (x_0 real). Denote the quotient in the division of $P_n(x)$ by $x - x_0$ by $P_{n-1}(x)$, so that

$$P_{n-1}(x) = s_0 x^{n-1} + s_1 x^{n-2} + \ldots + s_{n-1}.$$

The relationship (5.2) may be written as

$$P_n(x) = P_{n-1}(x)(x - x_0) + s_n.$$

Denote the quotient in the division of $P_{n-1}(x)$ by $x - x_0$ by

$$P_{n-2}(x) = s_0^{(1)} x^{n-2} + s_1^{(1)} x^{n-3} + \ldots + s_{n-2}^{(1)}.$$

One has that

$$P_{n-1}(x) = P_{n-2}(x)(x - x_0) + s_{n-1}^{(1)},$$

which means that the numbers $s_j^{(1)}$ may be obtained from the numbers s_j by means of the recurrence relations

$$s_j^{(1)} = s_j + s_{j-1}^{(1)} x_0, \quad j = 0, 1, 2, \ldots, n-1, \quad s_{-1}^{(1)} = 0.$$

Continuing in this way, one obtains the polynomials

$$P_n(x), P_{n-1}(x), P_{n-2}(x), \ldots, P_1(x), P_0(x), \tag{5.4}$$

and the triangular array of numbers

$$\left. \begin{array}{cccccc}
a_0 & a_1 & \cdots & a_{n-2} & a_{n-1} & a_n \\
s_0 & s_1 & \cdots & s_{n-2} & s_{n-1} & s_n \\
s_0^{(1)} & s_1^{(1)} & \cdots & s_{n-2}^{(1)} & s_{n-1}^{(1)} & \\
\cdots & \cdots & \cdots & \cdots & \cdots & \\
s_0^{(n-1)} & s_1^{(n-1)} & & & & \\
s_0^{(n)} & & & & &
\end{array} \right\} \tag{5.5}$$

Setting

$$a_j = s_j^{(-1)}, \qquad s_j = s_j^{(0)}, \qquad j = 0, 1, 2, \ldots, n,$$

one may confirm that the numbers $s_j^{(k)}$ on the $(k+2)$nd line are determined from the numbers $s_j^{(k-1)}$ on the $(k+1)$st line by means of the recurrence relations

$$\left. \begin{array}{l}
s_0^{(k)} = s_0^{(k-1)}, \\
s_1^{(k)} = s_1^{(k-1)} + s_0^{(k)} x_0, \\
s_2^{(k)} = s_2^{(k-1)} + s_1^{(k)} x_0, \qquad k = 0, 1, 2, \ldots, n. \\
\cdots \cdots \cdots \cdots \cdots \cdots \cdots \cdots \cdots \cdots \\
s_{n-k}^{(k)} = s_{n-k}^{(k)} + s_{n-k-1}^{(k)} x_0,
\end{array} \right\} \tag{5.6}$$

For $k = 0$, the relationships (5.6) coincide with (5.3).
The polynomials (5.4) are connected by the relationships

$$\left. \begin{array}{l}
P_n(x) = P_{n-1}(x)(x-x_0) + s_n, \\
P_{n-1}(x) = P_{n-2}(x)(x-x_0) + s_{n-1}^{(1)}, \\
\cdots \cdots \cdots \cdots \cdots \cdots \cdots \cdots \cdots \\
P_2(x) = P_1(x)(x-x_0) + s_2^{(n-2)}, \\
P_1(x) = P_0(x)(x-x_0) + s_1^{(n-1)}, \\
P_0(x) = s_0^{(n)},
\end{array} \right\} \tag{5.7}$$

from which we obtain the equation

$$P_n(x) = s_n + s_{n-1}^{(1)}(x-x_0) + s_{n-2}^{(2)}(x-x_0)^2 + \\
+ \ldots + s_1^{(n-1)}(x-x_0)^{n-1} + s_0^{(n)}(x-x_0)^n. \tag{5.8}$$

By virtue of the uniqueness of Taylor series expansions, we obtain from equation (5.8) that

$$P_n(x_0) = s_n, \quad \frac{P_n'(x_0)}{1!} = s_{n-1}^{(1)},$$

$$\frac{P_n''(x_0)}{2!} = s_{n-2}^{(2)}, \ldots, \quad \frac{P_n^{(n)}(x_0)}{n!} = s_0^{(n)}. \tag{5.9}$$

Therefore, in order to evaluate the derivatives of the polynomial $P_n(x)$ at $x = x_0$, it is sufficient to construct the array of numbers (5.5) by means of the recurrence relations (5.6). The desired values of the derivatives of the polynomial $P_n(x)$ are obtained from equations (5.9).

Equations (5.7) may be used to check the validity of the calculation. Setting $x = 1$ in them, we obtain the checking equations:

$$P_n(1) - S(1-x_0) = s_n,$$
$$S - S^{(1)}(1-x_0) = s_{n-1}^{(1)},$$
$$\cdot \quad \cdot \quad \cdot \quad \cdot \quad \cdot \quad \cdot \quad \cdot \quad \cdot \quad \cdot \quad \cdot \quad \cdot \quad \cdot \quad \cdot \quad \cdot$$
$$S^{(n-2)} - S^{(n-1)}(1-x_0) = s_1^{(n-1)},$$
$$S^{(n-1)} = s_0^{(n)},$$

where

$$S^{(j)} = s_0^{(j)} + s_1^{(j)} + \ldots + s_{n-j-1}^{(j)}, \quad j = 0, 1, \ldots, n,$$
$$S^{(0)} = S, \quad S^{(n)} = 0.$$

One may also obtain checking equations by setting $x = -1$ in (5.7).

Example 2. We evaluate the polynomial

$$P_4(x) = x^4 + 3.8491x^3 - 4.9126x^2 + 1.8312x - 2.3844$$

and its derivatives for $x = 1.2134$. We construct the array (5.5):

1	3.8491	−4.9126	1.8312	−2.3844	−0.6167	
1	5.0625	1.2302	3.3239	1.6488	10.6166	1.6489
1	6.2759	8.8454	14.0569		16.1213	14.0569
1	7.4893	17.9329			8.4893	17.9329
1	8.7027				1	8.7027
1					0	1

On the basis of (5.9), we obtain that

$$P_4(1.2134) = 1.6488; \quad P_4'(1.2134) = 14.0569;$$

$$\frac{P_4''(1.2134)}{2!} = 17.9329; \qquad \frac{P_4'''(1.2134)}{3!} = 8.7027;$$

$$\frac{P_4^{(IV)}(1.2134)}{4!} = 1.$$

The control numbers occupy the last two columns of the table. In the next-to-last column, the numbers

$$P_4(1), S, S^{(1)}, S^{(2)}, S^{(3)}, S^{(4)}$$

are entered. The entries in the last column are the numbers

$$P_4(1) - S(1 - x_0), \ S - S^{(1)}(1 - x_0), \ \ldots$$

We see that $P_4(1) - S(1 - x_0)$ differs from s_4 by one unit in the fourth decimal place, while the other numbers in the last column coincide with $s_3^{(1)}, s_2^{(2)}, s_1^{(3)}, s_0^{(4)}$, so that the evaluation is acceptable to four decimal places. Finally, the checking is performed not at the end of all the computation, but sequentially: The table is filled out by rows, including the control numbers.

We now proceed to describe a method for evaluating the polynomial (5.1) and its first derivative for $x = x_0$, where x_0 is a complex number:

$$x_0 = u_0 + iv_0.$$

The quadratic trinomial

$$x^2 + px + q,$$

where

$$p = -2u_0, \qquad q = u_0^2 + v_0^2,$$

has x_0 and $\bar{x}_0$ as its roots. The complex conjugate of the number x_0 is denoted by $\bar{x}_0$: $\bar{x}_0 = u_0 - iv_0$.

We divide the polynomial $P_n(x)$ by $x^2 + px + q$, and obtain the relationship

$$a_0 x^n + a_1 x^{n-1} + \ldots + a_n =$$
$$= (b_0 x^{n-2} + b_1 x^{n-3} + \ldots + b_{n-2})(x^2 + px + q) + b_{n-1}(x + p) + b_n.$$
$$(5.10)$$

The remainder in the division, a linear function, is written in a special form in order to obtain unique recurrence relations to define the numbers b_j.

Equating the coefficients of like powers of x in the relationship (5.10):

$$a_0 = b_0$$
$$a_1 = b_1 + pb_0,$$
$$a_2 = b_2 + pb_1 + qb_0,$$
$$\cdots\cdots\cdots\cdots\cdots\cdots$$
$$a_{n-1} = b_{n-1} + pb_{n-2} + qb_{n-3},$$
$$a_n = b_n + pb_{n-1} + qb_{n-2}.$$

We solve these equations for $b_0, b_1, \ldots, b_n$, and obtain the recurrence relations

$$\left. \begin{aligned} b_0 &= a_0, \\ b_1 &= a_1 - pb_0, \\ b_2 &= a_2 - pb_1 - qb_0, \\ &\cdots\cdots\cdots\cdots\cdots\cdots \\ b_{n-1} &= a_{n-1} - pb_{n-2} - qb_{n-3}, \\ b_n &= a_n - pb_{n-1} - qb_{n-2}, \end{aligned} \right\} \tag{5.11}$$

by means of which the numbers $b_0, b_1, \ldots, b_n$ are obtained sequentially. It is seen from the relationship (5.10) that

$$P_n(x_0) = b_{n-1}(x_0 + p) + b_n = B_1 + iB_2 = b_n - u_0 b_{n-1} + iv_0 b_{n-1}. \tag{5.12}$$

Together with $P_n(x_0)$, we obtain the coefficients of the quotient

$$Q(x) = b_0 x^{n-2} + b_1 x^{n-3} + \ldots + b_{n-2}, \tag{5.13}$$

in the division of $P_n(x)$ by $x^2 + px + q$.

In using a desk calculator for the computation, the numbers b_j may be formed cumulatively.

Equation (5.10) may be used to check the validity of the calculation.

Relationship (5.10) may be written in the form

$$P_n(x) = Q(x)(x^2 + px + q) + b_{n-1}(x + p) + b_n,$$

and differentiation of both sides with respect to x gives:

$$P'_n(x) = Q'(x)(x^2 + px + q) + Q(x)(2x + p) + b_{n-1}.$$

We set $x = x_0$ in this equation, and obtain

$$P'_n(x_0) = Q(x_0) \cdot i2v_0 + b_{n-1}. \tag{5.14}$$

In order to employ equation (5.14) for the evaluation of the derivative $P_n'(x)$ at $x = x_0$, we note that we may find the value $Q(x_0)$ of the polynomial (5.13) by calculating the numbers $c_0, c_1, \ldots, c_{n-2}$ by means of the relationships

$$\left.\begin{aligned}
c_0 &= b_0, \\
c_1 &= b_1 - pc_0, \\
c_2 &= b_2 - pc_1 - qc_0, \\
&\cdots \cdots \cdots \cdots \cdots \cdots \\
c_{n-2} &= b_{n-2} - pc_{n-3} - qc_{n-4}.
\end{aligned}\right\} \tag{5.15}$$

On the basis of (5.12), we have that

$$Q(x_0) = C_1 + iC_2 = c_{n-2} - u_0 c_{n-3} + iv_0 c_{n-3}. \tag{5.16}$$

Equation (5.14) may now be written as:

$$P_n'(x_0) = D_1 + iD_2 = -2v_0 C_2 + b_{n-1} + i2v_0 C_1. \tag{5.17}$$

The recording of the evaluation of a polynomial and its first derivative may be laid out according to the following diagram:

u_0	v_0	a_0	a_1	$\ldots$	a_{n-2}	a_{n-1}	a_n		
p	q	b_0	b_1	$\ldots$	b_{n-2}	b_{n-1}	b_n	B_1	B_2
		c_0	c_1	$\ldots$	c_{n-2}	C_1	C_2	D_1	D_2

Here, the pairs of numbers B_1, B_2; C_1, C_2; D_1, D_2 are defined by formulas (5.12), (5.16), and (5.17), respectively.

Example 3. We evaluate the polynomial

$$P_4(x) = x^4 - 2.7x^3 + 4x^2 - 3.3x + 1$$

and its first derivative at $x_0 = 0.52 + 1.16i$. The results of the calculation are written according to the diagram indicated earlier:

0.52	1.16	1	−2.7	4	−3.3	1		
−1.04	1.616	1	−1.66	0.6576	0.06646	0.00644	−0.02812	0.07709
		1	−0.62	−1.2808	−1.2808	−0.7192	1.73500	−2.97146

We obtain

$$P_4(x_0) = -0.02812+0.07709i,$$
$$P_4'(x_0) = 1.73500-2.97146i.$$

We now give another method for evaluating the polynomial (5.1) and all of its derivatives for $x = x_0 = u_0+iv_0$. We set

$$x = u+iv,$$

where u and v are real. We expand $P_n(x)$ in a Taylor series about u in powers of iv,

$$P_n(u+iv) = P_n(u)+ivP_n'(u)-\frac{v^2}{2}P_n''(u)-\frac{iv^3}{3!}P_n'''(u)+$$

$$+\frac{v^4}{4!}P_n^{(\mathrm{IV})}(u)+ \ldots +\frac{i^n v^n}{n!}P_n^{(n)}(u).$$

Separating the real and imaginary parts, we obtain

$$P_n(u+iv) = X(u,v)+iY(u,v),$$

where

$$X(u,v) = P_n(u)-\frac{P_n''(u)}{2!}v^2+\frac{P_n^{(\mathrm{IV})}(u)}{4!}v^4- \ldots,$$

$$Y(u,v) = \frac{P_n'(u)}{1!}v-\frac{P_n'''(u)}{3!}v^3+\frac{P_n^{(\mathrm{V})}(u)}{5!}v^5- \ldots.$$

(5.18)

If n is even, then $X(u_0,v)$ is a polynomial of degree n in v, and $Y(u_0,v)$ is a polynomial of degree $n-1$ in v. If n is odd, then $X(u_0,v)$ is a polynomial of degree $n-1$ in v, and $Y(u_0,v)$ is a polynomial of degree n in v. The coefficients of the polynomials $X(u_0,v)$ and $Y(u_0,v)$ are easy to find. It is sufficient to construct the array of numbers (5.5) from the coefficients of the polynomial $P_n(x)$ and the real number u_0.

After we find the coefficients of the polynomials $X(u_0,v)$ and $Y(u_0,v)$, we may evaluate these polynomials and their derivatives with respect to v at $v = v_0$:

$$X(u_0,v_0), \quad \frac{1}{1!}\frac{\partial X(u_0,v_0)}{\partial v}, \quad \frac{1}{2!}\frac{\partial^2 X(u_0,v_0)}{\partial v^2}, \ldots,$$

$$Y(u_0,v_0), \quad \frac{1}{1!}\frac{\partial Y(u_0,v_0)}{\partial v}, \quad \frac{1}{2!}\frac{\partial^2 Y(u_0,v_0)}{\partial v^2}, \ldots,$$

from the construction of the arrays (5.5) for $X(u_0,v)$ and $Y(u_0,v)$.

It is easy to see that the following equations hold:

$$P_n(x_0) = X(u_0, v_0) + iY(u_0, v_0),$$

$$\frac{i}{1!} P'_n(x_0) = \frac{1}{1!} \left[\frac{\partial X(u_0, v_0)}{\partial v} + i \frac{\partial Y(u_0, v_0)}{\partial v} \right],$$

$$-\frac{1}{2!} P''_n(x_0) = \frac{1}{2!} \left[\frac{\partial^2 X(u_0, v_0)}{\partial v^2} + i \frac{\partial^2 Y(u_0, v_0)}{\partial v^2} \right], \tag{5.19}$$

$$\cdots\cdots\cdots\cdots\cdots\cdots\cdots\cdots\cdots\cdots\cdots\cdots$$

$$\frac{i^n}{n!} P_n^{(n)}(x_0) = \frac{1}{n!} \left[\frac{\partial^n X(u_0, v_0)}{\partial v^n} + i \frac{\partial^n Y(u_0, v_0)}{\partial v^n} \right].$$

Thus, the evaluation of a polynomial and all of its derivatives for complex values of its argument may be reduced to the evaluation of three polynomials and their derivatives for real arguments by the method (5.5).

Example 4. We evaluate the polynomial

$$P_4(x) = x^4 - 2.7x^3 + 4x^2 - 3.3x + 1$$

and its derivatives for $x = 0.52 + 1.16i$.
We construct the array (5.5) from the coefficients of the polynomial and the number $u = 0.52$:

```
1   -2.7    4          -3.3      1
1   -2.18   2.8664     -1.80947  0.05908
1   -1.66   2.00320    -0.76781
1   -1.14   1.4104
1   -0.62
1
```

We write down the polynomials (5.18),

$$X(0.52, v) = 0.05908 - 1.4104v^2 + v^4,$$
$$Y(0.52, v) = -0.76781v + 0.62v^3.$$

The array (5.5) is constructed for the polynomial $X(0.52, v)$ and the number $v = 1.16$:

```
1   0       -1.4140    0         0.05908
1   1.16    -0.06480   -0.07517  -0.02812
1   2.32    2.6264     2.97145
1   3.48    6.6632
1   4.64
1
```

and also for the polynomial $Y(0.52, v)$ and $v = 1.16$:

0.62	0	-0.76781	0
0.62	0.7192	0.06646	0.07709
0.62	1.4834	1.73500	
0.62	2.1576		
0.62			

We write down equations (5.19),

$$P_4(x_0) = -0.02812 + 0.07709i,$$
$$iP_4'(x_0) = 2.97145 + 1.73500i$$
$$-\tfrac{1}{2}P_4''(x_0) = 6.6632 + 2.1576i,$$
$$-\tfrac{i}{6}P_4'''(x_0) = 4.64 + 0.62i,$$
$$\tfrac{1}{24}P_4^{(IV)}(x_0) = 1,$$

from which we obtain

$$P_4(x_0) = -0.02812 + 0.07709i,$$
$$P_4'(x_0) = 1.73500 - 2.97145i,$$
$$P_4''(x_0) = -13.3264 - 4.3152i,$$
$$P_4'''(x_0) = -3.72 + 27.84i,$$
$$P_4^{(IV)}(x_0) = 24.$$

6. Newton's method

Newton's method is a very effective procedure for solving algebraic and transcendental equations. Its principal merit lies in the fact that it furnishes rapid convergence for a relatively simple computational process. Newton's method is widely applicable, as is the method of iterations. It is capable of solving a large class of functional equations.

For consideration in the present case of algebraic and transcendental equations, of course, a method must be useful for the determination of both real and complex solutions of equations.

We proceed to describe Newton's method. Suppose that the equation

$$P(x) = 0 \tag{6.1}$$

is given, together with an initial approximation x_0 to a solution of the equation. We shall assume that the function $P(x)$ is differentiable in a neighborhood of x_0 in the complex plane. If the function $P(x)$ is real, and one is concerned with finding a real solution, then one may in fact assume that x_0 is real, and it is sufficient to require that $P(x)$ be differentiable in a neighborhood of x_0 on the real axis.

In the neighborhood of x_0, we replace equation (6.1) by the linear equation

$$P(x_0) + P'(x_0)(x - x_0) = 0, \tag{6.2}$$

the left side of which consists of the first two terms of the Taylor series expansion of the function $P(x)$. The solution

$$x_1 = x_0 - \frac{P(x_0)}{P'(x_0)}$$

of equation (6.2) may be taken as a new approximation to the solution of equation (6.1). The successive approximations

$$x_{n+1} = x_n - \frac{P(x_n)}{P'(x_n)}, \qquad n = 0, 1, 2, \ldots, \tag{6.3}$$

are defined analogously. We assume that

$$P'(x_n) \neq 0, \qquad n = 0, 1, 2, \ldots.$$

Sometimes, it is expedient to apply the modified form of Newton's method, for which the successive approximations are defined by the formula

$$x'_{n+1} = x'_n - \frac{P(x'_n)}{P'(x_0)}, \qquad x_0 = x'_0, \qquad n = 0, 1, 2, \ldots. \tag{6.4}$$

The computation using formula (6.4) is simpler than by formula (6.3), since it is not necessary to evaluate the derivative $P'(x)$ at each step. On the other hand, the modified method converges significantly slower than the ordinary form of Newton's method, as will be seen later.

We indicate a geometric interpretation of Newton's method for the case that $P(x)$, x_0, and the solution being sought are real. Geometrically, one step of Newton's method consists of replacing the curve $y = P(x)$ by the straight line

$$y = P(x_0) + P'(x_0)(x - x_0),$$

which is tangent to the curve at the point $x = x_0$. Because of this, Newton's method is also called the tangent method. The approximation

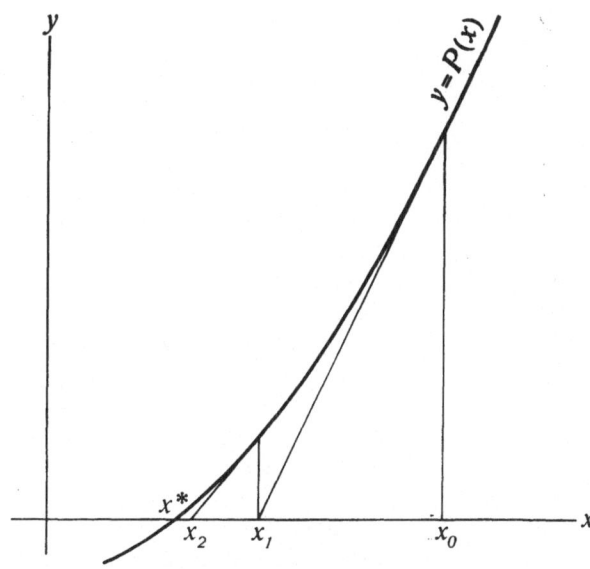

Fig. 7

x_{n+1} is the point of intersection of the tangent line to the curve $y = P(x)$ at the point $x = x_0$ with the x-axis (see fig. 7).

The approximation x'_{n+1} of the modified form of Newton's method is the point of intersection of the line drawn through the point $(x'_n, P(x'_n))$ with inclination $P'(x_0)$ to the x-axis. This line corresponds with the tangent line to the curve $y = P(x)$ only at the first step, for the determination of x'_1 (see fig. 8).

We shall not consider Newton's method for general functional equations, but limit ourselves to the case of systems of equations. Thus, suppose we have the system of equations

$$\left.\begin{array}{l} P_1(\xi_1, \xi_2, \ldots, \xi_k) = 0, \\ \cdot \cdot \cdot \cdot \cdot \cdot \cdot \cdot \cdot \cdot \cdot \cdot \cdot \\ P_k(\xi_1, \xi_2, \ldots, \xi_k) = 0, \end{array}\right\} \tag{6.5}$$

and an initial approximation $x_0 = (\xi_1^{(0)}, \xi_2^{(0)}, \ldots, \xi_k^{(0)})$ to its solution. We shall assume that the functions $P_j(\xi_1, \ldots, \xi_k)$ are real and have continuous partial derivatives in a neighborhood of the initial approximation x_0, which we shall assume to be real, together with the solution sought.

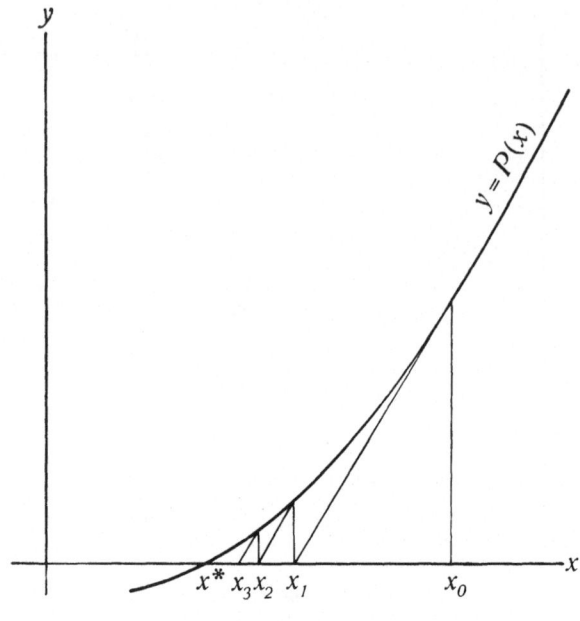

Fig. 8

We shall write the system (6.5) as the single equation

$$P(x) = 0, \tag{6.6}$$

where $P(x)$ is an operator which transforms the column vector

$$x = \begin{bmatrix} \xi_1 \\ \vdots \\ \xi_k \end{bmatrix}$$

into the column vector

$$P(x) = \begin{bmatrix} P_1(\xi_1, \ldots, \xi_k) \\ \cdots\cdots\cdots \\ P_k(\xi_1, \ldots, \xi_k) \end{bmatrix}.$$

The zero on the right-hand side of equation (6.6) denotes the k-dimensional column vector with zero components.

We introduce into consideration the Jacobian matrix of the left-hand side of the system (6.5),

$$P'(x) = \begin{pmatrix} \dfrac{\partial P_1(\xi_1, \ldots, \xi_k)}{\partial \xi_1} & \cdots & \dfrac{\partial P_1(\xi_1, \ldots, \xi_k)}{\partial \xi_k} \\ \cdots & \cdots & \cdots \\ \dfrac{\partial P_k(\xi_1, \ldots, \xi_k)}{\partial \xi_1} & \cdots & \dfrac{\partial P_k(\xi_1, \ldots, \xi_k)}{\partial \xi_k} \end{pmatrix}.$$

We replace system (6.6) by the linear algebraic system

$$P(x_0) + P'(x_0)(x - x_0) = 0. \tag{6.7}$$

We shall assume that the determinant of the matrix $P'(x_0)$ is not equal to zero. Then, the system (6.7) is solvable, and its solution, which is the vector x_1, may be taken as the first approximation to the solution of the system (6.5).

The approximation x_{n+1} is defined by the system

$$P(x_n) + P'(x_n)(x - x_n) = 0. \tag{6.8}$$

The system (6.8) may be written out in detail as:

$$\left. \begin{aligned} P_1(\xi_1^{(n)}, \ldots, \xi_k^{(n)}) + \sum_{j=1}^{k} \frac{\partial P_1(\xi_1^{(n)}, \ldots, \xi_k^{(n)})}{\partial \xi_j} (\xi_j + \xi_j^{(n)}) &= 0, \\ \cdots \cdots \cdots \cdots \cdots \cdots \cdots \cdots \cdots \cdots \\ P_k(\xi_1^{(n)}, \ldots, \xi_k^{(n)}) + \sum_{j=1}^{k} \frac{\partial P_k(\xi_1^{(n)}, \ldots, \xi_k^{(n)})}{\partial \xi_j} (\xi_j - \xi_j^{(n)}) &= 0. \end{aligned} \right\}$$

The successive approximations for system (6.5) by the modified form of Newton's method are defined by the linear algebraic systems

$$P(x_n') + P'(x_0)(x - x_n') = 0. \tag{6.9}$$

If the system (6.5) is linear, then systems (6.8) and (6.9) coincide with it, so that Newton's method does not apply to the solution of linear systems (or linear equations of more general form). We note, however, that the method of iterations is applicable to linear algebraic systems.

We mention the fact that Newton's method for the solution of systems (and more general functional equations) may be considered to be an iterative method which results from writing the system (6.5) in the form

$$x = x - [P'(x)]^{-1} P(x).$$

Here, one assumes that the matrix $P'(x)$ is nonsingular in some region which contains the initial approximation x_0.

7. Theorems on the convergence of Newton's method

We shall prove theorems concerning the ordinary and modified forms of Newton's method for the equation

$$P(x) = 0, \tag{7.1}$$

where the function $P(x)$ is assumed to be differentiable in some neighborhood of the initial approximation x_0 in the complex plane (and, consequently, analytic in this neighborhood).

In the proofs, we shall make use of the following assertion, concerning the convergence of Newton's method for the quadratic equation

$$at^2 + bt + c = 0,$$

where the coefficients a, b, c are real numbers, and $b^2 - 4ac \geq 0$. The sequence $\{t_n\}$ for the ordinary Newton's method, and the sequence $\{t_n'\}$ for the modified form $(n = 0, 1, 2, \ldots; t_0' = t_0)$ converge if the initial approximation t_0 is taken to be any real number different from $-b/2a$, the root of the derivative of the quadratic trinomial. Here, $\{t_n\}$ and $\{t_n'\}$ are monotone approximations to a solution of the equation (the one nearer to t_0) if t_0 is chosen outside of the interval which has the solutions of the equation considered as its end-points. This assertion is perfectly obvious geometrically (see figure 9, where the beginning of the sequence $\{t_n\}$ is shown), and, of course, may be easily proved analytically.

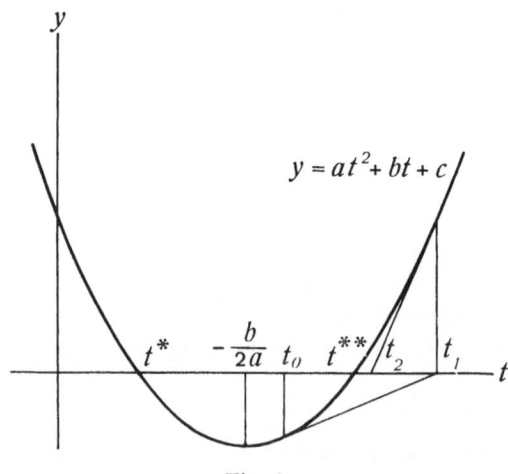

Fig. 9

Theorem 1 (on the convergence of Newton's method). Suppose that these conditions are satisfied for equation (7.1) and the initial approximation x_0:

1) $P'(x_0) \neq 0$, and

$$\frac{1}{|P'(x_0)|} \leqq B; \tag{7.2}$$

2) There holds the inequality

$$\left| \frac{P(x_0)}{P'(x_0)} \right| \leqq \eta; \tag{7.3}$$

3) For any x in the disk

$$|x - x_0| \leqq \delta, \tag{7.4}$$

the second derivative is bounded in modulus:

$$|P''(x)| \leqq K; \tag{7.5}$$

4) The numbers B, η, K are subject to the condition

$$h = BK\eta \leqq \tfrac{1}{2}. \tag{7.6}$$

If

$$\delta \geqq \frac{1 - \sqrt{1 - 2h}}{h} \eta, \tag{7.7}$$

then equation (7.1) has a solution x^* in the disk (7.4), to which the successive approximations

$$x_{n+1} = x_n - \frac{P(x_n)}{P'(x_n)}, \tag{7.8}$$

converge, for which the error estimates

$$|x_n - x^*| \leqq t^* - t_n \tag{7.9}$$

hold, where the t_n are the successive approximations by Newton's method for the quadratic equation

$$f(t) \equiv \tfrac{1}{2}Kt^2 - \frac{1}{B}t + \frac{\eta}{B} = 0, \tag{7.10}$$

with $t_0 = 0$, and t^* is the smaller positive solution of equation (7.10).

Proof. The discriminant of the quadratic trinomial $f(t)$ is equal to

$$\frac{1}{B^2} - 4 \cdot \tfrac{1}{2} K \frac{\eta}{B} = \frac{1}{B^2}(1-2h),$$

and, by virtue of (7.6), is non-negative. Consequently, equation (7.10) has real solutions, and, also, both solutions are easily seen to be positive. The sequence of approximations $\{t_n\}$ ($t_0 = 0, n = 0, 1, 2, \ldots$) obtained by the ordinary form of Newton's method for equation (7.10) converges to the smaller positive solution

$$t^* = \frac{1-\sqrt{1-2h}}{h}\,\eta,$$

of this equation, and

$$t_0 < t_1 < t_2 < \ldots..$$

We note the inequality

$$1 \leqq \frac{1-\sqrt{1-2h}}{h} \leqq 2 \qquad \text{for} \quad 0 \leqq h \leqq \tfrac{1}{2}, \tag{7.11}$$

which follows at once from the equation

$$\frac{1-\sqrt{1-2h}}{h} = \frac{2}{1+\sqrt{1-2h}}.$$

We now prove that

$$|x_{k+1}-x_k| \leqq t_{k+1}-t_k, \tag{7.12}$$

for $k = 0, 1, 2, \ldots$. We shall use mathematical induction. Setting $n = 0$ in (7.8) and using inequality (7.3), we obtain

$$|x_1-x_0| = \left|\frac{P(x_0)}{P'(x_0)}\right| \leqq \eta;$$

since $t_1 = \eta$, consequently

$$|x_1-x_0| \leqq t_1-t_0.$$

Therefore, inequality (7.12) is established for $k = 0$.
It follows, in particular, that x_1 lies in the disk (7.4), so that by virtue

of (7.11) and (7.7),

$$|x_1 - x_0| \leqq \eta \leqq \frac{1 - \sqrt{1 - 2h}}{h} \eta \leqq \delta.$$

We shall now assume that inequality (7.12) is true for $k = 0, 1, \ldots, n-1$, and prove its validity for $k = n$. It follows from the inductive hypothesis that $x_1, \ldots, x_n$ belong to the disk (7.4). In fact,

$$|x_0 - x_k| \leqq |x_0 - x_1| + |x_1 - x_2| + \ldots + |x_{k-1} - x_k| \leqq$$

$$\leqq t_1 + (t_2 - t_1) + \ldots + (t_k - t_{k-1}) \leqq \frac{1 - \sqrt{1 - 2h}}{h} \eta \leqq \delta,$$

for $k = 1, 2, \ldots, n$. We also obtain that

$$|x_0 - x_k| \leqq t_k, \qquad k = 1, 2, \ldots, n. \tag{7.13}$$

On the basis of the definition of the approximation x_n, we have that

$$P(x_n) = P(x_n) - P(x_{n-1}) - P'(x_{n-1})(x_n - x_{n-1}).$$

The term standing on the right-hand side may be represented by the remainder term in the expansion of $P(x)$ about x_{n-1} by Taylor's formula. Writing the integral form of the remainder term, we obtain

$$P(x_n) = \int_{x_{n-1}}^{x_n} P''(z)(x_n - z) \, dz, \tag{7.14}$$

where the integral is taken along the straight line segment $\overline{x_{n-1} x_n}$ joining the points x_{n-1} and x_n. We introduce the change of variable

$$z = x_{n-1} + t(x_n - x_{n-1})$$

in the integral, and obtain

$$P(x_n) = \int_0^1 P''(x_{n-1} + t(x_n - x_{n-1}))(x_n - x_{n-1})^2 (1 - t) \, dt.$$

Since the points x_{n-1} and x_n belong to the disk (7.4), then the segment $\overline{x_{n-1} x_n}$ also belongs to it, and, by condition (7.5),

$$|P''(x_{n-1} + t(x_n - x_{n-1}))| \leqq K.$$

We now estimate the last integral above:

$$|P(x_n)| \leqq K |x_n - x_{n-1}|^2 \int_0^1 (1 - t) \, dt = \tfrac{1}{2} K |x_n - x_{n-1}|^2.$$

Using inequality (7.12) for $k = n-1$, we obtain that

$$|P(x_n)| \leq \tfrac{1}{2}K(t_n - t_{n-1})^2. \tag{7.15}$$

A relationship analogous to (7.14) holds for $f(t_n)$:

$$f(t_n) = \int_{t_{n-1}}^{t_n} f''(t)(t_n - t)\,dt,$$

from which we obtain immediately that

$$f(t_n) = \tfrac{1}{2}K(t_n - t_{n-1})^2,$$

in view of the equation $f''(t) = K$. Comparing the last equation with inequality (7.15), we have that

$$|P(x_n)| \leq f(t_n). \tag{7.16}$$

We shall prove that

$$|P'(x_n)| \geq |f'(t_n)|. \tag{7.17}$$

From the relationship

$$P'(x_n) - P'(x_0) = \int_{x_0}^{x_n} P''(z)\,dz,$$

where the integral is taken along the segment with ends x_0 and x_n belonging to the disk (7.4), we obtain by virtue of the inequality $|P''(z)| \leq K$ that

$$|P'(x_n) - P'(x_0)| \leq K|x_0 - x_n| \leq Kt_n. \tag{7.18}$$

Here, inequality (7.13) has been used.
We have that

$$|P'(x_n)| \geq |P'(x_0)| - |P'(x_n) - P'(x_0)|.$$

Using condition (7.2) and inequality (7.18), we obtain that

$$|P'(x_n)| \geq \frac{1}{B} - Kt_n = |f'(t_n)|,$$

and inequality (7.17) is proved.
Now, it is easy to establish inequality (7.12) for $k = n$. We have

$$|x_{n+1} - x_n| = \left| \frac{P(x_n)}{P'(x_n)} \right|.$$

Since, by (7.16) and (7.17),

$$\left|\frac{P(x_n)}{P'(x_n)}\right| \leq \left|\frac{f(t_n)}{f'(t_n)}\right| = t_{+1} - t_n,$$

we obtain that

$$|x_{n+1} - x_n| \leq t_{n+1} - t_n.$$

It follows from inequality (7.12) that

$$|x_{n+p} - x_n| \leq |x_{n+p} - x_{n+p-1}| + \ldots + |x_{n+1} - x_n| \leq t_{n+p} - t_n, \quad (7.19)$$

for $p = 1, 2, \ldots$. Since the sequence $\{t_n\}$ converges, it follows from this that the sequence $\{x_n\}$ converges. Obviously, the limit x^* of this sequence lies in the disk (7.4) and is a solution of equation (7.1). In order to convince oneself of the latter, it is sufficient to pass to the limit as $n \to \infty$ in inequality (7.16), and take into account the continuity of $P(x)$.

We obtain the error estimate (7.9) by passing to the limit as $p \to \infty$ in inequality (7.19). The theorem is proved in its entirety.

Theorem 2 (on the uniqueness of the solution). Suppose that the hypotheses of theorem 1 are satisfied. For $h < \frac{1}{2}$, the solution x^* of equation (7.1) is unique in the disk (7.4) if

$$\delta < t^{**} = \frac{1 + \sqrt{1 - 2h}}{h} \eta. \qquad (7.20)$$

For $h = \frac{1}{2}$, the uniqueness of x^* in the disk (7.4) holds if

$$\delta = t^{**} = \frac{1 + \sqrt{1 - 2h}}{h} \eta = 2\eta. \qquad (7.21)$$

Proof. Suppose that $h < \frac{1}{2}$. In this case, equation (7.10) has two distinct solutions, t^* and t^{**}. We shall prove that equation (7.1) has a unique solution x^* in the disk (7.4), $|x - x_0| < \delta$.

Suppose that $\tilde{x}$ is some solution of equation (7.1) in the disk (7.4). In view of inequality (7.20),

$$|\tilde{x} - x_0| = \theta t^{**}, \qquad (7.22)$$

where $0 \leq \theta < 1$.

We have that

$$x_1 - \tilde{x} = \frac{1}{P'(x_0)} [P(\tilde{x}) - P(x_0) - P'(x_0)(\tilde{x} - x_0)].$$

The expression in square brackets may be represented by the remainder term in the expansion of $P(x)$ by Taylor's formula. Conducting the argument as in theorem 1, and using the fact that $\tilde{x}$ lies in the disk (7.4), we obtain the estimate

$$|x_1 - \tilde{x}| \leq \frac{1}{|P'(x_0)|} \tfrac{1}{2} K |\tilde{x} - x_0|^2.$$

On the basis of inequalities (7.17) and (7.22), we have that

$$|x_1 - \tilde{x}| \leq \frac{1}{|f'(t_0)|} \tfrac{1}{2} K \theta^2 t^{**2}.$$

Now, obviously,

$$\frac{1}{f'(t_0)} \tfrac{1}{2} K t^{**2} = \frac{1}{f'(t_0)} [f(t^{**}) - f(t_0) - f'(t_0)(t^{**} - t_0)] =$$

$$= -t^{**} + t_0 - \frac{f(t_0)}{f'(t_0)} = t_1 - t^{**},$$

therefore,

$$|x_1 - \tilde{x}| \leq \theta^2 (t^{**} - t_1).$$

In the same way, we obtain that

$$|x_n - \tilde{x}| \leq \theta^{2^n} (t^{**} - t_n). \tag{7.23}$$

Since $0 \leq \theta < 1$, then $|x_n - \tilde{x}| \to 0$ as $n \to \infty$, and $\tilde{x} = x^*$.
If $h = \tfrac{1}{2}$, then equation (7.21) holds, so that θ may be equal to one in (7.22). However, in this case, $t^* = t^{**}$, and $t_n \to t^*$, so that we obtain $|x_n - \tilde{x}| \to 0$ again from (7.23). The theorem is proved.

Theorem 3 (on the convergence of the modified form of Newton's method). Suppose that the hypotheses of theorem 1 are satisfied for equation (7.1), $P(x) = 0$, and the initial approximation x_0. Then, the sequence of approximations

$$x'_{n+1} = x'_n - \frac{P(x'_n)}{P'(x_0)}, \qquad x_0 = x'_0, \qquad n = 0, 1, 2, \ldots, \tag{7.24}$$

converges to a solution x^* of equation (7.1), and, moreover, the error estimate

$$|x'_n - x^*| \leq t^* - t'_n, \tag{7.25}$$

where $\{t'_n\}$ is the sequence of approximations of the modified form of Newton's method for the quadratic equation (7.10), where $t'_0 = 0$, and t^* is the smaller positive root of equation (7.10).

Proof: The sequence of approximations t'_n of the modified form of Newton's method for equation (7.10) converges to t^*, and is monotone increasing:

$$t'_0 = 0 < t'_1 < t'_2 < \ldots$$

We shall establish the inequality

$$|x'_{k+1} - x'_k| \leqq t'_{k+1} - t'_k. \tag{7.26}$$

For $k = 0$, this inequality has already been established in the proof of theorem 1, since $x'_1 = x_1$, $t'_1 = t_1$.

We shall assume that the inequality is true for $k = 0, 1, 2, \ldots, n-1$, and prove that it is valid for $k = n$. As a consequence of the assumption, it follows that

$$|x_0 - x'_k| \leqq t'_k, \qquad k = 1, 2, \ldots, n, \tag{7.27}$$

in particular, x'_k lies in the disk (7.4).

On the basis of the definition of the sequence of approximations (7.24), we have that

$$x'_{n+1} - x'_n = -\frac{1}{P'(x_0)} [P(x'_n) - P(x'_{n-1}) - P'(x_0)(x'_n - x'_{n-1})].$$

If we add and subtract $P'(x'_{n-1})(x'_n - x'_{n-1})$, the quantity in square brackets may be expressed in terms of the remainder in the expansion of $P(x)$ about x'_{n-1} by Taylor's formula:

$$x'_{n+1} - x'_n = -\frac{1}{P'(x_0)} [P(x'_n) - P(x'_{n-1}) - P'(x'_{n-1})(x'_n - x'_{n-1})] -$$

$$-\frac{1}{P'(x_0)} (x'_n - x'_{n-1})[P'(x'_{n-1}) - P'(x_0)]. \tag{7.28}$$

Here, the quantities in square brackets may be represented in the form of the integrals

$$P(x'_n) - P(x'_{n-1}) - P'(x'_{n-1})(x'_n - x'_{n-1}) = \int_{x'_{n-1}}^{x'_n} P''(z)(x'_n - z)dz,$$

$$P'(x'_{n-1}) - P'(x_0) = \int_{x_0}^{x'_{n-1}} P''(z)dz.$$

Since x'_n and x'_{n-1} lie in the disk (7.4) by virtue of (7.27), the inequality $|P''(x)| \leq K$ may be used in the estimation of the integrals, and we obtain that

$$|P(x'_n) - P(x'_{n-1}) - P'(x'_{n-1})(x'_n - x'_{n-1})| \leq \tfrac{1}{2}K|x'_n - x'_{n-1}|^2,$$
$$|P'(x'_{n-1}) - P'(x_0)| \leq K|x'_{n-1} - x_0|.$$

We estimate the left side of the relationship (7.28) by taking into account the preceding inequalities,

$$|x'_{n+1} - x'_n| \leq \frac{1}{|P'(x_0)|} \left[\tfrac{1}{2}K|x'_n - x'_{n-1}|^2 + |x'_n - x'_{n-1}|K|x'_{n-1} - x_0| \right].$$

Now, we employ inequalities (7.26) and (7.27) for $k = n-1$, and obtain

$$|x'_{n+1} - x'_n| \leq \frac{1}{|P'(x_0)|} \left[\tfrac{1}{2}K(t'_n - t'_{n-1})^2 + (t'_n - t'_{n-1})Kt'_{n-1} \right] =$$

$$= \frac{1}{|P'(x_0)|} \cdot \tfrac{1}{2}K(t'^2_n - t'^2_{n-1}).$$

Taking condition (7.2) into account, we may write the last inequality in the form:

$$|x'_{n+1} - x'_n| \leq \tfrac{1}{2}BK(t'^2_n - t'^2_{n-1}). \tag{7.29}$$

From the definition of t'_k, we obtain that

$$t'_k = t'_{k-1} - \frac{f(t'_{k-1})}{f'(t_0)} = t'_{k-1} + B\left[\tfrac{1}{2}Kt'^2_{k-1} - \frac{1}{B}t'_{k-1} + \frac{\eta}{B} \right] =$$

$$= \tfrac{1}{2}BKt'^2_{k-1} + \eta.$$

We write this equation for $k = n+1$ and $k = n$:

$$t'_{n-1} = \tfrac{1}{2}BKt'^2_n + \eta,$$
$$t'_n = \tfrac{1}{2}BKt'^2_{n-1} + \eta,$$

and by subtracting the second equation term-by-term from the first, we obtain

$$t'_{n+1} - t'_n = \tfrac{1}{2}BK(t'^2_n - t'^2_{n-1}).$$

It follows from (7.29) and the previous inequality that inequality (7.26) is valid for $k = n$.

We obtain from inequality (7.26) that

$$|x'_{n+p} - x'_n| \leq t'_{n+p} - t'_n, \tag{7.30}$$

from which the existence of $\lim_{n \to \infty} x'_n$ follows. By passing to the limit as $n \to \infty$ in (7.24), one convinces oneself that the foregoing limit is a solution of equation (7.1), and, by the theorem on uniqueness,

$$\lim_{n \to \infty} x'_n = x^*.$$

We obtain the error estimate (7.25) by passing to the limit as $p \to \infty$ in inequality (7.30). The theorem is proved.

We make some supplementary remarks to the proofs of the theorems. For δ, the radius of the disk (7.4), one may take

$$\delta = 2\eta, \tag{7.31}$$

since condition (7.7) is fulfilled for this δ, by virtue of inequality (7.11). In the case that $P(x)$ and the initial approximation x_0 are real, theorems 1, 2, and 3, as well as their proofs, are preserved if the existence of $P'(x)$ and $P''(x)$ and the satisfaction of inequality (7.5), $|P''(x)| \leq K$, are required on the interval $|x - x_0| \leq \delta$ on the real axis, instead of in the disk (7.4).

We remark that all inequalities obtained in theorems 1, 2, and 3 become exact equalities for the quadratic equation (7.10), $f(t) = 0$. In particular, the error estimates (7.9) and (7.25) are exact.

In order to use the error estimate (7.9) [or (7.25)], it is necessary to solve equation (7.10)

$$f(t) \equiv \tfrac{1}{2}Kt^2 - \frac{1}{B}t + \frac{\eta}{B} = 0,$$

and find the approximation t_n (or t'_n) by Newton's method for equation (7.10).

We introduce the change of variables $t = \eta\tau$ into equation (7.10), and obtain

$$f(\eta\tau) = \frac{\eta}{B}(\tfrac{1}{2}h\tau^2 - \tau + 1) = \frac{\eta}{B}\phi(\tau).$$

We now consider the quadratic equation

$$\phi(\tau) \equiv \tfrac{1}{2}h\tau^2 - \tau + 1 = 0. \tag{7.32}$$

We denote the sequences of approximations by Newton's methods for this equation by $\{\tau_n\}$ and $\{\tau'_n\}$, $\tau_0 = \tau'_0 = 0$. It is easy to verify that

$$t_n = \eta\tau_n; \qquad t'_n = \eta\tau'_n, \qquad n = 0, 1, 2, \ldots. \tag{7.33}$$

Since equation (7.32) depends only on the single parameter h, then one may construct a table of the quantities $\tau^* - \tau_n$ and $\tau^* - \tau'_n$ for $0 \le h \le \frac{1}{2}$ and $n = 0, 1, 2, \ldots$.

The values of $\tau^* - \tau_n$ are entered in table 5 for equally spaced values of the argument h in steps of 0.05, and for $n = 0, 1, 2, 3, 4, 5$. Table 5 is useful for estimating the error $|x_n - x^*|$. In fact, by virtue of (7.33), the estimate (7.9) may be written as

$$|x_n - x^*| \le \eta(\tau^* - \tau_n). \tag{7.34}$$

Table 5

h \ n	0	1	2	3	4	5
0.05	1.026	$0.263 \cdot 10^{-1}$	$0.183 \cdot 10^{-4}$	$0.877 \cdot 10^{-11}$	$0.203 \cdot 10^{-23}$	
0.10	1.056	$0.557 \cdot 10^{-1}$	$0.173 \cdot 10^{-3}$	$0.166 \cdot 10^{-8}$	$0.155 \cdot 10^{-18}$	
0.15	1.089	$0.889 \cdot 10^{-1}$	$0.698 \cdot 10^{-3}$	$0.436 \cdot 10^{-7}$	$0.171 \cdot 10^{-15}$	
0.20	1.127	0.127	$0.202 \cdot 10^{-2}$	$0.525 \cdot 10^{-6}$	$0.356 \cdot 10^{-13}$	
0.25	1.172	0.172	$0.491 \cdot 10^{-2}$	$0.425 \cdot 10^{-5}$	$0.319 \cdot 10^{-11}$	$0.180 \cdot 10^{-23}$
0.30	1.225	0.225	$0.109 \cdot 10^{-1}$	$0.278 \cdot 10^{-4}$	$0.184 \cdot 10^{-9}$	$0.802 \cdot 10^{-20}$
0.35	1.292	0.292	$0.230 \cdot 10^{-1}$	$0.166 \cdot 10^{-3}$	$0.885 \cdot 10^{-8}$	$0.250 \cdot 10^{-16}$
0.40	1.382	0.382	$0.486 \cdot 10^{-1}$	$0.101 \cdot 10^{-2}$	$0.459 \cdot 10^{-6}$	$0.942 \cdot 10^{-13}$
0.45	1.519	0.519	0.110	$0.749 \cdot 10^{-2}$	$0.395 \cdot 10^{-4}$	$0.111 \cdot 10^{-8}$
0.50	2	1	0.5	0.25	0.125	$0.625 \cdot 10^{-1}$

For the modified form of Newton's method, we obtain in place of (7.25) the estimate

$$|x'_n - x^*| \le \eta(\tau^* - \tau'_n). \tag{7.35}$$

We shall explain the character of the rate of convergence of $\tau^* - \tau_n$ to zero as $n \to \infty$. One has that

$$\phi(\tau_n) = \phi(\tau_n) - \phi(\tau_{n-1}) - \phi'(\tau_{n-1})(\tau_n - \tau_{n-1}) =$$
$$= \tfrac{1}{2}\phi''(\xi)(\tau_n - \tau_{n-1})^2 = \tfrac{1}{2}h(\tau_n - \tau_{n-1})^2,$$
$$\phi'(\tau_n) = h\tau_n - 1.$$

From these relationships, one obtains that

$$\tau_{n+1}-\tau_n = -\frac{\phi(\tau_n)}{\phi'(\tau_n)} = \tfrac{1}{2}\frac{h}{1-h\tau_n}(\tau_n-\tau_{n-1})^2. \tag{7.36}$$

Since $\tau_0 = 0$, $\tau_1 = 1$, and $h \leqq \tfrac{1}{2}$, then (7.36) for $n = 1$ gives

$$\tau_2-\tau_1 = \tfrac{1}{2}\frac{h}{1-h} \leqq \tfrac{1}{2}.$$

Obviously,

$$\tau_2 = \tau_1+(\tau_2-\tau_1) \leqq 1+\tfrac{1}{2} = \tfrac{3}{2}.$$

Again from the relationship (7.36), one obtains for $n = 2$ that

$$\tau_3-\tau_2 = \tfrac{1}{2}\frac{h}{1-h_2}(\tau_2-\tau_1)^2 \leqq \tfrac{1}{2}\frac{h}{1-h\frac{3}{2}}(\tfrac{1}{2})^2 \leqq \tfrac{1}{4},$$

and, consequently,

$$\tau_3 = \tau_2+(\tau_3-\tau_2) \leqq \tfrac{3}{2}+\tfrac{1}{4} = \tfrac{7}{4}.$$

Proceeding in the same way, one obtains for any $n = 0, 1, 2, \ldots$ that

$$\tau_n \leqq 2-2^{1-n}. \tag{7.37}$$

The relationship

$$\tau_n-\tau_n^* = \frac{1}{\phi'(\tau_{n-1})}\left[\phi(\tau^*)-\phi(\tau_{n-1})-\phi'(\tau_{n-1})(\tau^*-\tau_{n-1})\right],$$

which is a consequence of the definition of τ_n and the equation $\phi(\tau_n) = 0$, may be represented in the following manner on the basis of Taylor's theorem:

$$\tau^*-\tau_n = \frac{1}{1-h\tau_{n-1}}\tfrac{1}{2}h(\tau_{n-1}-\tau^*)^2.$$

Evidently, by virtue of (7.37), an upper bound for the first factor is:

$$\frac{1}{1-h\tau_{n-1}} \leqq \frac{1}{1-\tfrac{1}{2}(2-2^{2-n})} = 2^{n-1},$$

and thus

$$\tau^*-\tau_n \leqq 2^{n-2}h(\tau_{n-1}-\tau^*)^2. \tag{7.38}$$

Since, by (7.11),

$$\tau^* = \frac{1-\sqrt{1-2h}}{h} \leqq 2,$$

then, for $n = 1$, one obtains from (7.38) that

$$\tau^* - \tau_1 \leqq 2h.$$

Taking this last inequality into account, one obtains again from (7.38) for $n = 2$ that

$$\tau^* - \tau_2 \leqq h(2h)^2 = \tfrac{1}{2}(2h)^3.$$

By continuing the estimation, one obtains that

$$\tau^* - \tau_n \leqq \frac{1}{2^{n-1}}(2h)^{2n-1}. \tag{7.39}$$

This is the inequality which we wanted to establish. It shows that the difference $\tau^* - \tau_n$ tends to zero very rapidly for $2h < 1$: Generally speaking, the error is squared in passing from n to $n+1$. In this case, it is said the convergence is *quadratic*.

We shall now show that if $h < \tfrac{1}{2}$, then $\tau^* - \tau_n'$ goes to zero with the speed of a geometric progression as $n \to \infty$. To this end, we note that τ_n' and τ_{n+1}' are related by the equation

$$\tau_{n+1}' = 1 + \tfrac{1}{2}h\tau_n'^2.$$

The relationship

$$\tau^* = 1 + \tfrac{1}{2}h\tau^{*2}$$

also holds for the root τ^*. Subtracting the first equation term-by-term from the second, we obtain that

$$\tau^* - \tau_{n+1}' = \tfrac{1}{2}h(\tau^* + \tau_n')(\tau^* - \tau_n').$$

Since $\tau_n' < \tau^*$, we get from this that

$$\tau^* - \tau_{n+1}' \leqq h\tau^*(\tau^* - \tau_n'). \tag{7.40}$$

However, we have that

$$h\tau^* = 1 - \sqrt{1-2h} = q,$$

from which $0 < q < 1$ if $0 < h < \tfrac{1}{2}$. The assertion is proved.

Now, from inequality (7.40) we obtain the desired inequality

$$\tau^* - \tau'_{n+1} \leqq q^n(\tau^* - \tau_1) \tag{7.41}$$

where $q = 1 - \sqrt{1 - 2h} < 1$ for $0 < h < \frac{1}{2}$.

Newton's method for general functional equations $P(x) = 0$, where $P(x)$ is a nonlinear operator which is twice differentiable (for example, in the sense of Fréchet), and which maps a Banach (complete normed linear) space X into a Banach space Y, was first proposed and studied in papers by L. V. Kantorovič [10, 11, 12] (see also [13]). In particular, theorems 1, 2, and 3 of this section were established for general functional equations in his papers. We remark that the proofs of the theorems presented in this section are general: They carry over to the case of functional equations almost verbatim [11].

We state, without proof, L. V. Kantorovič's theorem on the convergence of Newton's method for the nonlinear system of equations (6.5):

$$\left.\begin{array}{c} P_1(\xi_1, \xi_2, \ldots, \xi_k) = 0, \\ \cdots\cdots\cdots\cdots\cdots \\ P_k(\xi_1, \xi_2, \ldots, \xi_k) = 0. \end{array}\right\} \tag{7.42}$$

We shall assume that the initial approximation to the solution of the system (7.42),

$$x_0 = (\xi_1^{(0)}, \xi_2^{(0)}, \ldots, \xi_k^{(0)}), \tag{7.43}$$

is a vector with real components. In a neighborhood of the initial approximation (7.43) to be specified later, it is assumed that the functions $P_j(\xi_1, \xi_2, \ldots, \xi_k)$, $j = 1, 2, \ldots, k$, are real and twice continuously differentiable.

In order to make the formulation of the theorem more compact, use will be made of the concept of the *norm* of a matrix. Suppose that $A = (a_{ij})$ is some matrix. The norm of the matrix A will be defined to be the number

$$\|A\| = \max_{1 \leqq i \leqq k} \sum_{j=1}^{k} |a_{ij}|.$$

The Jacobian matrix of the left side of the system (7.42) will be denoted by $P'(x)$, as in § 6, so that the element of the matrix $P'(x)$ with indices i and j is

$$\{P'(x)\}_{ij} = \frac{\partial P_i(\xi_1, \xi_2, \ldots, \xi_k)}{\partial \xi_j}.$$

Theorem 4. Suppose that the system (7.42) and the initial approximation (7.43) to its solution satisfy these conditions:

1) The Jacobian matrix $P'(x_0)$ of the left side of system (7.42) is non-singular at the initial approximation, and a bound $\|[P'(x_0)]^{-1}\| \leq B$ is known for the norm of its inverse;

2) At the initial approximation, the left side of the system statisfies the inequality

$$\max_{1 \leq i \leq k} |P_i(\xi_1^{(0)}, \xi_2^{(0)}, \ldots, \xi_k^{(0)})| \leq \eta;$$

3) For all points in the k-dimensional cube,

$$|\xi_i - \xi_i^{(0)}| \leq 2B\eta, \qquad i = 1, 2, \ldots, k, \tag{7.44}$$

the inequality

$$\left| \frac{\partial^2 P_i(\xi_1, \xi_2, \ldots, \xi_k)}{\partial \xi_p \partial \xi_q} \right| \leq L, \qquad i, p, q = 1, 2, \ldots, k,$$

holds;

4) The constants B, η, L satisfy the condition

$$h = B^2 \eta L k^2 \leq \tfrac{1}{2}.$$

Then, the system (7.42) has a solution $x^* = (\xi_1^*, \xi_2^*, \ldots, \xi_k^*)$ in the cube

$$|\xi_i - \xi_i^{(0)}| \leq \frac{1 - \sqrt{1 - 2h}}{h} B\eta, \qquad i = 1, 2, \ldots, k,$$

which is unique in the cube (7.44), and to which converge the approximations

$$x_n = (\xi_1^{(n)}, \xi_2^{(n)}, \ldots, \xi_k^{(n)}), \qquad n = 0, 1, 2, \ldots,$$

of the ordinary form of Newton's method defined by the system (6.8), for which the error estimate

$$\max_{1 \leq i \leq k} |\xi_i^{(k)} - \xi_i^*| \leq \frac{1}{2^{n-1}} (2h)^{2n-1} B\eta$$

holds. The sequence of approximations x_n' of the modified form of Newton's method defined by the system (6.9) also converges to the solution x^*.

8. Remarks on the practical application of Newton's method

The calculation of a solution of the equation $P(x) = 0$ by Newton's method may be laid out according to the following diagram:

n	0	1	2
x_n	x_0	$x_1 = x_0 + \Delta x_0$	
$P(x_n)$	$P(x_0)$		
$P'(x_n)$	$P'(x_0)$		
$\Delta x_n = \dfrac{-P(x_n)}{P'(x_n)}$	Δx_0		

If the calculation of a complex solution is being pursued, it is convenient to employ a diagram in which only real numbers are entered:

n	0		1	2
$x_n = u_n + iv_n$	u_0	v_0	$u_1 = u_0 + \Delta u_0$	$v_1 = v_0 + \Delta v_0$
$P(x_n) = a_n + ib_n$	a_0	b_0		
$P'(x_n) = c_n + id_n$	c_0	d_0		
$P(x_n)\overline{P'(x_n)}$	$a_0 c_0 + b_0 d_0$	$b_0 c_0 - a_0 d_0$		
$P'(x_n)\overline{P'(x_n)}$	$c_0^2 + d_0^2$			
$\Delta x_n = -\dfrac{P\overline{P'}}{P'\overline{P'}}$	$\Delta u_0 =$ $= -\dfrac{a_0 c_0 + b_0 d_0}{c_0^2 + d_0^2}$	$\Delta v_0 =$ $= -\dfrac{b_0 c_0 - a_0 d_0}{c_0^2 + d_0^2}$		

The computational pattern may be simplified in the case of the modified form of Newton's method.

It is necessary to continue the computation until two successive approximations x_k and x_{k+1} coincide to the required number of places. Before calculating, it is useful to convince oneself that the sequence $\{x_n\}$ (or $\{x_n'\}$) will converge to the solution desired. For this purpose, it is sufficient to verify that the hypotheses of theorem 1 of § 7 are fulfilled.

If $P(x)$ is a polynomial, then the calculation of $P(x_n)$ and $P'(x_n)$ may be carried out according to the methods in § 5 for x_n real or complex. Here, it is not necessary to actually differentiate the polynomial.

We remark that some books on computational methods (see, for example, [33], p. 106) make the unfounded assertion that Newton's method is not appropriate for the calculation of complex solutions of algebraic equations, or that it is less efficient than factorization methods.

We shall indicate a sufficient condition for the convergence of Newton's method for a real equation $P(x) = 0$ and real x_0. Obviously, Newton's method will only lead to a real solution of the equation in this case. We shall assume that an interval $[a, b]$ exists for which

$$P(a)P(b) < 0,$$

and that

$$P'(x) \quad \text{and} \quad P''(x)$$

do not change sign on $[a, b]$. If, for example,

$$P(b)P''(b) > 0,$$

then for the initial values $x_0 = x_0' = b$, the sequences $\{x_n\}$ and $\{x_n'\}$

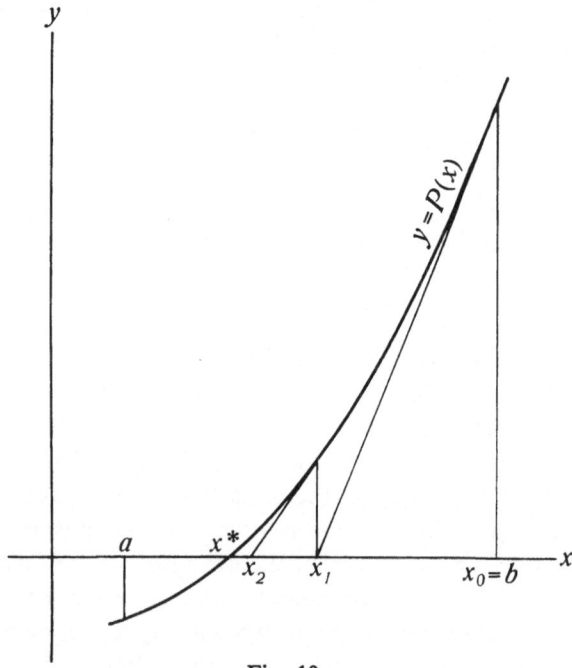

Fig. 10

converge to the unique solution x^* of the equation $P(x) = 0$ in the interval $[a, b]$. This proposition is obvious geometrically. The case that $P'(x) > 0$, $P''(x) > 0$ on $[a, b]$, and $P(b)P''(b) > 0$ is depicted in fig. 10. We mention what we shall call the combined method of tangents and secants, which is applicable under the conditions indicated above. We shall limit ourselves to a geometrical discussion of the method (see fig. 11).

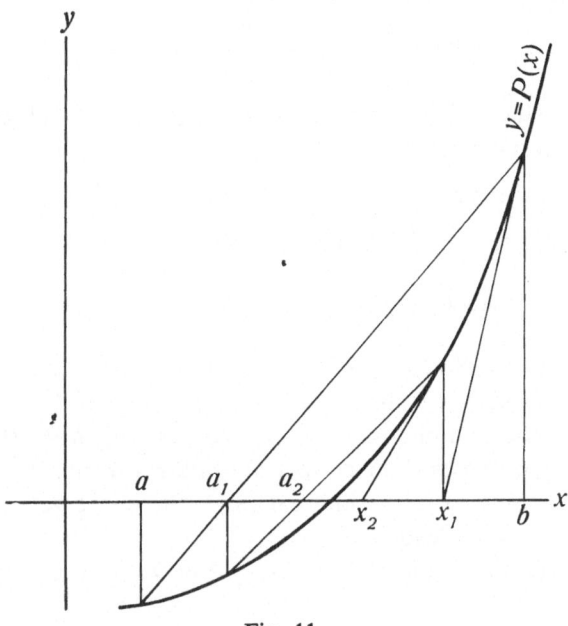

Fig. 11

At the first step, one obtains the approximation a_1 by the method of secants, and the approximation x_1 by Newton's method. In exactly the same way, the second step furnishes a_2 and x_2, etc. Clearly, the solution x^* lies between a_n and x_n. From the figure,

$$a_1 < a_2 < x^* < x_2 < x_1.$$

Consequently, the combined method permits one to obtain an estimate of the error $|x_n - x^*|$.

We give some numerical examples.

Example 1. We shall compute the smallest positive solution of the equation

$$P(x) \equiv x - \tan x = 0$$

by Newton's method. The initial approximation $x_0 = 4.5$ was obtained in § 1. Obviously, $P'(x) = -\tan^2 x$. The five place mathematical tables by B. I. Segal and K. A. Semendjaev [24] were used for the calculations. We shall apply theorem 1 of § 7. By computation,

$$P(x_0) = -0.1373, \quad P'(x_0) = -21.505;$$

and one may choose for η and B the values

$$\left| \frac{P(x_0)}{P'(x_0)} \right| < \eta = 0.0064, \quad B = \frac{1}{21.5} \cong 0.047.$$

The absolute value of $P''(x)$ is to be estimated on the interval $|x - x_0| \leqq 2\eta$; in our case, on the interval

$$|x - 4.5| \leqq 0.0128. \tag{8.1}$$

It is not difficult to verify that the function

$$P''(x) = -2 \tan x \frac{1}{\cos^2 x}$$

attains its value on (8.1) which is greatest in absolute value at the right end of the interval, that is, for $x = 4.5128$. This value of $P''(x)$ is equal to 251.51, so that we may choose $K = 252$. The quantity $h = BK\eta - 0.076 < \frac{1}{2}$, and all hypotheses of theorem 1 are fulfilled. From this follow the existence of a solution in the interval (8.1), the convergence of the sequence $\{x_n\}$, and error estimates.

For error estimation, table 5 will be used. For $h = 0.1$, $\tau^* - \tau_1 = 0.056$, so that

$$|x_1 - x^*| < 0.0064 \cdot 0.056 < 0.00036.$$

Therefore, x_1 differs from the solution by no more than four units in the fourth decimal place. Analogously, for x_2 we obtain that

$$|x_2 - x^*| < 0.12 \cdot 10^{-5},$$

that is, x_2 differs from the solution by no more than two units in the sixth decimal place.

It should be noted that in the calculation, we do not obtain x_2, but rather some $\tilde{x}_2$ is obtained instead (see §3). The error in $\tilde{x}_2$ (as an approximation to x^*) is not obliged to obey the last inequality, as we used five-place tables, and carried out the calculations to five significant figures.

The results of the calculations are shown in table 6.

Table 6

n	0	1	2	3
x_n	4.5	4.4936	4.4934	4.4934
$\tan x_n$	4.6373	4.4975	4.4932	
$P'(x_n) = -\tan^2 x_n$	-21.505	-20.228	-20.189	
$P(x_n) = x_n - \tan x_n$	-0.1373	-0.0041	0.0002	
$\Delta x_n = -\dfrac{P(x_n)}{P'(x_n)}$	-0.0064	-0.0002	0.0000	

Example 2. We shall calculate a complex solution of the equation

$$P(x) \equiv x^4 - 2.7x^3 + 4x^2 - 3.3x + 1 = 0,$$

choosing $x_0 = 0.52 + 1.16i$ as the initial approximation.
First, we shall verify the satisfaction of the hypotheses of theorem 1 of § 7.
Calculation yields

$$\left| \frac{P(x_0)}{P'(x_0)} \right| = 0.02385, \qquad \frac{1}{|P'(x_0)|} = 0.29062,$$

so that one may choose

$$\eta = 0.024, \qquad B = 0.3.$$

Now, we shall bound $|P''(x)|$ from above in the disk $|x - x_0| \leq 2\eta$, that is, in our case, the disk

$$|x - x_0| \leq 0.048. \tag{8.2}$$

For this purpose, we use the values of the derivatives of $P(x)$, computed as in § 5 (example 4):

$$P''(x_0) = -13.3264 - 4.3152i,$$
$$P'''(x_0) = -3.72 + 27.84i,$$
$$P^{(IV)}(x_0) = 24.$$

The moduli of the derivatives are equal to

$$|P''(x_0)| = 14.0076, |P'''(x_0)| = 28.087, |P^{(IV)}(x_0)| = 24.$$

The exact value of the modulus will not be given, but rather, an upper bound for it. We expand $P''(x)$ into a Taylor series

$$P''(x) = P''(x_0) + P'''(x_0)(x - x_0) + \tfrac{1}{2}P^{(IV)}(x_0)(x - x_0)^2,$$

from which we obtain the inequality

$$|P''(x)| \leq 14.0076 + 28.078 \cdot 0.048 + 12 \cdot 0.048^2 = 15.383,$$

for x in the disk (8.2). Consequently, we may take $K = 16$.
We have that

$$h = BK\eta = 0.3 \cdot 16 \cdot 0.024 = 0.12 < 0.5.$$

All of the hypotheses of the theorem are satisfied, so that there exists a solution of the equation in the neighborhood of x_0 to which the sequence of approximations x_n converges. We shall estimate the error of the approximation x_2. With the aid of table 5, we obtain that $\tau^* - \tau_2 = 0.21 \cdot 10^{-3}$ (for $h = 0.12$), from which

$$|x_2 - x^*| \leq 0.024 \cdot 0.21 \cdot 10^{-3} < 0.6 \cdot 10^{-5}. \tag{8.3}$$

It is apparent from the estimate (8.3) that it would be of no value to calculate further approximations to five decimal places.

Table 7

		1	−2.7	4	−3.3	1		
0.52	1.16	1	−1.66	0.6576	0.06646	0.00644	−0.02812	0.07709
−1.04	1.616	1	−0.62	−1.6032	−1.2808	−0.7192	1.73500	−2.97146
0.54347	1.15576	1	−1.61306	0.61556	0.00020	−0.00385	−0.00396	0.00023
−1.08694	1.63114	1	−0.52612	−1.58744	−1.30151	−0.60807	1.40577	−3.00847
0.54404	1.15681	1	−1.61192	0.61191	−0.00001	0.00001	0.00002	−0.00001
−1.08808	1.63419							

The calculations are displayed in tables 7 and 8. In table 7, the values of $P(x_n)$ and $P'(x_n)$ are calculated according to the method given in § 5.

Table 8

n	0		1		2	
x_n	0.52	1.16	0.54347	1.15576	0.54404	1.15681
$P(x_n)$	−0.02812	0.07709	−0.00396	0.00023	0.00002	−0.00001
$P'(x_n)$	1.73500	−2.97146	1.40577	−3.00847		
$P\overline{P}'$	−0.27786	0.05019	−0.00626	−0.01159		
$P'\overline{P}'$	11.840		11.027			
Δx_n	0.02347	−0.00424	0.00057	0.00105		

Example 3. We seek the solution of the system

$$P(\xi, \eta) \equiv \xi^3 + \eta^3 - 4 = 0, \\ Q(\xi, \eta) \equiv \xi^4 + \eta^2 - 3 = 0, \Big\} \qquad (8.4)$$

which is located in the first quadrant of the coordinate plane. The initial approximation

$$\xi_0 = 1, \qquad \eta_0 = 1.4,$$

we obtained in § 1 (example 5).

In order to obtain the first approximation (ξ_1, η_1) by Newton's method [see (6.7)], it is necessary to solve the linear algebraic system:

$$P(\xi_0, \eta_0) + \frac{\partial P(\xi_0, \eta_0)}{\partial \xi} (\xi - \xi_0) + \frac{\partial P(\xi_0, \eta_0)}{\partial \eta} (\eta - \eta_0) = 0, \\ Q(\xi_0, \eta_0) + \frac{\partial Q(\xi_0, \eta_0)}{\partial \xi} (\xi - \xi_0) + \frac{\partial Q(\xi_0, \eta_0)}{\partial \eta} (\eta - \eta_0) = 0. \Bigg\} \qquad (8.5)$$

This system (and, likewise, the systems defining subsequent approximations) may be solved by means of determinants. We remark that the numerical solution of linear algebraic systems by means of determinants is convenient only for systems of two equations. In the case of systems with a larger number of equations, considerably more efficient methods, for example, Gaussian elimination, are used.

The results of the computations are displayed in tables 9 and 10. The coefficients and right-hand sides of the linear algebraic systems from which the pairs $(\xi_1 - \xi_0, \eta_1 - \eta_0)$, $(\xi_2 - \xi_1, \eta_2 - \eta_1)$, ... are determined to give the successive approximations (ξ_1, η_1), (ξ_2, η_2), ... are entered in the first two rows of table 10. The determinants occupy the third row, and the values of the pairs obtained are entered in the fourth row.

9. Lobačevskiĭ's method *

In contrast to the methods considered previously, which may be applied to transcendental as well as algebraic equations, Lobačevskiĭ's method is only applicable to finding roots of algebraic equations

$$P(x) \equiv a_0 x^n + a_1 x^{n-1} + \ldots + a_{n-1} x + a_n = 0. \qquad (9.1)$$

---

* Translator's note: This method is almost universally called Graeffe's root-squaring method in the Western literature.

Table 9

m	ξ_0^m	η_0^m	ξ_1^m	η_1^m	ξ_2^m	η_2^m	ξ_3^m	η_3^m
1	1	1.4	0.96815	1.45979	0.96764	1.45716	0.96764	1.45715
2	1	1.96	0.93731	2.13099	0.93633	2.12332		2.12329
3	1	2.744	0.90746	3.11080	0.90603	3.09402		3.09395
4	1		0.87855		0.87671			

Table 10

	$\left(\dfrac{\partial}{\partial\xi}\right)_0$	$\left(\dfrac{\partial}{\partial\eta}\right)_0$	$-(\ )_0$	$\left(\dfrac{\partial}{\partial\xi}\right)_1$	$\left(\dfrac{\partial}{\partial\eta}\right)_1$	$-(\ )_1$	$\left(\dfrac{\partial}{\partial\xi}\right)_2$	$\left(\dfrac{\partial}{\partial\eta}\right)_2$	$-(\ )_2$	$-(\ )_3$
P	3	5.88	0.256	2.81193	6.39297	−0.01826	2.80899	6.36996	−0.00005	0.00002
	0.4816	−0.904	−15.12	0.00768	0.03946	−14.996	0.00005	0.00010	−14.899	
Q	4	2.8	0.04	3.62984	2.91958	−0.00954	3.62412	2.91432	−0.00003	0.00000
	−0.03185	0.05979		−0.00051	−0.00263		0.00000	−0.00001		

It will be assumed that the coefficients $a_0, a_1, \ldots, a_n$ are real, and that $a_0 \neq 0$.

In order to apply Lobačevskiĭ's method, it is not required that an initial approximation be known, so that this method may be used to find such initial approximations.

We shall explain the method which makes it possible to construct a polynomial of degree n with roots which are the squares of the roots of equation (9.1). We denote the solutions of equation (9.1) by $x_1, x_2, \ldots, x_n$, and obtain that

$$P(x) = a_0(x-x_1)(x-x_2)\ldots(x-x_n). \tag{9.2}$$

Consider the polynomial

$$P^*(x) = a_0 x^n - a_1 x^{n-1} + a_2 x^{n-2} - \ldots + (-1)^n a_n. \tag{9.3}$$

Clearly, the roots of this polynomial are

$$-x_1, -x_2, \ldots, -x_n,$$

so that

$$P^*(x) = a_0(x+x_1)(x+x_2)\ldots(x+x_n). \tag{9.4}$$

It follows from (9.2) and (9.4) that

$$P(x)P^*(x) = a_0^2(x^2-x_1^2)(x^2-x_2^2)\ldots(x^2-x_n^2).$$

We introduce the change of variable $x^2 = y$. Then, we obtain the required polynomial of degree n in y:

$$\tilde{P}_1(y) = P(\sqrt{y})P^*(\sqrt{y}) = a_0^2(y-x_1^2)(y-x_2^2)\ldots(y-x_n^2), \tag{9.5}$$

the roots of which are the squares of the roots of the polynomial (9.1). The polynomial $\tilde{P}_1(y)$ may be written in the form *

$$\tilde{P}_1(y) = a_0^{(1)}y^n - a_1^{(1)}y^{n-1} + a_2^{(1)}y^{n-2} - \ldots + (-1)^n a_n^{(1)},$$

and its coefficients obtained by making use of the relationship

$$a_0^{(1)}y^n - a_1^{(1)}y^{n-1} + a_2^{(1)}y^{n-2} - \ldots + (-1)^n a_n^{(1)} =$$
$$= [a_0(\sqrt{y})^n + a_1(\sqrt{y})^{n-1} + a_2(\sqrt{y})^{n-2} + \ldots + a_{n-1}\sqrt{y} + a_n] \times$$
$$\times [a_0(\sqrt{y})^n - a_1(\sqrt{y})^{n-1} + a_2(\sqrt{y})^{n-2} - \ldots + (-1)^{n-1}a_{n-1}\sqrt{y} + a_n],$$

---

* The coefficients of the polynomial $\tilde{P}_1(y)$ are denoted by $a_0^{(1)}, -a_1^{(1)}, \ldots, (-1)^n a_n^{(1)}$ so that a uniform system for calculating the numbers $a_0^{(1)}, a_1^{(1)}, \ldots, a_n^{(1)}$ may be obtained.

which is a consequence of (9.5), (9.1), and (9.3). One obtains that

$$a_0^{(1)} = a_0^2,$$
$$a_1^{(1)} = a_1^2 - 2a_0 a_2,$$
$$a_2^{(1)} = a_2^2 - 2a_1 a_3 + 2a_0 a_4,$$
$$a_3^{(1)} = a_3^2 - 2a_2 a_4 + 2a_1 a_5 - 2a_0 a_6, \tag{9.6}$$
$$\cdots \cdots \cdots \cdots \cdots \cdots$$
$$a_{n-1}^{(1)} = a_{n-1}^2 - 2a_{n-2} a_n,$$
$$a_n^{(1)} = a_n^2.$$

Obviously, the polynomial

$$P_1(y) = a_0^{(1)} y^n + a_1^{(1)} y^{n-1} + a_2^{(1)} y^{n-2} + \ldots + a_n^{(1)}$$

has the property that its roots are equal to the negatives of the squares of the solutions of equation (9.1). In order to convince oneself of this, it is sufficient to compare the polynomials $P_1(y)$ and $\tilde{P}_1(y)$.

Applying the process described to the polynomial $P_1(x)$, one obtains the polynomial $P_2(x)$, which has the roots

$$-x_1^4, \ -x_2^4, \ldots, \ -x_n^4.$$

Continuing in the same way, at the kth step, one obtains the polynomial

$$P_k(x) = a_0^{(k)} x^n + a_1^{(k)} x^{n-1} + \ldots + a_n^{(k)}, \tag{9.7}$$

the roots of which are

$$-x_1^m, \ -x_2^m, \ldots, \ -x_n^m,$$

where $m = 2^k$.

The following relationships hold,

$$\left. \begin{array}{r} x_1^m + x_2^m + \ldots + x_n^m = a_1^{(k)}/a_0^{(k)}, \\ x_1^m x_2^m + x_1^m x_3^m + \ldots + x_{n-1}^m x_n^m = a_2^{(k)}/a_0^{(k)}, \\ x_1^m x_2^m x_3^m + \ldots + x_{n-2}^m x_{n-1}^m x_n^m = a_3^{(k)}/a_0^{(k)}, \\ \cdots \cdots \cdots \cdots \cdots \cdots \\ x_1^m x_2^m \ldots x_n^m = a_n^{(k)}/a_0^{(k)}, \end{array} \right\} \tag{9.8}$$

as may be obtained by equating the coefficients of like powers of x in the identity

$$a_0^{(k)} x^n + a_1^{(k)} x^{n-1} + \ldots + a_n^{(k)} = a_0^{(k)}(x + x_1^m)(x + x_2^m) \ldots (x + x_n^m).$$

Suppose now that the solutions of equation (9.1) are real and distinct in absolute value. We shall assume that they are numbered in such a way as to satisfy the inequality

$$|x_1| > |x_2| > |x_3| > \ldots > |x_n|. \tag{9.9}$$

Clearly, by virtue of inequality (9.9), the left sides of equations (9.8) are dominated by the first term for sufficiently large k, and we obtain the approximate equations

$$\left.\begin{aligned}
x_1^m &\cong a_1^{(k)}/a_0^{(k)}, \\
x_1^m x_2^m &\cong a_2^{(k)}/a_0^{(k)}, \\
x_1^m x_2^m x_3^m &\cong a_3^{(k)}/a_0^{(k)}, \\
&\cdots\cdots\cdots \\
x_1^m x_2^m \ldots x_n^m &\cong a_n^{(k)} a_0^{(k)},
\end{aligned}\right\} \tag{9.10}$$

from which we obtain sequentially that

$$\left.\begin{aligned}
x_1^m &\cong a_1^{(k)}/a_0^{(k)}, \\
x_2^m &\cong a_2^{(k)}/a_1^{(k)}, \\
&\cdots\cdots\cdots \\
x_n^m &\cong a_n^{(k)}/a_{n-1}^{(k)}.
\end{aligned}\right\} \tag{9.11}$$

Equations (9.11) permit one to determine the absolute values $|x_1|$, $|x_2|, \ldots, |x_n|$ of the solutions of equation (9.1). The signs of the solutions may be determined by substitution into the equation.

Thus, the basic labor in the application of Lobačevskiĭ's method consists of the calculation of the coefficients

$$a_0^{(k)}, a_1^{(k)}, a_2^{(k)}, \ldots, a_n^{(k)}, \qquad k = 1, 2, 3, \ldots.$$

It is convenient to lay out the calculation in the form:

k	$a_0^{(k)}$	$a_1^{(k)}$	$a_2^{(k)}$	$\ldots$	$a_{n-1}^{(k)}$	$a_n^{(k)}$
0	a_0	a_1	a_2	$\ldots$	a_{n-1}	a_n
	a_0^2	a_1^2 $-2a_0a_2$	a_2^2 $-2a_1a_3$ $2a_0a_4$		a_{n-1}^2 $-2a_{n-2}a_n$	a_n^2
1	$a_0^{(1)}$	$a_1^{(1)}$	$a_2^{(1)}$	$\ldots$	$a_{n-1}^{(1)}$	$a_n^{(1)}$
	$a_0^{(1)^2}$	$a_1^{(1)^2}$ $-2a_0^{(1)}a_2^{(1)}$	$a_2^{(1)^2}$ $-2a_1^{(1)}a_3^{(1)}$ $2a_0^{(1)}a_4^{(1)}$	$a_{n-1}^{(1)^2}$ $-2a_{n-2}^{(1)}a_n^{(1)}$		$a_n^{(1)^2}$
2	$a_0^{(2)}$	$a_1^{(2)}$	$a_2^{(2)}$	$\ldots$	$a_{n-1}^{(2)}$	$a_1^{(2)}$

It still remains to discuss the question of the value of k at which to stop the computation, in other words, how to find out if equations (9.10) hold to an acceptable degree of accuracy. In order to determine this, one assumes that equations (9.10) hold to the required degree of accuracy for some $k = k_0$:

$$x_1^{m_0} = a_1^{(k_0)}/a_0^{(k_0)}, \quad x_1^{m_0}x_2^{m_0} = a_2^{(k_0)}/a_0^{(k_0)}, \quad x_1^{m_0}x_2^{m_0}x_3^{m_0} = a_3^{(k_0)}/a_0^{(k_0)}, \cdots. \tag{9.12}$$

Here, $m_0 = 2^{k_0}$. For $k = k_0 + 1$, equations (9.10) are even more accurate.

$$x_1^{2m_0} = a_1^{(k_0+1)}/a_0^{(k_0+1)}, \quad x_1^{2m_0}x_2^{2m_0} = a_2^{(k_0+1)}/a_0^{(k_0+1)},$$
$$x_1^{2m_0}x_2^{2m_0}x_3^{2m_0} = a_3^{(k_0+1)}/a_0^{(k_0+1)}, \ldots \tag{9.13}$$

Since $a_0^{(k_0+1)} = [a_0^{(k_0)}]^2$, we obtain from (9.12) and (9.13) that

$$a_1^{(k_0+1)} = [a_1^{(k_0)}]^2, \quad a_2^{(k_0+1)} = [a_2^{(k_0)}]^2, \quad a_3^{(k_0+1)} = [a_3^{(k_0)}]^2, \ldots$$

Consequently, the computation may be stopped when one obtains a polynomial $P_{k_0+1}(x)$ with coefficients which are, with acceptable accuracy the squares of the coefficients of the polynomial $P_{k_0}(x)$ obtained at the previous step.

We consider now the more general case that all of the solutions of equation (9.1) are real, but that some of them may be equal (or nearly equal) in absolute value. Suppose, for example, that the two solutions x_2 and x_3 are equal in absolute value, so that

$$|x_1| > |x_2| = |x_3| > |x_4| > \ldots > |x_n|. \tag{9.14}$$

It follows from the relationships (9.8) and (9.14) that for sufficiently large $m = 2^k$, the following approximate equations are valid:

$$\left.\begin{aligned}
x_1^m &\cong a_1^{(k)}/a_0^{(k)}, \\
2x_1^m x_2^m &\cong a_2^{(k)}/a_0^{(k)}, \\
x_1^m x_2^{2m} &\cong a_3^{(k)}/a_0^{(k)}, \\
x_1^m x_2^{2m} x_4^m &\cong a_4^{(k)}/a_0^{(k)}, \\
&\cdots\cdots\cdots\cdots\cdots \\
x_1^m x_2^{2m} x_4^m \ldots x_n^m &\cong a_n^{(k)}/a_0^{(k)}.
\end{aligned}\right\} \tag{9.15}$$

From these, we obtain that

$$x_1^m \cong a_1^{(k)}/a_0^{(k)}, \quad x_2^{2m} \cong a_3^{(k)}/a_1^{(k)}, \quad x_4^{(m)} \cong a_4^{(k)}/a_3^{(k)}, \ldots$$

The absolute values of the solutions may be determined from these equations.

In order to find out to which value of k to carry out the computation, we assume that equations (9.15) are satisfied to an acceptable degree of accuracy for $k = k_0$. As in the previous case, by writing these equations for $k = k_0 + 1$, we convince ourselves of the validity of the equations

$$a_i^{(k_0+1)} = [a_i^{(k_0)}]^2, \quad i = 0, 1, 3, 4, \ldots, n, \quad a_2^{(k_0+1)} = \tfrac{1}{2}[a_2^{(k_0)}]^2.$$

Satisfaction of these equations serves as a criterion for stopping the computation.

Suppose that equation (9.1) has the complex solutions

$$x_2 = re^{i\phi}, \quad x_3 = re^{-i\phi},$$

and that the inequality

$$|x_1| > r > |x_4| > \ldots > |x_n| \tag{9.16}$$

holds. The solutions $x_1, x_4, x_5, \ldots, x_n$ are assumed to be real. By virtue of the equations

$$x_2^m + x_3^m = 2r^m \cos m\phi, \quad x_2^m x_3^m = r^{2m},$$

the relationships (9.8) may be written as follows:

$$\left.\begin{aligned}
x_1^m + 2r^m \cos m\phi + x_4^m + \ldots + x_n^m &= a_1^{(k)}/a_0^{(k)}, \\
2x_1^m r^m \cos m\phi + x_1^m x_4^m + \ldots &= a_2^{(k)}/a_0^{(k)}, \\
x_1^m r^{2m} + x_1^m x_2^m x_4^m + \ldots &= a_3^{(k)}/a_0^{(k)}, \\
x_1^m r^{2m} x_4^m + \ldots &= a_4^{(k)}/a_0^{(k)}, \\
\cdots\cdots\cdots\cdots\cdots\cdots\cdots & \\
x_1^m r^{2m} x_4^m \ldots x_n^m &= a_n^{(k)}/a_0^{(k)}.
\end{aligned}\right\} \tag{9.17}$$

For $m = 2^k$ sufficiently large, the first term on the left side dominates in every equation but the second, and we obtain the approximate equations

$$\left.\begin{aligned}
x_1^m &\cong a_1^{(k)}/a_0^{(k)}, \\
x_1^m r^{2m} &\cong a_3^{(k)}/a_0^{(k)}, \\
x_1^m r^{2m} x_4^m &\cong a_4^{(k)}/a_0^{(k)}, \\
\cdots\cdots\cdots\cdots\cdots & \\
x_1^m r^{2m} x_4^m \ldots x_n^m &\cong a_n^{(k)}/a_0^{(k)}.
\end{aligned}\right\}$$

From these, we get the relationships

$$x_1^m \cong a_1^{(k)}/a_0^{(k)}, \qquad r^{2m} \cong a_3^{(k)}/a_1^{(k)},$$

$$x_4^m \cong a_4^{(k)}/a_3^{(k)}, \ldots, \qquad x_n^m \cong a_n^{(k)}/a_n^{(k)}/a_{n-1}^{(k)},$$

which permit the determination of the absolute values $|x_1|, |x_4|, \ldots, |x_n|$
of the real solutions, and the modulus $r = |x_2| = |x_3|$ of the complex
solutions.

Following exactly the same line of reasoning as in the case of real solutions, we convince ourselves that the computation may be stopped when
the relationships

$$a_1^{(k_0+1)} = [a_1^{(k_0)}]^2, \qquad a_3^{(k_0+1)} = [a_3^{(k_0)}]^2, \ldots, \qquad a_n^{(k_0+1)} = [a_n^{(k_0)}]^2$$

are satisfied, that is, when all of the coefficients of $P_{k_0+1}(x)$ except $a_2^{k_0+1}$
are equal to the squares of the corresponding coefficients of $P_{k_0}(x)$ to an
acceptable degree of accuracy. The behavior of the coefficient $a_2^{(k)}$ is
irregular, because of the presence of the term

$$2x_1^m r^m \cos m\phi$$

in the second equation of (9.17). It may change sign as k varies, a circumstance which indicates computationally that there are complex solutions.
We have obtained $|x_1|, |x_4|, \ldots, |x_n|$, and r. Assume that $x_1, x_4, \ldots, x_n$
have been determined (for example, by substitution into the equation).
In order to obtain x_2, the relationship

$$x_1 + x_2 + x_3 + x_4 + \ldots + x_n = -a_1/a_0,$$

or

$$x_1 + 2r \cos \phi + x_4 + \ldots + x_n = -a_1/a_0,$$

is used, from which $\cos \phi$ is determined. After calculating $\sin \phi =$
$= \sqrt{1 - \cos^2 \phi}$, one obtains that

$$x_2 = r(\cos \phi + i \sin \phi).$$

We have limited our discussion above to the most important cases which
might be encountered in the calculation of solutions by Lobačevskiĭ's
method. More complete investigations of Lobačevskiĭ's method may
be found, for example, in the books [16] and [9]. Generally speaking,

Lobačevskiĭ's method permits finding all of the solutions of an equation. We consider some numerical examples.

Example 1. We shall find the solutions of the equation

$$x^4 - 3.6x^3 + 4.2x^2 - 1.8x + 0.2 = 0.$$

The basic calculations are presented in table 11. In the table, each step of the computation occupies two lines.

Table 11

k		$a_0^{(k)}$	$a_1^{(k)}$	$a_2^{(k)}$	$a_3^{(k)}$	$a_4^{(k)}$
0	1	1	−3.6	4.2	−1.8	0.2
	2		−7.2	8.4	−3.6	0.4
1	1	1	4.56	5.08	1.56	0.04
	2		9.12	10.16	3.12	0.08
2	1	1	$0.10634 \cdot 10^2$	$0.11659 \cdot 10^2$	$0.20272 \cdot 10$	0.0016
	2		$0.21268 \cdot 10^2$	$0.23318 \cdot 10^2$	$0.40544 \cdot 10$	0.0032
3	1	1	$0.89764 \cdot 10^2$	$0.92821 \cdot 10^2$	$0.40722 \cdot 10$	$0.256 \cdot 10^{-5}$
	2		$0.17953 \cdot 10^3$	$0.18564 \cdot 10^3$	$0.81444 \cdot 10$	$0.512 \cdot 10^{-5}$
4	1	1	$0.78719 \cdot 10^4$	$0.78847 \cdot 10^4$	$0.16582 \cdot 10^2$	$0.65536 \cdot 10^{-11}$
	2		$0.15744 \cdot 10^5$	$0.15769 \cdot 10^5$	$0.33164 \cdot 10^2$	$0.13107 \cdot 10^{-10}$
5	1	1	$0.61951 \cdot 10^8$	$0.61907 \cdot 10^8$	$0.27496 \cdot 10^8$	$0.42950 \cdot 10^{-22}$
	2		$0.12390 \cdot 10^9$	$0.12381 \cdot 10^9$	$0.54992 \cdot 10^8$	$0.85900 \cdot 10^{-22}$
6	1	1	$0.38379 \cdot 10^{16}$	$0.38325 \cdot 10^{16}$	$0.75603 \cdot 10^5$	$0.18447 \cdot 10^{-44}$

The coefficients of the polynomial are written on the first line (corresponding to $k = 0$). The first step of the computation is begun by filling out the second line, on which the coefficients of the polynomial are doubled. This is done so that the coefficients $a_i^{(1)}$ may be calculated by accumulating

$$a_i^{(1)} = a_i^2 - (2a_{i-1}) \times a_{i+1} + (2a_{i-2}) \times a_{i+2} - \ldots.$$

The coefficients $a_i^{(1)}$ are written on the third line (corresponding to $k = 1$). The computation is stopped at the sixth step, because the equations

$$a_i^{(6)} = [a_i^{(5)}]^2, \qquad i = 0, 1, 2, 3, 4,$$

are satisfied to five significant figures. It follows from these equations that all of the solutions are real. The absolute values $|x_i|$ of the solutions are found from the equations

$$x_i^{64} = a_i^{(6)}/a_{i-1}^{(6)}, \qquad i = 1, 2, 3, 4.$$

As an example, we obtain $|x_1|$. We have

$$x_1^{64} = 0.38379 \cdot 10^{16},$$

from which

$$64 \log |x_1| = 15 + \log 3.8379. \tag{9.18}$$

Using a five-place table of logarithms, we find that

$$\log 3.8379 = 0.58409.$$

From (9.18), we obtain

$$\log |x_1| = 0.24350.$$

Again with the aid of the table of logarithms, we find that

$$|x_1| = 1.7519.$$

It may be seen from the original equation that none of the solutions can be negative, since the coefficients of the odd powers of x are negative. Thus, $x_1 = 1.7519$.

The other solutions may be found in an analogous fashion. They turn out to be

$$x_2 = 0.99999, \qquad x_3 = 0.68033, \qquad x_4 = 0.16780.$$

To verify the validity of the results of the calculation, the values found for the solutions may be substituted into the left side of the equation. To do this, it is necessary to use Horner's procedure.

Example 2. We shall calculate the solutions of the equation

$$x^4 + 0.75x^3 + 0.625x^2 - 0.375x - 0.125 = 0.$$

The calculations are presented in table 12.

Table 12

k		$a_0^{(k)}$	$a_1^{(k)}$	$a_2^{(k)}$	$a_3^{(k)}$	$a_4^{(k)}$
0	1	0.75	0.625	-0.375	-0.125
	2	1.5	1.25	-0.75	-0.25
1	1	-0.6875	0.70312	0.29688	$0.15625 \cdot 10^{-1}$
	2	$-0.13750 \cdot 10$	$0.14062 \cdot 10$	0.59376	$0.31250 \cdot 10^{-1}$
2	1	-0.93358	0.93384	$0.66166 \cdot 10^{-1}$	$0.24414 \cdot 10^{-3}$
	2	$-0.18672 \cdot 10$	$0.18677 \cdot 10$	0.13233	$0.48828 \cdot 10^{-3}$
3	1	-0.99611	0.99609	$0.39220 \cdot 10^{-2}$	$0.59604 \cdot 10^{-7}$
	2	$-0.19922 \cdot 10$	$0.19922 \cdot 10$	$0.78440 \cdot 10^{-2}$	$0.11921 \cdot 10^{-6}$
4	1	-0.99994	$0.10000 \cdot 10$	$0.15263 \cdot 10^{-4}$	$0.35526 \cdot 10^{-14}$
	2	$-0.19999 \cdot 10$	$0.20000 \cdot 10$	$0.30740 \cdot 10^{-4}$	$0.71052 \cdot 10^{-14}$
5	1	$-0.10001 \cdot 10$	$0.10000 \cdot 10$	$0.23296 \cdot 10^{-9}$	$0.12621 \cdot 10^{-28}$

The behavior of the coefficient $a_1^{(k)}$ is irregular, but all of the other coefficients obtained at the fifth step are the squares of the corresponding coefficients obtained at the fourth step. From this, we conclude that the equation has two complex solutions x_1 and x_2, and two real solutions x_3 and x_4, for which

$$|x_1| = |x_2| > |x_3| > |x_4|.$$

The modulus $|x_1|$ of the complex solutions, and the absolute values $|x_3|$ and $|x_4|$ of the real solutions are found from the relationships

$$|x_1|^{64} = a_2^{(5)}/a_0^{(5)}, \qquad x_3^{32} = a_3^{(5)}/a_2^{(5)}, \qquad x_4^{32} = a_4^{(5)}/a_3^{(5)}.$$

As a result of the calculation, we obtain that

$$|x_1| = 1.00000, \qquad x_3 = 0.50001, \qquad x_4 = -0.24999.$$

From the relationship

$$2|x_1| \cos \phi + x_3 + x_4 = -\frac{a_1}{a_0},$$

we find that $\cos \phi = -0.50001$, and

$$x_{1,2} = -0.50001 \pm 0.86602i.$$

10. Factorization methods

The methods considered in this section are applicable to finding solutions of the algebraic equation

$$P(x) \equiv a_0 x^n + a_1 x^{n-1} + a_2 x^{n-2} + \ldots + a_n = 0. \tag{10.1}$$

They permit one to find polynomials of degree $m < n$ which are divisors of the polynomial $P(x)$. We shall consider the highly important and frequent case that $m = 2$, in which one seeks divisors of second degree. It will be assumed that $a_0, a_1, \ldots, a_n$ are real numbers.

One well-known method for factorization is the method of penultimate remainders, due to Shih-nge Lin. We now turn to the exposition of the method of penultimate remainders. Suppose that $x^2 + p_0 x + q_0$ is an initial approximation to one of the second degree divisors of the polynomial $P(x)$. Then, carry out the division of $P(x)$ by $x^2 + p_0 x + q_0$. Since this quadratic trinomial is not an exact divisor, we shall obtain a remainder as a result of the division which will be some linear function. However, we shall not carry out the division to the end, but stop at the penultimate remainder, which, generally speaking, will be a quadratic trinomial

$$ax^2 + bx + c.$$

We divide the coefficients of this quadratic trinomial by a, assuming that $a \neq 0$. We obtain what is called the reduced penultimate remainder,

$$x^2 + p_1 x + q_1.$$

In the same way, the reduced penultimate remainder $x^2 + p_2 x + q_2$ may be constructed from $P(x)$ and $x^2 + p_1 x + q_1$, and so forth.

If the limits

$$\lim_{n \to \infty} p_n = p^* \quad \text{and} \quad \lim_{n \to \infty} q_n = q^* \tag{10.2}$$

exist, then the quadratic trinomial

$$x^2 + p^* x + q^*$$

is a divisor of the polynomial $P(x)$, as we shall see.

We shall show how to compute the coefficients of the penultimate remainder. One uses the relationship (5.10),

$$a_0 x^n + a_1 x^{n-1} + \ldots + a_n =$$
$$= (b_0^{(0)} x^{n-2} + b_1^{(0)} x^{n-3} + \ldots + b_{n-2}^{(0)})(x^2 + p_0 x + q_0) + b_{n-1}^{(0)}(x + p_0) +$$
$$+ b_n^{(0)}.$$

Obviously, the penultimate remainder in the division of $P(x)$ by $x^2 + p_0 x + q_0$ is equal to

$$b^{(0)}_{n-2}(x^2 + p_0 x + q_0) + b^{(0)}_{n-1}(x + p_0) + b^{(0)}_n =$$
$$= b^{(0)}_{n-2} x^2 + (p_0 b^{(0)}_{n-2} + b^{(0)}_{n-1})x + b^{(0)}_n + p_0 b^{(0)}_{n-1} + q_0 b^{(0)}_{n-2},$$

which, if the last recurrence relation of (5.11) is used, is equal to

$$b^{(0)}_{n-2} x^2 + (p_0 b^{(0)}_{n-2} + b^{(0)}_{n-1})x + a_n.$$

Therefore, the coefficients p_1 and q_1 of the reduced penultimate remainder may be calculated by the formulas

$$p_1 = p_0 + b^{(0)}_{n-1}/b^{(0)}_{n-2}, \quad q_1 = a_n/b^{(0)}_{n-2}.$$

It is assumed that $b^{(0)}_{n-2} \neq 0$. The numbers $b^{(0)}_j$ $(j = 0, 1, 2, \ldots, n-1)$ are defined by the sequence of recurrence relationships (5.11),

$$b^{(0)}_j = a_j - p_0 b^{(0)}_{j-1} - q_0 b^{(0)}_{j-2}, \quad j = 0, 1, 2, \ldots, n-1,$$
$$b_{-1} = b_{-2} = 0.$$

The numbers $b^{(k)}_j$ $(j = 0, 1, 2, \ldots, n-1)$ are defined in exactly the same way by the coefficients p_k and q_k, and lead to the coefficients

$$p_{k+1} = p_k + b^{(k)}_{n-1}/b^{(k)}_{n-2}, \quad q_k = a_n/b^{(k)}_{n-2}. \tag{10.3}$$

It is convenient to record the results of the computation in the following table:

k	p_k	$q_k = \dfrac{a_n}{b^{(k)}_{n-2}}$	a_0	a_1	$\ldots$	a_{n-2}	a_{n-1}	$\dfrac{b^{(k)}_{n-1}}{b^{(k)}_{n-2}}$
0	p_0	q_0	$b^{(0)}_0$	$b^{(0)}_1$	$\ldots$	$b^{(0)}_{n-2}$	$b^{(0)}_{n-1}$	$\dfrac{b^{(0)}_{n-1}}{b^{(0)}_{n-2}}$
1	p_1	q_1	$\ldots$	$\ldots$	$\ldots$	$\ldots$	$\ldots$	$\ldots$

The computation is continued until the equations

$$p_{k_0} = p_{k_0+1}, \quad q_{k_0} = q_{k_0+1},$$

are satisfied to the required degree of accuracy. Solving the quadratic equation

$$x^2 + p_{k_0} x + q_{k_0} = 0,$$

we obtain two solutions of equation (10.1).

In order to construct the initial approximation $x^2 + p_0 x + q_0$ to a divisor of the polynomial $P(x)$, it is necessary to know initial approximations to two solutions of equation (10.1). If initial approximations $x_0^{(1)}$ and $x_0^{(2)}$ to real solutions are known, then we obtain

$$p_0 = -x_0^{(1)} - x_0^{(2)}, \qquad q_0 = x_0^{(1)} x_0^{(2)}. \tag{10.4}$$

If an initial approximation $x_0 = u_0 + iv_0$ to a complex solution is known, then $\bar{x}_0 = u_0 - iv_0$ is an initial approximation to the complex conjugate solution, and one may take

$$p_0 = -2u_0, \qquad q_0 = u_0^2 + v_0^2. \tag{10.5}$$

The following approach may be taken to factorizations methods. The remainder in the division of the polynomial $P(x)$ by the quadratic trinomial $x^2 + px + q$ may be represented as the linear function

$$b_{n-1} x + b_n + p b_{n-1}, \tag{10.6}$$

where the b_j are defined by the recurrence relations (5.11),

$$b_j = a_j - p b_{j-1} - q b_{j-2}, \qquad j = 0, 1, 2, \ldots, n,$$
$$b_{-1} = b_{-2} = 0. \tag{10.7}$$

The quantities b_j are continuously differentiable functions of the variables p and q (since they are polynomials in p and q).

In order for the quadratic trinomial $x^2 + px + q$ to be a divisor of the polynomial $P(x)$, it is necessary and sufficient that the linear function (10.6) vanish identically, or, what is the same, that

$$b_n + p b_{n-1} = 0, \qquad b_{n-1} = 0. \tag{10.8}$$

Thus, the problem of obtaining a divisor of second degree of the polynomial $P(x)$ is equivalent to obtaining a solution (p^*, q^*) of the system (10.8). Obviously, system (10.8) is equivalent to the system

$$b_n = 0, \qquad b_{n-1} = 0. \tag{10.9}$$

Factorization methods are some variants of the method of iterations for solving the system (10.8) or (10.9). For example, the method of penultimate remainders is the method of iterations for solving system (10.8) corresponding to writing the system in the form

$$p = p + \frac{b_{n-1}}{b_{n-2}}, \qquad q = q + \frac{b_n + p b_{n-1}}{b_{n-2}}.$$

Indeed, the successive approximations of the method of iterations for this system may be calculated by the formulas:

$$p_{k+1} = p_k + \frac{b_{n-1}^{(k)}}{b_{n-2}^{(k)}}, \qquad q_{k+1} = q_k + \frac{b_n^{(k)} + p_k b_{n-1}^{(k)}}{b_{n-2}^{(k)}}, \qquad (10.10)$$

$$k = 0, 1, 2, \ldots.$$

Here, the index k on $b_{n-j}^{(k)}$, $j = 0$, 1, 2, denotes that the functions b_j are calculated for the values $p = p_k$, $q = q_k$, in other words, that the values $b_j^{(k)}$ are defined by the relationships (10.7) for $p = p_k$, $q = q_k$. Obviously, the relationships (10.10) coincide with relationships (10.3), from which our assertion about the method of penultimate remainders follows.

Consideration of the method of penultimate remainders as an iteration method makes evident the assertion made earlier that if the relationships (10.2) are satisfied, then $x^2 + p^* x + q^*$ is a divisor of the polynomial $P(x)$. Indeed, if the relationships (10.2) hold, the p^* and q^* satisfy system (10.8).

Other methods of factoring polynomials will be indicated. If system (10.9) is written in the form

$$p = p + b_n, \qquad q = q + b_{n-1},$$

then we obtain a method for which the approximations (p_{k+1}, q_{k+1}) are defined by the formulas

$$p_{k+1} = p_k + b_n^{(k)}, \qquad q_{k+1} = q_k + b_{n-1}^{(k)}, \qquad k = 0, 1, 2, \ldots. \qquad (10.11)$$

A defect of the method of penultimate remainders and the method defined by formulas (10.11) lies in the fact that these methods may diverge even in cases that the initial approximation is as close as one pleases to a divisor of the polynomial. This defect is absent in the methods which are obtained by applying the ordinary and modified forms of Newton's method to the solution of the system (10.9).

We shall need the first order partial derivatives of the functions b_j with respect to p and q. Differentiating both sides of the relationships (10.7) with respect to p and q, we obtain

$$
\left.\begin{array}{l}
\dfrac{\partial b_j}{\partial p} = -b_{j-1} - p\,\dfrac{\partial b_{j-1}}{\partial p} - q\,\dfrac{\partial b_{j-2}}{\partial p}, \\[3mm]
\dfrac{\partial b_j}{\partial q} = -b_{j-2} - p\,\dfrac{\partial b_{j-1}}{\partial q} - q\,\dfrac{\partial b_{j-2}}{\partial q}, \\[3mm]
j = 0, 1, 2, \ldots, n, \qquad b_{-1} = b_{-2} = 0.
\end{array}\right\} \tag{10.12}
$$

The numbers c_j, $j = 0, 1, 2, \ldots, n-1$ are defined by means of the recurrence relations

$$
\begin{array}{l}
c_j = b_j - p c_{j-1} - q c_{j-2}, \\[2mm]
j = 0, 1, 2, \ldots, n-1, \qquad c_{-1} = c_{-2} = 0.
\end{array} \tag{10.13}
$$

The numbers c_j are obtained from the numbers b_j in the same way as the numbers b_j are obtained from the numbers a_j.

Relationships (10.12) may be written in the form

$$
-\left(\frac{\partial b_j}{\partial p}\right) = b_{j-1} - p\left(-\frac{\partial b_{j-1}}{\partial p}\right) - q\left(-\frac{\partial b_{j-2}}{\partial p}\right),
$$

$$
-\left(\frac{\partial b_j}{\partial q}\right) = b_{j-2} - p\left(-\frac{\partial b_{j-1}}{\partial q}\right) - q\left(-\frac{\partial b_{j-2}}{\partial q}\right),
$$

and, by comparing these with (10.13), we conclude that

$$
\begin{array}{l}
\dfrac{\partial b_j}{\partial p} = -c_{j-1}, \qquad \dfrac{\partial b_j}{\partial q} = -c_{j-2}, \\[3mm]
j = 0, 1, 2, \ldots, n, \qquad c_{-1} = c_{-2} = 0.
\end{array} \tag{10.14}
$$

We shall now apply Newton's method to the solution of the system (10.9). By virtue of (10.14), the Jacobian matrix of the left side of the system is:

$$
C = \begin{pmatrix} \dfrac{\partial b_n}{\partial p} & \dfrac{\partial b_n}{\partial q} \\[3mm] \dfrac{\partial b_{n-1}}{\partial p} & \dfrac{\partial b_{n-1}}{\partial q} \end{pmatrix} = -\begin{pmatrix} c_{n-1} & c_{n-2} \\ c_{n-2} & c_{n-3} \end{pmatrix}, \tag{10.15}
$$

and thus the procedure for determining $(\Delta p_k, \Delta q_k)$ by Newton's method [see (6.8)] is:

$$
-\begin{pmatrix} c_{n-1}^{(k)} & c_{n-2}^{(k)} \\ c_{n-2}^{(k)} & c_{n-3}^{(k)} \end{pmatrix} \begin{pmatrix} \Delta p_k \\ \Delta q_k \end{pmatrix} = -\begin{pmatrix} b_n^{(k)} \\ b_{n-1}^{(k)} \end{pmatrix}. \tag{10.16}
$$

Here, the $c_{n-j}^{(k)}, j = 1, 2, 3$, are defined by the recurrence relations (10.13) which result from setting $p = p_k$, $q = q_k$.

We shall assume that the determinants of the matrices

$$C_k = - \begin{pmatrix} c_{n-1}^{(k)} & c_{n-2}^{(k)} \\ c_{n-2}^{(k)} & c_{n-3}^{(k)} \end{pmatrix} \qquad (k = 0, 1, 2, \ldots)$$

are different from zero. They will be denoted by $\varDelta_k$. It is not difficult to write system (10.16) as

$$\begin{aligned} \varDelta p_k &= \alpha_{11}^{(k)} b_n^{(k)} + \alpha_{12}^{(k)} b_{n-1}^{(k)}, \\ \varDelta q_k &= \alpha_{21}^{(k)} b_n^{(k)} + \alpha_{22}^{(k)} n_{n-1}^{(k)}, \end{aligned} \tag{10.17}$$

where

$$\alpha_{11}^{(k)} = \frac{c_{n-3}^{(k)}}{\varDelta_k}, \qquad \alpha_{12}^{(k)} = \alpha_{21}^{(k)} = -\frac{c_{n-2}^{(k)}}{\varDelta_k}, \qquad \alpha_{22}^{(k)} = \frac{c_{n-1}^{(k)}}{\varDelta_k}. \tag{10.18}$$

In order to find $(\varDelta p_k, \varDelta q_k)$ by formulas (10.17), it is necessary to calculate the two sequences of numbers

$$b_0^{(k)}, b_1^{(k)}, \ldots, b_{n-1}^{(k)}, b_n^{(k)},$$
$$c_0^{(k)}, c_1^{(k)}, \ldots, c_{n-1}^{(k)},$$

by means of the relationships (10.7) and (10.13).

In order to simplify the calculation, the modified form of Newton's method may be used. In this case, it is necessary to calculate only the single sequence of numbers

$$c_0^{(0)}, c_1^{(0)}, \ldots, c_{n-1}^{(0)}.$$

The formulas for determining $(\varDelta p_k', \varDelta q_k')$ are:

$$\begin{aligned} \varDelta p_k' &= \alpha_{11}^{(0)} b_n'^{(k)} + \alpha_{12}^{(0)} b_{n-1}'^{(k)}, \\ \varDelta q_k' &= \alpha_{21}^{(0)} b_n'^{(k)} + \alpha_{22}^{(0)} b_{n-1}'^{(k)}. \end{aligned} \tag{10.19}$$

Here, the numbers $\alpha_{ij}^{(0)}$ $(i, j = 1, 2)$ are determined from formulas (10.18) for $k = 0$; the $b_j'^{(k)}$, $(j = 0, 1, 2, \ldots, n)$ are determined from the recurrence relations (10.7) which result from setting

$$p = p_k', \qquad q = q_k' \qquad (p_0' = p_0, q_0' = q_0).$$

We have already indicated that the methods defined by formulas (10.17) and (10.19) will converge if the initial approximation (p_0, q_0) is taken sufficiently close to the coefficients of a quadratic trinomial $x^2 + p^*x + q^*$ which divides the polynomial $P(x)$. Of course, it is assumed that the Jacobian matrix (10.15) is nonsingular in some neighborhood of (p_0, q_0). The assertion about convergence is a consequence of theorem 4 of § 7.

Example 1. We shall calculate the solutions of the equation

$$x^4 - 6.4x^3 + 14x^2 - 11.5x + 3 = 0 \qquad (10.20)$$

by the method of penultimate remainders. We take the initial approximations

$$x_0^{(1)} = 0.5, \qquad x_0^{(2)} = 1,$$

to the solutions of equations (10.20). On the basis of (10.4), we choose

$$p_0 = -1.5, \quad q_0 = 0.5.$$

The results of the calculation are displayed in table 13.

Table 13

k	p_k	q_k	1	-6.4	14	-11.5	Δp_k
0	-1.5	0.5	1	-4.9	6.15	0.175	0.02846
1	-1.47154	0.48780	1	-4.92846	6.25977	0.11560	0.01847
2	-1.45307	0.47925	1	-4.94693	6.33251	0.07240	0.01143
3	-1.44164	0.47375	1	-4.95836	6.37808	0.04392	0.00689
4	-1.43475	0.47036	1	-4.96525	6.40575	0.02610	0.00407
5	-1.43068	0.46833	1	-4.96932	6.42216	0.01534	0.00239
6	-1.42829	0.46713	1	-4.97171	6.43183	0.00895	0.00139
7	-1.42690	0.46643	1	-4.97310	6.43745	0.00520	0.00081
8	-1.42609	0.46602	1	-4.97391	6.44074	0.00302	0.00047
9	-1.42562	0.46578	1	-4.97438	6.44264	0.00172	0.00027
10	-1.42535	0.46565	1	-4.97465	6.44373	0.00102	0.00016
11	-1.42519	0.46557	1	-4.97481	6.44438	0.00059	0.00009
12	-1.42510	0.46552	1	-4.97490	6.44475	0.00033	0.00005
13	-1.42505	0.46550	1	-4.97495	6.44495	0.00022	0.00003
14	-1.42502	0.46548	1	-4.97498	6.44507	0.00011	0.00002
15	-1.42500	0.46547	1	-4.97500	6.44516	0.00007	0.00001
16	-1.42499	0.46547	1	-4.97501	6.44519	0.00005	0.00001
17	-1.42498	0.46546	1	-4.97502	6.44524	0.00001	0.00000
18	-1.42498	0.46546					

It is seen from the table that

$$x^2 - 1.42498x + 0.46546$$

is the desired divisor of the polynomial $P(x)$. The roots of this quadratic trinomial,

$$x_1 = 0.50710, \qquad x_2 = 0.91787,$$

are solutions of equation (10.20). The other two solutions of equation (10.20) are found from the quadratic equation

$$x^2 - 4.97502x + 6.44524 = 0.$$

They turn out to be equal to

$$x_{3,4} = 2.48751 \pm 0.50748i.$$

Example 2. The equation

$$x^4 + 0.75x^3 + 0.625x^2 - 0.375x - 0.125 = 0, \tag{10.21}$$

was solved in § 9 by Lobačevskiĭ's method. One of its solutions turned out to be equal to

$$-0.50001 + 0.86602i.$$

We shall solve this equation by the method defined by formulas (10.17), taking

$$x_0 = -0.55 + 0.9i$$

as an initial approximation to a solution. The coefficients of the initial approximation to the quadratic trinomial sought are found from formulas (10.15) to be

$$p_0 = 1.1, \qquad q_0 = 1.1125. \tag{10.22}$$

The calculations are displayed in table 14. In the table, each step of the calculation occupies two lines. In particular, the numbers $b_j^{(k)}$, $j = 0, 1, 2, 3, 4$ are written on the first line, and the numbers $c_j^{(k)}$, $j = 0, 1, 2, 3$ are written on the second line. We see that $\Delta p_3 = \Delta q_3 = 0$ to five decimal places. Actually, $p_3 = 1$, $q_3 = 1$ are the exact coefficients of the divisor $x^2 + x + 1$ of the polynomial (10.21). The roots of the divisor are

$$x_{1,2} = -0.5 \pm 0.86602i.$$

The other two solutions of equation (10.21) are the roots of the quadratic trinomial

$$x^2 - 0.25x - 0.125,$$

and they are equal to

$$x_3 = 0.5, \qquad x_4 = -0.25.$$

The results of the solution of equation (10.21) by the factorization method based on using the modified form of Newton's method are displayed in table 15. The quantities $\Delta p'_k$ and $\Delta q'_k$ are calculated by formulas (10.19), which, in this case, are written as:

$$\Delta p'_k = 0.70332 b'^{(k)}_n + 0.18432 b'^{(k)}_{n-1},$$
$$\Delta q'_k = 0.18432 b'^{(k)}_n - 0.64135 b'^{(k)}_{n-1}.$$

The numbers $\alpha^{(0)}_{ij}$, $i, j = 1, 2$, are taken from table 14.

We remark that the method of penultimate remainders for equation (10.22) does not converge to the divisor $x^2 + x + 1$ in spite of the fact that (p_0, q_0) is close to $(1, 1)$. One may convince oneself of its divergence from the first steps, if one actually carries out the calculations.

The cause of the divergence of the method of penultimate remainders for the divisor $x^2 + x + 1$ will be given below.

Table 14

k	p_k	q_k	1	0.75	0.625	-0.375	-0.125
0	1.1	1.1125	1	-0.35	-0.1025	0.12712	-0.15080
			1	-1.45	0.38	1.32224	
1	1.01737	1.00318	1	-0.26737	-0.10617	0.00123	-0.01974
			1	-1.28474	0.19771	1.08891	
2	0.99990	0.99953	1	-0.24990	-0.12465	-0.00058	0.00017
			1	-1.24980	0.12550	1.12315	
3	1.00000	1.00000	1	-0.25000	-0.12500	0.00000	0.00000

k	Δ_k	$\alpha^{(k)}_{11}$	$\alpha^{(k)}_{12}$	$\alpha^{(k)}_{22}$	Δp_k	Δq_k
0	-2.06165	0.70332	0.18432	-0.64135	-0.08263	-0.10932
1	-1.43806	0.89338	0.13748	-0.75721	-0.01747	-0.00365
2	-1.41946	0.88048	0.08841	-0.79125	0.00010	0.00047
3						

Table 15

		1	0.75	0.625	−0.375	−0.125		
k	p'_k	q'_k	$b'^{(k)}_0$	$b'^{(k)}_1$	$b'^{(k)}_2$	$b'^{(k)}_3$	$b'^{(k)}_4$	$\Delta p'_k$	$\Delta q'_k$
0	1.1	1.1125	1	−0.35	−0.1025	0.12712	−0.15080	−0.08263	−0.10932
1	1.01737	1.00318	1	−0.26737	−0.10617	0.00123	−0.01974	−0.01366	−0.00443
2	1.00371	0.99875	1	−0.25371	−0.11910	−0.00207	−0.00397	−0.00317	0.00060
3	1.00054	0.99935	1	−0.25054	−0.12367	−0.00089	−0.00052	−0.00053	0.00047
4	1.00001	0.99982	1	−0.25001	−0.12481	−0.00022	0.00001	−0.00003	0.00014
5	0.99998	0.99996	1	−1.24998	−0.12498	−0.00005	0.00002	0.00000	0.00003
6	0.99998	0.99999	1	−0.24998	−0.12501	−0.00001	0.00002	0.00001	0.00001
7	0.99999	1.00000	1	−0.24999	−0.12501	0.00000	0.00001	0.00001	0.00000
8	1.00000	1.00000	1	−0.25000	−0.12500	0.00000	0.00000		

We have seen that the method of penultimate remainders is the method of iterations for solving the nonlinear system

$$
\left.
\begin{aligned}
p &= \phi(p, q) \equiv p + \frac{b_{n-1}}{b_{n-2}}, \\
q &= \psi(p, q) \equiv \frac{a_n}{b_{n-2}}.
\end{aligned}
\right\} \tag{10.23}
$$

In a small neighborhood of (p^*, q^*), where p^* and q^* are the coefficients of an exact divisor of the polynomial, the nonlinear system (10.23) is close to the linear algebraic system

$$
\left.
\begin{aligned}
p &= \phi(p^*, q^*) + \frac{\partial \phi(p^*, q^*)}{\partial p}(p - p^*) + \frac{\partial \phi(p^*, q^*)}{\partial q}(q - q^*), \\
q &= \psi(p^*, q^*) + \frac{\partial \psi(p^*, q^*)}{\partial p}(p - p^*) + \frac{\partial \psi(p^*, q^*)}{\partial q}(q - q^*).
\end{aligned}
\right\} \tag{10.24}
$$

It follows from this that for $p_0 - p^*$, $q_0 - q^*$ sufficiently small, the character of the convergence of the method of iterations for the system (10.23), generally speaking, is determined by the character of the convergence of the method of iterations for the system (10.24).

It is well-known that for the convergence of the method of iterations for the linear algebraic system

$$
x = Ax + b,
$$

it is necessary and sufficient that all of the characteristic values of the matrix A be less than one in modulus. The matrix A of system (10.24) has the form

$$A = \begin{pmatrix} \dfrac{\partial \phi(p^*, q^*)}{\partial p} & \dfrac{\partial \phi(p^*, q^*)}{\partial q} \\ \dfrac{\partial \psi(p^*, q^*)}{\partial p} & \dfrac{\partial \psi(p^*, q^*)}{\partial q} \end{pmatrix}.$$

With the aid of formulas (10.14), we find that

$$\frac{\partial \phi}{\partial p} = 1 + \frac{-c_{n-2} b_{n-2} + b_{n-1} c_{n-2}}{b_{n-2}^2}, \quad \frac{\partial \phi}{\partial q} = \frac{-c_{n-3} b_{n-2} + b_{n-1} c_{n-4}}{b_{n-2}^2},$$

$$\left.\frac{\partial \psi}{\partial p} = \frac{a_n c_{n-3}}{b_{n-2}^2}, \quad \frac{\partial \psi}{\partial q} = \frac{a_n c_{n-4}}{b_{n-2}^2}.\right\} \tag{10.25}$$

Returning to our example, we calculate the quantities b_j, c_j by formulas (10.7) and (10.13) for $p = p^* = 1$, $q = q^* = 1$:

j	0	1	2	3	4
a_j	1	0.75	0.625	−0.375	−0.125
b_j	1	−0.25	−0.125	0	0
c_j	1	−1.25	0.125	1.125	

and substitute them into the right-hand sides of (10.25). We obtain

$$A = \begin{pmatrix} 2 & -10 \\ 10 & -8 \end{pmatrix}.$$

The characteristic values of the matrix A are the solutions of the equation

$$\begin{vmatrix} 2-\lambda & -10 \\ 10 & -8-\lambda \end{vmatrix} = 0,$$

which are equal to

$$\lambda_{1,2} = 3 \pm 5i\sqrt{3}.$$

We see that $|\lambda_{1,2}|$ are considerably larger than one. It is because of this fact that the method of penultimate remainders does not converge to the divisor $x^2 + x + 1$.

EXERCISES FOR CHAPTER I

§ 1

1. Find initial approximations to the two real solutions of the equation

$$x - \ln x - 2 = 0.$$

2. Find initial approximations to the real and complex solutions of

$$x^3 - 9x^2 - 9x - 15 = 0.$$

3. Find an initial approximation to the solution of the system

$$\xi^2 + \eta^2 = 1,$$
$$\eta = \tan \xi,$$

which lies in the first quadrant.

§ 2

1. Calculate the real solution of

$$x^3 - 9x^2 - 9x - 15 = 0$$

to five decimal places by the secant method.

2. Calculate $\sqrt{2}$ to ten decimal places by the secant method.

3. Calculate the positive solution of

$$x^4 - 11x^2 - 6x + 10 = 0$$

which is least in magnitude to four decimal places by the secant method.

§ 3

1. Show that the method of iterations converges to a solution of

$$x = \tfrac{1}{10}(x^3 + x + 1)$$

for $x_0 = 0$. Give an estimate for the radius δ of the disk in which Theorem 1 guarantees uniqueness of this solution.

2. Give an estimate for the error obtained in the solution of the equation in Problem 1 by the method of iterations, assuming that all numbers calculated are rounded to four decimal places.

3. Show that the method of iterations will not converge to the real solution of the equation

$$x^3 - 9x^2 - 9x - 15 = 0.$$

§ 4

1. Apply the method of iterations to the system

$$\xi_1 = 0.5\xi_2 + 0.1\xi_1^2,$$
$$\xi_2 = 0.2\xi_1 - 0.5,$$

with the initial approximation $(\xi_1^{(0)}, \xi_2^{(0)}) = (0, 0)$. How many iterations would be required for five decimal place accuracy, assuming that all calculations are performed exactly?

2. Show that the method of iterations will always converge to the solution of the linear system

$$\xi_1 = a_{11}\xi_1 + a_{12}\xi_2 + \ldots + a_{1n}\xi_n + b_1,$$
$$\xi_2 = a_{21}\xi_1 + a_{22}\xi_2 + \ldots + a_{2n}\xi_n + b_2,$$
$$\cdots \cdots \cdots \cdots \cdots \cdots \cdots$$
$$\xi_n = a_{n1}\xi_1 + a_{n2}\xi_2 + \ldots + a_{nn}\xi_n + b_n,$$

if

$$\max_{(i)} \sum_{j=1}^{n} |a_{ij}| < 1, \qquad i = 1, \ldots, n,$$

independently of the choice of $(\xi_1^{(0)}, \xi_2^{(0)}, \ldots, \xi_n^{(0)})$.

3. Investigate the application of the method of iterations to finding the solution of the system

$$\xi_1^2 + \xi_2^2 = 1,$$
$$\xi_2 = \tan \xi_1,$$

in the first quadrant of the (ξ_1, ξ_2)-plane.

§ 5

1. Evaluate the polynomial

$$P_3(x) = x^3 + 6.3x^2 + 11.23x + 0.061$$

and its derivatives at $x = -0.005$. Retain five decimal places.

2. By the method of (5.15), evaluate

$$P_5(x) = x^5 - x^4 - 11x^3 + 5x^2 + 7x - 1$$

and its first derivative at $x_0 = 1.1 - 0.6i$. Retain five places, and check the computation.

3. Evaluate

$$P_4(x) = x^4 + 11x^2 + 10x + 50$$

and all of its derivatives at $x_0 = 0.5 + 0.7i$. Retain five decimal places, and check the computation.

§ 6

1. Calculate the two real solutions of the equation

$$x - \ln x - 2 = 0$$

by Newton's method, terminating the computation when successive approximations agree to five decimal places.

2. Calculate the real solution of

$$x^3 - 9x^2 - 9x - 15 = 0$$

to five decimal places by Newton's method, taking $x_0 = 10.0$.
Compare with the secant method. Repeat, using the modified form of Newton's method, and compare with the above.

3. Calculate a solution of the system

$$\xi_1 = 0.5\xi_2 + 0.1\xi_1^2,$$
$$\xi_2 = 0.2\xi_1 - 0.5,$$

by Newton's method, taking

$$(\xi_1^{(0)}, \xi_2^{(0)}) = (0, 0).$$

Terminate the computation when successive approximations agree to five decimal places. Compare with the method of iterations.

§ 7

1. Calculate t_1, t_2, t_3 for $t_0 = 0$ and

$$f(t) \equiv \tfrac{1}{2}Kt^2 - \frac{1}{B}t + \frac{\eta}{B} = 0$$

by Newton's method.

2. For $t_0 = 0$, calculate t_1', t_2', t_3' by the modified form of Newton's method applied to the equation in Problem 1.

3. Construct a table of $\tau^* - \tau_n'$ in the same form as Table 5.

§ 8

1. Calculate the complex solution of

$$x^3 - 9x^2 - 9x - 15 = 0$$

which lies in the upper half-plane by Newton's method to a guaranteed accuracy of four decimal places.

2. Apply the combined method of tangents and secants to find the real solution of the equation

$$x^5 - 2 = 0,$$

obtaining an accuracy of six decimal places.

3. Solve the system

$$\xi_1 = 0.5\xi_2 + 0.2\xi_1^2,$$
$$\xi_2 = 0.2\xi_1 - 0.5,$$

$(\xi_1^{(0)}, \xi_2^{(0)}) = (0, 0)$ by the modified form of Newton's method to five decimal place accuracy.

§ 9

1. Obtain the solutions of

$$x^4 - 28x^2 + 24x + 12 = 0$$

by Lobačevskiï's method.

2. Obtain the solutions of

$$x^4 + 10x^2 + 12x + 40 = 0$$

by Lobačevskiï's method.

3. Solve for the root of largest magnitude of

$$x^4 - 12x^3 + 11x^2 - 6x + 1 = 0$$

by Lobačevskiï's method.

§ 10

1. Investigate finding the quadratic divisor of

$$P(x) = x^3 - 9x^2 - 9x - 15$$

with real coefficients by (a) the method of penultimate remainders, (b) the method of iterations as given by (10.11).

2. Find the quadratic divisor of

$$P(x) = x^3 - 9x^2 - 9x - 15$$

with real coefficients by (a) Newton's method, (b) the modified form of Newton's method.

3. Find a quadratic divisor of

$$P(x) = x^4 - 25x^2 + 12x + 18$$

with real coefficients.

Chapter II

ALGEBRAIC INTERPOLATION

1. Introduction

Many numerical methods are based on the idea of replacing the functions which appear in the formulation of the problem by simpler functions which are close to the given ones in some sense. For example, to solve the equation $\phi(x) = 0$ by Newton's method, in the neighborhood of the initial approximation x_0 to the solution, the function $\phi(x)$ is replaced by the linear function

$$\phi_0(x) \equiv \phi(x_0) + \phi'(x_0)(x - x_0).$$

The solution of the equation $\phi_0(x) = 0$ is taken as the next approximation to the solution of the equation $\phi(x) = 0$.
Simpson's Formula

$$\int_a^b \phi(x)dx \cong \frac{b-a}{6}\left[\phi(a) + 4\phi\left(\frac{a+b}{2}\right) + \phi(b)\right]$$

for the approximate calculation of integrals is based on the replacement of the function $\phi(x)$ in the integrand by the polynomial $P(x)$ of second (or lower) degree, which at

$$x = a, \frac{a+b}{2}, b$$

has the same values as $\phi(x)$. The integral

$$\int_a^b P(x)dx = \frac{b-a}{6}\left[\phi(a) + 4\phi\left(\frac{a+b}{2}\right) + \phi(b)\right]$$

is taken as an approximation to the value of the integral $\int_a^b \phi(x)dx$. In both examples, polynomials were taken as the approximating functions. The basis for taking polynomials as approximations to continuous functions is the property of the denseness of polynomials in the class of continuous functions, which follows from the theorem of Weierstrass: If $f(x)$ is a continuous function on a closed finite interval [a, b], then for any $\varepsilon > 0$, one may find a polynomial $P(x)$ such that

$$|f(x) - P(x)| < \varepsilon$$

for all x on [a, b].

Another important property of polynomials is that they are functions which are simple in nature. In order to compute the value of a polynomial, it is only necessary to perform a finite number of the arithmetic operations: addition, subtraction, and multiplication. Derivatives and (indefinite) intergrals of polynomials are also polynomials.

The theorem of Weierstrass does not give a way for constructing the approximating polynomial, it only asserts that it is possible in principle to construct such a polynomial. There exist various methods of approximation of functions by polynomials. Among the numerical methods of wide applicability is the process of interpolation, the essence of which is the following.

From [a, b], one chooses the $n+1$ distinct points

$$x_0, x_1, \ldots, x_n,$$

called the basic points (or knots) of interpolation, and constructs the polynomial of degree not greater than n

$$P(x) = a_0 x^n + a_1 x^{n-1} + \ldots + a_{n-1} x + a_n,$$

with values at the points $x_0, x_1, \ldots, x_n$ which coincide with the values of the continuous function $f(x)$ defined on [a, b]. As we shall see later, such a polynomial exists and is unique. It may be taken as an approximation to $f(x)$. This is called interpolation of the values of the function, which was what was discussed in the second example above.

Sometimes it is required at the knots of interpolation that not only the values $f(x)$ and $P(x)$ coincide, but also the values of their successive derivates of some order. This type of interpolation was dealt with in the first example.

The basic facts of the theory of polynomial interpolation, or what is called algebraic interpolation, will be considered in this chapter.

2. Finite differences

Suppose that values of some function $y = f(x)$ are known for equally spaced values of the argument $x_k = x_0 + kh$, $k = 0, 1, 2, \ldots, N, h > 0$:

$$y_0 = f(x_0), \; y_1 = f(x_1), \; y_2 = f(x_2), \ldots, y_N = f(x_N).$$

We shall say that this gives a *table of the function* $f(x)$ for the initial value of the argument x_0, step h, and final value of the argument x_N. The event that a table of the function $f(x)$ is given from $x = x_0$ to $x = x_N$ with step h can be denoted by:

$$x = x_0(h)x_N.$$

The numbers

$$\Delta y_0 = y_1 - y_0, \; \Delta y_1 = y_2 - y_1, \; \Delta y_2 = y_3 - y_2, \ldots$$

are called finite differences of the first order of the function $f(x)$. In an analogous way, one may define finite differences of the second order

$$\Delta^2 y_0 = \Delta y_1 - \Delta y_0, \; \Delta^2 y_1 = \Delta y_2 - \Delta y_1, \ldots$$

Suppose that finite differences of order k

$$\Delta^k y_0, \; \Delta^k y_1, \; \Delta^k y_2, \; \ldots$$

are already known. Then, differences of order $k+1$ may be defined by the relationships

$$\Delta^{k+1} y_0 = \Delta^k y_1 - \Delta^k y_0, \; \Delta^{k+1} y_1 = \Delta^k y_2 - \Delta^k y_1, \ldots$$

The table of values of a function and the finite differences of them may be written in the following manner:

x	y	Δy	$\Delta^2 y$	$\Delta^3 y$	$\Delta^4 y$
x_0	y_0				
		Δy_0			
x_1	y_1		$\Delta^2 y_0$		
		Δy_1		$\Delta^3 y_0$	
x_2	y_2		$\Delta^2 y_1$		$\Delta^4 y_0$
		Δy_2		$\Delta^3 y_1$	
x_3	y_3		$\Delta^2 y_2$		$\Delta^4 y_1$
		Δy_3		$\Delta^3 y_2$	
x_4	y_4		$\Delta^2 y_3$		
		Δy_4			
x_5	y_5				

Thus, all differences of even order are written on the same horizontal

lines as the arguments, and all odd differences are written on intermediate lines.

Sometimes, the table of finite differences is written in the form:

x	y	Δy	$\Delta^2 y$	$\Delta^3 y$	$\Delta^4 y$
x_0	y_0	Δy_0	$\Delta^2 y_0$	$\Delta^3 y_0$	$\Delta^4 y_0$
x_1	y_1	Δy_1	$\Delta^2 y_1$	$\Delta^3 y_1$	$\Delta^4 y_1$
x_2	y_2	Δy_2	$\Delta^2 y_2$	$\Delta^3 y_2$	
x_3	y_3	Δy_3	$\Delta^2 y_3$		
x_4	y_4	Δy_4			
x_5	y_5				

Example 1. We shall construct the table of finite differences for the
function $y = \sin x$, $x = 0(0.1)0.7$. The values of $\sin x$ are taken from
[24]. The results of the calculations are shown in Table 16. In the table
of differences the zeros which precede the first significant digit are not
entered. This is done in order to make the table more compact, and not
to waste time writing down zeros.

Table 16

x	y	Δy	$\Delta^2 y$	$\Delta^3 y$	$\Delta^4 y$
0	0				
		9983			
0.1	0.09983		-99		
		9884		-100	
0.2	0.19867		-199		4
		9685		-96	
0.3	0.29552		-295		2
		9390		-94	
0.4	0.38942		-389		3
		9001		-91	
0.5	0.47943		-480		8
		8521		-83	
0.6	0.56464		-563		
		7958			
0.7	0.64422				

The finite differences of functions which are defined in the form of a table
play rôles similar to those played by the derivatives of functions which
depend continuously on an argument. The basic properties of finite

differences which are analogous to the corresponding properties of derivatives are indicated below.

1. The finite difference of the sum of two functions $F(x) = f(x) + g(x)$ is equal to the sum of the finite differences of the addends $f(x)$ and $g(x)$.

2. If $f(x) = Af(x)$, where A is a constant, then

$$\Delta F(x_l) = A \Delta f(x_l).$$

It is clear that properties 1 and 2 hold for differences of arbitrary order. The finite difference of a polynomial of degree n is a polynomial of degree $n-1$. For the polynomial $P(x) = x^k$, this assertion is obvious:

$$(x+h)^k - x^k = khx^{k-1} + \frac{k(k-1)}{2} h^2 x^{k-2} + \dots$$

On the basis of properties 1 and 2, this assertion is obvious for arbitrary polynomials.

From this, the following property follows.

3. The finite difference of order n of a polynomial of degree n is equal to a constant, and, consequently, all differences of higher order are equal to zero.

The difference of order k may be expressed directly in terms of values of the function. One has that

$$\Delta y_l = y_{l+1} - y_l,$$

$$\Delta^2 y_l = \Delta y_{l+1} - \Delta y_l = (y_{l+2} - y_{l+1}) - (y_{l+1} - y_l) =$$
$$= y_{l+2} - 2y_{l+1} + y_l.$$

It may be established by induction that

$$\Delta^k y_l = y_{k+l} - \frac{k}{1!} y_{k+l-1} + \frac{k(k-1)}{2!} y_{k+l-2} - \dots + (-1)^k y_l.$$

If the shift operator E is introduced, which applied to a function increases its argument by the amount h:

$$Ef(x) = f(x+h), \quad Ey_m = y_{m+1},$$

then the previous formula may be written in the following way:

$$\Delta^k y_l = (E-1)^k y_l.$$

This symbolic formula may be understood as follows. The expansion of

$(E-1)^k$ by Newton's binomial formula and multiplication by y_l results in the polynomial

$$E^k y_l - \frac{k}{1!} E^{k-1} y_l + \frac{k(k-1)}{2!} E^{k-2} y_l - \ldots + (-1)^k y_l.$$

In order to obtain the formula for $\Delta^k y_l$ from this, it is necessary to employ the equations

$$E^m y_l = y_{l+m}, \quad m = 0, 1, 2, \ldots$$

It is also easy to show the validity of the formula

$$y_{k+l} = y_l + \frac{k}{1!} \Delta y_l + \frac{k(k-1)}{2!} \Delta^2 y_l + \ldots + \Delta^k y_l,$$

which gives a representation of the values of a function in terms of finite differences. This formula may be written compactly in the symbolic form

$$y_{k+l} = (1+\Delta)^k y_l.$$

In the tables of functions with which one comes to deal in approximate calculations, the values given are in error, which ordinarily does not exceed half a unit in the last place retained. Thus, it is of interest to explain how the errors in the value of the function influence the finite differences. We consider the simple case that in some portion of the table, there is an error in only one of the function values. The influence of this error on the finite differences may be discerned from the table presented below, where it is assumed that there is an error ε in the value given for y_l.

x	y	Δy	$\Delta^2 y$	$\Delta^3 y$	$\Delta^4 y$
x_{l-3}	y_{l-3}				
		Δy_{l-3}			
x_{l-2}	y_{l-2}		$\Delta^2 y_{l-3}$		$\Delta^4 y_{l-4} + \varepsilon$
		Δy_{l-2}		$\Delta^3 y_{l-3} + \varepsilon$	
x_{l-1}	y_{l-1}		$\Delta^2 y_{l-2} + \varepsilon$		$\Delta^4 y_{l-3} - 4\varepsilon$
		$\Delta y_{l-1} + \varepsilon$		$\Delta^3 y_{l-2} - 3\varepsilon$	
x_l	$y_l + \varepsilon$		$\Delta^2 y_{l-1} - 2\varepsilon$		$\Delta^4 y_{l-2} + 6\varepsilon$
		$\Delta y_l - \varepsilon$		$\Delta^3 y_{l-1} + 3\varepsilon$	
x_{l+1}	y_{l+1}		$\Delta^2 y_l + \varepsilon$		$\Delta^4 y_{l-1} - 4\varepsilon$
		Δy_{l+1}		$\Delta^3 y_l - \varepsilon$	
x_{l+2}	y_{l+2}		$\Delta^2 y_{l+1}$		$\Delta^4 y_l + \varepsilon$
		Δy_{l+2}			
x_{l+3}	y_{l+3}				

It is seen from the table that the error influences $k+1$ values of the kth difference, for which the coefficients of ε are the binomial coefficients

$$C_k^{(m)} = \frac{k(k-1)\ldots(k-m+1)}{m!}, \qquad m = 0, 1, 2, \ldots, k,$$

with alternating signs.

If the function is sufficiently smooth, that is, if it and its derivatives are approximated sufficiently well by polynomials of low degree, then it is clear that differences of some order may be constructed to an accuracy of one or two units in the last place. This fact may be used to control the validity of the calculation of function values. The table given above of the propagation of errors in finite differences permits one to discover an "isolated" error, and remove it.

Example 2. We shall cite an example which illustrates the process of finding an error in a table. Table 17 gives the finite differences of the polynomial.

$$P(x) = 70x^4 - 140x^3 + 90x^2 - 20x + 1, \quad x = 0(0.02)0.20.$$

Table 17

x	$P(x)$	$\Delta P(x)$	$\Delta^2 P(x)$	$\Delta^3 P(x)$	$\Delta^4 P(x)$
0	1	-36511			
0.02	0.63489		6544		
		-29967		-632	
0.04	0.33522		5912		27
		-24055		-605	
0.06	0.09467		5307		27
		-18748		-578	
0.08	-0.09281		4729		19
		-14019		-559	
0.10	-0.23300		4170		60
		-9849		-499	
0.12	-0.33149		3671		-23
		-6178		-522	
0.14	-0.39327		3149		58
		-3029		-464	
0.16	-0.42356		2685		23
		-344		-441	
0.18	-0.42700		2244		
		1900			
0.20	-0.40800				

If the values of $P(x)$ are given exactly, then the fourth difference will be constant by property 3. Since the values of $P(x)$ are given to five decimal places, the worst case of deviation of the fourth difference from its exact value is equal to

$$0.5 \cdot 10^{-5}(1+4+6+4+1) = 8 \cdot 10^{-5},$$

since the value of the fourth difference is influenced by five values of $P(x)$, for which the errors enter into the fourth difference with coefficients $C_4^{(k)}$ ($k = 0, 1, 2, 3, 4$). Thus the fourth differences in our case may not differ by more than about 16 units in the fifth place. However, in the table, the fourth differences oscillate strongly in value, and this indicates the presence of an error.

In the table, the largest distortion in the fourth difference occurs on the level $x = 0.12$. This shows that the error apparently is in the value of $P(x)$ for $x = 0.12$. It is possible to indicate the magnitude of the error. Denote the value of the fourth difference by α. Since the sum of the binomial coefficients with alternating signs is equal to zero, the sum of all the fourth differences (in our case there are seven of them) is approximately equal to 7α:

$$191 \cong 7\alpha,$$

from which $\alpha = 27$ units in the fifth decimal place. The value of the fourth derivative on the line $x = 0.12$ is equal to $x+6\varepsilon$, from which

$$27+6\varepsilon \cong -23$$

and $\varepsilon = -8$.

The corrected value of the polynomial equals

$$P(0.12) = -0.33149 + 0.00008 = -0.33141,$$

which differs from the true value -0.3314048 by half a unit in the fifth decimal place. The values of the fourth differences after correction (in terms of units in the fifth decimal place) are:

27;   27;   27;   28;   25;   26;   31.

3. Divided differences

Frequently, one has to deal with functions with values which are known for unequally spaced values of the argument. It is not possible to use finite differences for such, but in these cases what are called *difference ratios* or *divided differences*.

Suppose that for various values $x_0, x_1, \ldots, x_n$ of the argument, a table of values of the function $f(x)$ is given,

$$f(x_0), f(x_1), \ldots, f(x_n).$$

The values $x_0, x_1, \ldots, x_n$ are not assumed to be equally spaced, nor are they required to increase or decrease in a monotone fashion with increasing index.

The numbers

$$f(x_1, x_0) = \frac{f(x_1) - f(x_0)}{x_1 - x_0}, \qquad f(x_2, x_1) = \frac{f(x_2) - f(x_1)}{x_2 - x_1},$$

$$f(x_3, x_2) = \frac{f(x_3) - f(x_2)}{x_3 - x_2}, \ldots$$

are called divided differences of the first order. Geometrically, the divided difference of first order is the slope of a chord to the graph of the function $y = f(x)$.

Divided differences of second order are defined in terms of divided differences of first order

$$f(x_2, x_1, x_0) = \frac{f(x_2, x_1) - (x_1, x_0)}{x_2 - x_0},$$

$$f(x_3, x_2, x_1) = \frac{f(x_3, x_2) - f(x_2, x_1)}{x_3 - x_1}, \ldots$$

In an analogous fashion, one defines divided differences of third order

$$f(x_3, x_2, x_1, x_0)$$

$$= \frac{f(x_3, x_2, x_1) - f(x_2, x_1, x_0)}{x_3 - x_0}, \ldots \text{ and so forth.}$$

Tables of divided differences may be written in the same way as tables of finite differences.

The construction of a table of finite differences requires a greater outlay of computational labor than the construction of a table of finite differences.

x	Divided differences of order			
	0	1	2	3
x_0	$f(x_0)$			
		$f(x_1, x_0)$		
x_1	$f(x_1)$		$f(x_2, x_1, x_0)$	
		$f(x_2, x_1)$		$f(x_3, x_2, x_1, x_0)$
x_2	$f(x_2)$		$f(x_3, x_2, x_1)$	
		$f(x_3, x_2)$		$f(x_4, x_3, x_2, x_1)$
x_3	$f(x_3)$		$f(x_4, x_3, x_2)$	
		$f(x_4, x_3)$		
x_4	$f(x_4)$			

Example. We shall construct the divided differences of the polynomial

$$P(x) = x^4 + 2.5x^3 - 3x^2 + 2x + 1.8$$

for the system of arguments

$$x_0 = 1.2, \ x_1 = 1, \ x_2 = 1.3, \ x_3 = 1.5, \ x_4 = 1.1, \ x_5 = 1.8.$$

The calculations are recorded in Table 18.

Table 18

x	Divided differences of order				
	0	1	2	3	4
1.2	6.27360				
		9.86800			
1	4.30000		13.94000		
		11.26200		7.50000	
1.3	7.67860		16.19000		1.00000
		19.35700		7.40000	
1.5	11.55000		16.93000		1.00000
		15.97100		8.20000	
1.1	5.16160		21.03000		
		22.28000			
1.8	20.75760				

We note the basic properties of divided differences:

1. The divided difference of the sum $F(x) = f(x) + g(x)$ of functions is equal to the sum of the divided differences of the addends $f(x)$ and $g(x)$.

2. If $F(x) = Af(x)$, where A is a constant, then

$$F(x_{l+1}, x_l) = Af(x_{l+1}, x_l).$$

Properties 1 and 2 are valid for divided differences of any order.

3. *The divided differences of order n of a polynomial of degree n are equal to a constant, and, consequently, the divided differences of higher order are equal to zero.*

By virtue of what has been said earlier, it is sufficient to establish this property for the simple polynomial $P(x) = x^n$, and for this, in turn, it is sufficient to prove that the divided difference $P(x_1, x_0)$ of first order is a polynomial of degree $n-1$ in the arguments x_0, x_1. However, this is obvious:

$$P(x_1, x_0) = \frac{x_1^n - x_0^n}{x_1 - x_0} = x_1^{n-1} + x_1^{n-2}x_0 + \ldots + x_0^{n-1}.$$

One may express a divided difference of any order by means of values of the function $f(x)$. One has that

$$f(x_1, x_0) = \frac{f(x_1) - f(x_0)}{x_1 - x_0} = \frac{f(x_0)}{x_0 - x_1} + \frac{f(x_1)}{x_1 - x_0}.$$

Now, one may form the divided difference of second order:

$$f(x_2, x_1, x_0) = \frac{f(x_2, x_1) - f(x_1, x_0)}{x_2 - x_0} =$$

$$= \frac{1}{x_2 - x_0} \left\{ \left[\frac{f(x_1)}{x_1 - x_2} + \frac{f(x_2)}{x_2 - x_1} \right] - \left[\frac{f(x_0)}{x_0 - x_1} + \frac{f(x_1)}{x_1 - x_0} \right] \right\} =$$

$$= \frac{f(x_0)}{(x_0 - x_1)(x_0 - x_2)} + \frac{f(x_2)}{(x_2 - x_0)(x_2 - x_1)} +$$

$$+ \frac{f(x_1)}{x_2 - x_0} \left[\frac{1}{x_1 - x_2} - \frac{1}{x_1 - x_0} \right] =$$

$$= \frac{f(x_0)}{(x_0 - x_1)(x_0 - x_2)} + \frac{f(x_1)}{(x_1 - x_0)(x_1 - x_2)} + \frac{f(x_2)}{(x_2 - x_0)(x_2 - x_1)}.$$

It is not difficult to verify that the divided difference of order n may be represented in the form

$$f(x_n, x_{n-1}, \ldots, x_1, x_0) =$$

$$= \sum_{k=0}^{n} \frac{f(x_k)}{(x_k - x_0) \ldots (x_k - x_{k-1})(x_k - x_{k+1}) \ldots (x_k - x_n)}. \tag{3.1}$$

The special notation

$$\omega(x) = (x - x_0)(x - x_1) \ldots (x - x_n) \tag{3.2}$$

is introduced for the polynomial of degree $n+1$ which has the zeros $x_0, x_1, \ldots, x_n$. Obviously,

$$\omega'(x_k) = (x_k - x_0) \ldots (x_k - x_{k-1})(x_k - x_{k+1}) \ldots (x_k - x_n). \tag{3.3}$$

If relationship (3.3) is used, then formula (3.1) may be written in the more compact form:

$$f(x_n, x_{n-1}, \ldots, x_1, x_0) = \sum_{k=0}^{n} \frac{f(x_k)}{\omega'(x_k)}. \tag{3.4}$$

The following property of divided differences follows from formula 3.4.
4. *Divided differences are symmetric functions of their arguments, in other words, they are invariant under arbitrary interchanges of their arguments.*
A formula is now stated, which expresses the function value $f(x_k)$ by means of divided differences:

$$f(x_k) = f(x_0) + (x_k - x_0)f(x_0, x_1) + (x_k - x_0)(x_k - x_1)f(x_0, x_1, x_2) +$$

$$\ldots + (x_k - x_0)(x_k - x_1) \ldots (x_k - x_{k-1})f(x_0, x_1, \ldots, x_k). \tag{3.5}$$

For $k = 1$, formula (3.5) follows from the definition of the divided difference

$$f(x_1) = f(x_0) + (x_1 - x_0)f(x_0, x_1).$$

The method of mathematical induction will be employed. It will be assumed that formula (3.5) is valid for $k = n-1$, and its validity will be proved for $k = n$. By the inductive hypothesis,

$$f(x_n) = f(x_1) + (x_n - x_1)f(x_1, x_2) +$$

$$+ (x_n - x_1)(x_n - x_2)f(x_1, x_2, x_3) + \ldots$$

$$\ldots + (x_n - x_1)(x_n - x_2) \ldots (x_n - x_{n-1})f(x_1, x_2, \ldots, x_n). \tag{3.6}$$

By the definition of divided differences,

$$f(x_1, x_2, \ldots, x_k) = f(x_0, x_1, \ldots, x_{k-1}) +$$
$$+ (x_k - x_0) f(x_0, x_1, \ldots, x_k). \tag{3.7}$$

In (3.6), the right-hand side of (3.7) is substituted for $f(x_1, x_2, \ldots, x_k)$, $k = 1, 2, \ldots, n$. One obtains that

$$f(x_n) = f(x_0) + (x_1 - x_0) f(x_0, x_1) +$$
$$+ (x_n - x_1)\{f(x_0, x_1) + (x_2 - x_0) f(x_0, x_1, x_2)\} +$$
$$+ (x_n - x_1)(x_n - x_2)\{f(x_0, x_1, x_2) +$$
$$+ (x_3 - x_0) f(x_0, x_1, x_2, x_3)\} + \ldots$$
$$\ldots + (x_n - x_1)(x_n - x_2) \ldots (x_n - x_{n-1}) \times$$
$$\times \{f(x_0, x_1, \ldots, x_{n-1}) + (x_n - x_0) f(x_0, x_1, \ldots, x_n)\}.$$

Unifying the terms which contain the same divided difference, one obtains (3.5) for $k = n$.

We note one important fact which is a result of formula (3.5).

Consider the polynomial of degree not exceeding n,

$$P(x) = f(x_0) + (x - x_0) f(x_0, x_1) +$$
$$+ (x - x_0)(x - x_1) f(x_0, x_1, x_2) + \ldots$$
$$\ldots + (x - x_0)(x - x_1) \ldots (x - x_{n-1}) f(x_0, x_1, \ldots, x_n). \tag{3.8}$$

By virtue of formula (3.5), this polynomial is equal to $f(x_k)$ for $x = x_k$:

$$P(x_k) = f(x_k), \qquad k = 0, 1, 2, \ldots, n. \tag{3.9}$$

The connection between divided differences and differentiation is given by the following theorem.

Theorem 1. If $f(x)$ is differentiable n times on the interval [a, b], and the points $x_0, x_1, \ldots, x_n$ belong to [a, b], then there exists a point ξ, $a < \xi < b$, such that

$$f(x_0, x_1, \ldots, x_n) = \frac{f^{(n)}(\xi)}{n!}. \tag{3.10}$$

Proof. Construct the difference

$$\phi(x) = f(x) - P(x),$$

where $P(x)$ is the polynomial defined by the formula (3.8).

The function $\phi(x)$ is differentiable n times on [a, b] and vanishes at the

$n+1$ distinct points $x_0, x_1, \ldots, x_n$ of this interval. By Rolle's theorem, $\phi'(x)$ has at least n zeros interior to $[a, b]$. Applying Rolle's theorem to $\phi'(x)$, the second derivative $\phi''(x)$ has at least $n-1$ distinct zeros interior to $[a, b]$. Proceeding in the same way, one finds that $\phi^{(n)}(x)$ has at least one zero interior to $[a, b]$. Denote it by ξ, $a < \xi < b$.

In the derivative

$$\phi^{(n)}(x) = f^{(n)}(x) - n!\, f(x_0, x_1, \ldots, x_n)$$

set $x = \xi$. Since $\phi^{(n)}(\xi) = 0$, one obtains (3.10).

If the arguments x_k are equally spaced,

$$x_k = x_0 + kh, \ h > 0, \ k = 0, 1, 2, \ldots, N,$$

then $f(x)$ may be constructed by finite differences, and one obtains that

$$f(x_0, x_1) = \frac{f(x_1) - f(x_0)}{x_1 - x_0} = \frac{\Delta f(x_0)}{h},$$

$$f(x_1, x_2) = \frac{f(x_2) - f(x_1)}{x_2 - x_1} = \frac{\Delta f(x_1)}{h}.$$

From the definitions of divided differences and finite differences of the second order, one has that

$$f(x_0, x_1, x_2) = \frac{f(x_1, x_2) - f(x_0, x_1)}{x_2 - x_0} =$$

$$= \frac{1}{2h}\left[\frac{\Delta f(x_1)}{h} - \frac{\Delta f(x_0)}{h}\right] = \frac{1}{2!}\frac{\Delta^2 f(x_0)}{h^2}.$$

It is not difficult to show that for arbitrary n,

$$f(x_0, x_1, \ldots, x_n) = \frac{\Delta^n f(x_0)}{n!\, h^n}. \tag{3.11}$$

By comparison of formulas (3.10) and (3.11), the theorem which establishes the connection between finite differences and derivatives is obtained.

Theorem 2. If $f(x)$ is differentiable n times on $[x_0, x_0 + h]$, then there exists a point ξ interior to this interval such that

$$\Delta^n f(x_0) = h^n f^{(n)}(\xi). \tag{3.12}$$

It follows from formula (3.12) that $\Delta^n f(x_0)$ is a small quantity of order

h^n (if h is small), and that by decreasing the step h in the table by a factor of r leads to a decrease in the differences of order n by a factor of r^n.

4. The general problem of interpolation

In sufficient generality for our view, the problem of interpolation may be formulated in the following way. Suppose that the values of the function $f(x)$ are known at p_0 distinct points

$$x_1^{(0)}, x_2^{(0)}, \ldots, x_{p_0}^{(0)},$$

the values of its derivative $f'(x)$ are known at p_1 distinct points

$$x_1^{(1)}, x_2^{(1)}, \ldots, x_{p_1}^{(1)}$$

and so forth, the values of the derivative $f^{(m)}(x)$ of order m being known at p_m distinct points

$$x_1^{(m)}, x_2^{(m)}, \ldots, x_{p_m}^{(m)}$$

Therefore, there are

$$n = p_0 + p_1 + \ldots + p_m$$

facts given concerning the function $f(x)$. It is required to construct a function which approximates $f(x)$. One of the ways to solve this problem is to construct a generalized interpolating polynomial.

One chooses a system of linearly independent functions of simple form

$$\phi_1(x), \phi_2(x), \phi_3(x), \ldots, \phi_k(x), \ldots \tag{4.1}$$

and constructs a linear combination with constant coefficients of the first n functions of the system

$$S_n(x) = c_1 \phi_1(x) + c_2 \phi_2(x) + \ldots + c_n \phi_n(x). \tag{4.2}$$

One may call $S_n(x)$ a *generalized polynomial*. The constants $c_1, c_2, \ldots, c_m$ are chosen so that at the point $x_i^{(k)}$, the derivative of order k of the generalized polynomial $S_m(x)$ coincides with the kth derivative of the function $f(x)$:

$$S_n^{(k)}(x_i^{(k)}) = f^{(k)}(x_i^{(k)}), \tag{4.3}$$

$$k = 0, 1, 2, \ldots, m, \quad i = 1, 2, \ldots, p_k.$$

Equation (4.3) leads to the following linear algebraic system for the unknown coefficients $c_1, c_2, \ldots, c_m$ of the generalized polynomial (4.2):

$$
\left.
\begin{aligned}
&c_1\phi_1(x_1^{(0)})+c_2\phi_2(x_1^{(0)}) &&+ \ldots +c_n\phi_n(x_1^{(0)}) &&= f(x_1^{(0)}), \\
&c_1\phi_1(x_2^{(0)})+c_2\phi_2(x_2^{(0)}) &&+ \ldots +c_n\phi_n(x_2^{(0)}) &&= f(x_2^{(0)}), \\
& \cdots \cdots \cdots \cdots \cdots \cdots \cdots \cdots \cdots \cdots \cdots \cdots \\
&c_1\phi_1(x_{p_0}^{(0)})+c_2\phi_2(x_{p_0}^{(0)}) &&+ \ldots +c_n\phi_n(x_{p_0}^{(0)}) &&= f(x_{p_0}^{(0)}), \\
&c_1\phi_1'(x_1^{(1)})+c_2\phi_2'(x_1^{(1)}) &&+ \ldots +c_n\phi_n'(x_1^{(1)}) &&= f'(x_1^{(1)}), \\
&c_1'\phi_1(x_2^{(1)})+c_2\phi_2'(x_2^{(1)}) &&+ \ldots +c_n\phi_n'(x_2^{(1)}) &&= f'(x_2^{(1)}), \\
& \cdots \cdots \cdots \cdots \cdots \cdots \cdots \cdots \cdots \cdots \cdots \cdots \\
&c_1\phi_1'(x_{p_1}^{(1)})+c_2\phi_2'(x_{p_1}^{(1)}) &&+ \ldots +c_n\phi_n'(x_{p_1}^{(1)}) &&= f'(x_{p_1}^{(1)}), \\
& \cdots \cdots \cdots \cdots \cdots \cdots \cdots \cdots \cdots \cdots \cdots \cdots \\
&c_1\phi_1^{(m)}(x_1^{(m)})+c_2\phi_2^{(m)}(x_1^{(m)})+ \ldots +c_n\phi_n^{(m)}(x_1^{(m)}) &&= f^{(m)}(x_1^{(m)}), \\
&c_1\phi_1^{(m)}(x_2^{(m)})+c_2\phi_2^{(m)}(x_2^{(m)})+ \ldots +c_n\phi_n^{(m)}(x_2^{(m)}) &&= f^{(m)}(x_2^{(m)}), \\
& \cdots \cdots \cdots \cdots \cdots \cdots \cdots \cdots \cdots \cdots \cdots \cdots \\
&c_1\phi_1^{(m)}(x_{p_m}^{(m)})+c_2\phi_2^{(m)}(x_{p_m}^{(m)})+ \ldots +c_n\phi_n^{(m)}(x_{p_m}^{(m)}) &&= f^{(m)}(x_{p_m}^{(m)})
\end{aligned}
\right\} \quad (4.4)
$$

If the determinant of the system (4.4) is different from zero, then the system has a unique solution, and, consequently, a unique generalized polynomial (4.2) may be constructed which approximates the function $f(x)$:

$$ f(x) \cong S_n(x). $$

The error of this approximate equality depends on the properties of the function $f(x)$, on the system (4.1), and on the choice of the points $x_i^{(k)}$.

In what follows, we shall always deal with the system of functions

$$ \phi_k(x) = x^{k-1}, \qquad k = 1, 2, 3, \ldots $$

In this case, the generalized polynomial (4.2) turns out to be an ordinary polynomial, and the interpolation is said to be algebraic. Algebraic interpolation is frequently used in approximate calculations.

If the function $f(x)$ is periodic, with period 2π, then it is natural to use the trigonometric system

$$ 1, \cos x, \sin x, \cos 2x, \sin 2x, \ldots, \cos kx, \sin kx, \ldots $$

as the system $\{\phi_k(x)\}$. In this case, the generalized polynomial (4.2) is called a trigonometric polynomial, and the interpolation is called trigonometric.

We remark that the set of trigonometric polynomials is dense in the class of 2π-periodic continuous functions by the second theorem of Weierstrass.

5. Interpolation of function values

We shall consider the frequent and simple case of algebraic interpolation, in which at the basic points of interpolation

$$x_0, x_1, \ldots, x_n$$

there are given function values

$$f(x_0), f(x_1), \ldots, f(x_n).$$

The points are assumed to be distinct. It is required to construct the polynomial

$$P_n(x) = a_0 x^n + a_1 x^{n-1} + a_2 x^{n-2} + \ldots a_n, \tag{5.1}$$

which coincides with $f(x)$ at the points x_i:

$$P_n(x_i) = f(x_i), \qquad i = 0, 1, 2, \ldots, n. \tag{5.2}$$

Let M_k denote the point in the (x, y)-plane with coordinates $(x_k, f(x_k))$, $k = 0, 1, 2, \ldots, n$. Geometrically, condition (5.2) means that the graph of the polynomial $y = P_n(x)$ must pass through the points M_0, $M_1, \ldots, M_n$. The case $n = 3$ is depicted in figure 12. Since the graph of the curve defined by the equation $y = P_n(x)$ is a parabola (of degree n or lower), algebraic interpolation is sometimes called parabolic interpolation.

Conditions (5.2), which the polynomial (5.1) must satisfy, leads to a linear algebraic system for the unknown $a_0, a_1, \ldots, a_n$:

$$\left. \begin{aligned} a_0 x_0^n + a_1 x_0^{n-1} + a_2 x_0^{n-2} + \ldots a_n &= f(x_0), \\ a_0 x_1^n + a_1 x_1^{n-1} + a_2 x_1^{n-2} + \ldots a_n &= f(x_1), \\ \cdots \cdots \cdots \cdots \cdots \cdots \cdots \cdots \\ a_0 x_n^n + a_1 x_n^{n-1} + a_2 x_n^{n-2} + \ldots a_n &= f(x_n). \end{aligned} \right\} \tag{5.3}$$

The determinant of this system is called the Vandermonde determinant

$$
W_{n+1} = \begin{vmatrix} x_0^n & x_0^{n-1} & x_0^{n-2} & \dots & x_0 & 1 \\ x_1^n & x_1^{n-1} & x_1^{n-2} & \dots & x_1 & 1 \\ \cdot & \cdot & \cdot & \cdot & \cdot & \cdot \\ x_n^n & x_n^{n-1} & x_n^{n-2} & \dots & x_n & 1 \end{vmatrix}.
$$

It is different from zero, since the numbers $x_0, x_1, \dots, x_n$ are assumed to be distinct. It follows from this that the system (5.3) has a unique solution, and, consequently, the interpolating polynomial exists and is unique.

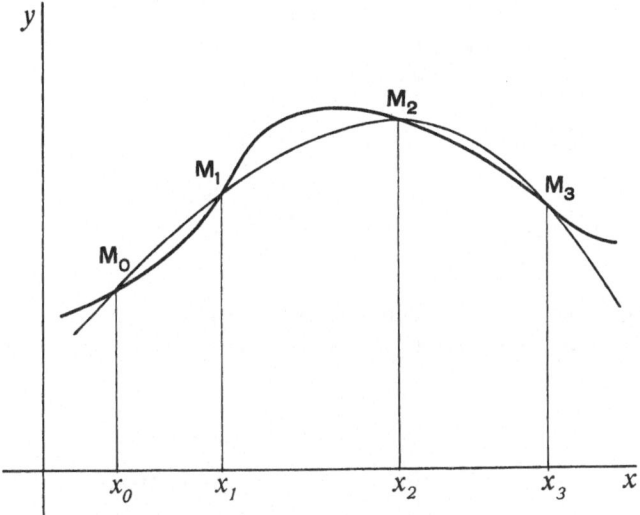

Fig. 12

We rewrite system (5.3) and adjoin equation (5.1) to it:

$$
\left.\begin{aligned}
a_0 x_0^n + a_1 x_0^{n-1} + \dots + a_n - 1 \cdot f(x_0) &= 0, \\
a_0 x_1^n + a_1 x_1^{n-1} + \dots + a_n - 1 \cdot f(x_1) &= 0, \\
\cdot\cdot\cdot\cdot\cdot\cdot\cdot\cdot\cdot\cdot\cdot\cdot\cdot\cdot\cdot\cdot\cdot\cdot\cdot\cdot\cdot\cdot\cdot\cdot\cdot\cdot& \\
a_0 x_n^n + a_1 x_n^{n-1} + \dots + a_n - 1 \cdot f(x_n) &= 0, \\
a_0 x^n + a_1 x^{n-1} + \dots + a_n - 1 \cdot P_n(x) &= 0.
\end{aligned}\right\}
$$

This system of $n+2$ equations may be considered to be a homogeneous linear algebraic system with $n+2$ "unknowns" $a_0, a_1, \dots, a_n, -1$. Then, its determinant is equal to zero:

$$\begin{vmatrix} x_0^n & x_0^{n-1} & \ldots & x_0 & 1 & f(x_0) \\ x_1^n & x_1^{n-1} & \ldots & x_1 & 1 & f(x_1) \\ \hdotsfor{6} \\ x_n^n & x_n^{n-1} & \ldots & x_n & 1 & f(x_n) \\ x^n & x^{n-1} & & x & 1 & P_n(x) \end{vmatrix} = 0.$$

Decomposing this determinant into the sum of two determinants

$$\begin{vmatrix} x_0^n & x_0^{n-1} & \ldots & x_0 & 1 & 0 \\ \hdotsfor{6} \\ x_n^n & x_n^{n-1} & \ldots & x_n & 1 & 0 \\ x^n & x^{n-1} & \ldots & x & 1 & P_n(x) \end{vmatrix} +$$

$$+ \begin{vmatrix} x_0^n & x_0^{n-1} & \ldots & x_0 & 1 & f(x_0) \\ \hdotsfor{6} \\ x_n^n & x_n^{n-1} & \ldots & x_n & 1 & f(x_n) \\ x^n & x^{n-1} & \ldots & x & 1 & 0 \end{vmatrix} = 0$$

and noting that the first determinant is equal to $P_n(x) \cdot W_{n+1}$, one obtains a representation of the interpolating polynomial in terms of determinants

$$P_n(x) = -\frac{1}{W_{n+1}} \begin{vmatrix} x_0^n & x_0^{n-1} & \ldots & x_0 & 1 & f(x_0) \\ x_1^n & x_1^{n-1} & \ldots & x_1 & 1 & f(x_1) \\ \hdotsfor{6} \\ x_n^n & x_n^{n-1} & \ldots & x_n & 1 & f(x_n) \\ x^n & x^{n-1} & \ldots & x & 1 & 0 \end{vmatrix}. \tag{5.4}$$

We shall give two further representations of $P_n(x)$, Lagrange's formula and Newton's formula, which in many cases are more convenient than the representation (5.4). We begin with the construction of the interpolating polynomial in the form of Lagrange.

We construct the polynomial $\omega_k(x)$, which is equal to one for $x = x_k$, and is equal to zero at the remaining knots $x_0, x_1, \ldots, x_{k-1}, x_{k+1}, \ldots, x_n$. Obviously,

$$\omega_k(x) = C(x-x_0)(x-x_1) \ldots (x-x_{k-1})(x-x_{k+1}) \ldots (x-x_n),$$

where the constant C is determined by the condition $\omega_k(x_k) = 1$. Thus,

$$\omega_k(x) = \frac{(x-x_0)(x-x_1)\ldots(x-x_{k-1})(x-x_{k+1})\ldots(x-x_n)}{(x_k-x_0)(x_k-x_1)\ldots(x_k-x_{k-1})(x_k-x_{k+1})\ldots(x_k-x_n)}.$$

It is clear that

$$P_n(x) = \sum_{k=0}^{n} \omega_k(x)f(x_k). \tag{5.5}$$

Actually, since $P_n(x)$ is a linear combination with constant coefficients of polynomials of degree n, it is a polynomial of degree $\leq n$ for which

$$P_n(x_i) = \sum_{k=0}^{n} \omega_k(x_i)f(x_k) = f(x_i).$$

One may use the notation (3.2),

$$\omega(x) = (x-x_0)(x-x_1)\ldots(x-x_n).$$

One has that

$$\omega_k(x) = \frac{\omega(x)}{(x-x_k)\omega'(x_k)},$$

and formula (5.5) may be written in the form

$$P_n(x) = \sum_{k=0}^{n} \frac{\omega(x)}{(x-x_k)\omega'(x_k)}f(x_k). \tag{5.6}$$

Formula (5.6) is called *Lagrange's interpolation formula.*

Newton's interpolation formula was essentially established in § 3. Formula (3.8) gives a representation of the interpolating polynomial in the form due to Newton:

$$P_n(x) = f(x_0)+(x-x_0)f(x_0,x_1)+(x-x_0)(x-x_1)f(x_0,x_1,x_2)+\ldots$$
$$\ldots +(x-x_0)(x-x_1)\ldots(x-x_{n-1})f(x_0,x_1,\ldots,x_n). \tag{5.7}$$

Newton's formula has an advantage in comparison with Lagrange's formula in the following respect. If the polynomial $P_n(x)$ has already been constructed, but does not give a sufficiently good approximation, then the need arises to add a new knot and construct $P_{n+1}(x)$. In both Newton's and Lagrange's formulas, the addition of a new knot adds one term, but in Newton's formula, the previous terms remain unaltered, while in Lagrange's formula all terms are changed.

The numerical evaluation of $P_n(x)$ for $x = \tilde{x}$ by means of Newton's formula proceeds in the following way. Construct a table of divided differences for the function $f(x)$ (for definiteness, we take $n = 4$).

x	Divided differences of order				
	0	1	2	3	4
x_0	$f(x_0)$				
		$f(x_0, x_1)$			
x_1	$f(x_1)$		$f(x_0, x_1, x_2)$		
		$f(x_1, x_2)$		$f(x_0, x_1, x_2, x_3)$	
x_2	$f(x_2)$		$f(x_1, x_2, x_3)$		$f(x_0, x_1, x_2, x_3, x_4)$
		$f(x_2, x_3)$		$f(x_1, x_2, x_3, x_4)$	
x_3	$f(x_3)$		$f(x_2, x_3, x_4)$		
		$f(x_3, x_4)$			
x_4	$f(x_4)$				

The divided differences which enter into Newton's interpolation formula are located on the upper diagonal line in the table. Their values are underlined in the table. One writes formula (5.7) for $x = \tilde{x}$ in the form

$$P_4(\tilde{x}) = \{[(f(x_0, x_1, x_2, x_3, x_4)(\tilde{x} - x_3) +$$
$$+ f(x_0, x_1, x_2, x_3))(\tilde{x} - x_2) +$$
$$+ f(x_0, x_1, x_2)](\tilde{x} - x_1) + f(x_0, x_1)\}(\tilde{x} - x_0) + f(x_0). \qquad (5.8)$$

It is seen from this that the calculation of $P_n(\tilde{x})$ may proceed according to a scheme which is analogous to Horner's method for the numerical evaluation of polynomials.

Example. A table of the function $f(x) = \sinh x$ is given:

x	0.40	0.55	0.65	0.80	0.90	1.05
$f(x)$	0.41075	0.57815	0.69675	0.88811	1.02652	1.25386

It is required to find an approximate value of $\sinh x$ for $x = \tilde{x} = 0.596$. The divided differences of $f(x)$ are calculated in Table 19. It is seen from the table that the divided differences of order four are constant, and thus one may replace $f(x)$ by the polynomial $P_4(x)$.

Table 19

x	Divided differences of order				
	0	1	2	3	4
0.40	0.41075				
		1.1160			
0.55	0.57815		0.2800		
		1.1860		0.197	
0.65	0.69675		0.3588		0.034
		1.2757		0.214	
0.80	0.88811		0.4336		0.034
		1.3841		0.231	
0.90	1.02652		0.5260		
		1.5156			
1.05	1.25386				

Choose $x_0 = 0.40$. In order to calculate $P_4(\tilde{x})$ by Newton's formula, one needs the numbers

$$\tilde{x} - x_3 = -0.204, \quad \tilde{x} - x_2 = -0.054$$
$$\tilde{x} - x_1 = 0.046, \quad \tilde{x} - x_0 = 0.196$$

The calculation of $P_4(\tilde{x})$ by formula (5.8) may be written in the same way as in the case of Horner's method:

0.034	0.197	0.2800	1.1160	0.41075
	-0.007	-0.0103	0.0124	0.22117
	0.190	0.2697	1.1284	0.63192

The last value on the third line is equal to $P_4(\tilde{x})$. Thus, $\sinh \tilde{x} \cong 0.63192$. The approximate value of $\sinh 0.596$ found coincides with that given in [24].

6. On the remainder term in interpolation

If the interpolated function $f(x)$ is a polynomial of degree n or less, then the interpolating polynomial $P_n(x)$ agrees with $f(x)$ exactly:

$$f(x) - P_n(x) \equiv 0.$$

This follows from the uniqueness of the interpolating polynomial.

If $f(x)$ is not a polynomial of degree less than or equal to n on the interval [a, b] which contains the points of interpolation $x_0, x_1, \ldots, x_n$, then the difference

$$R_n(f, x) = f(x) - P_n(x) \tag{6.1}$$

is equal to zero for $x = x_0, x_1, \ldots, x_n$, but is known not to be identically zero on [a, b]. The function $R_n(f, x)$ characterizes the accuracy of the approximation of the function $f(x)$ by the interpolating polynomial $P_n(x)$, and is called the *remainder term* or the *remainder of interpolation*. We shall prove a theorem on the representation of the remainder term in the form of Lagrange.

Theorem. If $f(x)$ is differentiable $n+1$ times on the interval [a, b] which contains the points of interpolation $x_0, x_1, \ldots, x_n$, then for any x in [a, b], there exists a point ξ, $a < \xi < b$, such that

$$R_n(f, x) = \frac{\omega(x)}{(n+1)!} f^{(n+1)}(\xi), \tag{6.2}$$

where

$$\omega(x) = (x - x_0)(x - x_1) \ldots (x - x_n).$$

Proof: Choose a point $x \in$ [a, b] which differs from all of the knots $x_0, x_1, \ldots, x_n$, and consider the function values

$$f(x_0), f(x_1), \ldots, f(x_n), f(x).$$

By formula (3.5) for $k = n+1$ ($x_{n+1} = x$), one has that

$$f(x) = f(x_0) + (x - x_0)f(x_0, x_1) + \ldots$$
$$\ldots + (x - x_0)(x - x_1) \ldots (x - x_{n-1})f(x_0, x_1, \ldots, x_n) +$$
$$+ \omega(x)f(x_0, x_1, \ldots, x_n, x).$$

The sum of the first $n+1$ terms on the right side agrees with the interpolating polynomial $P_n(x)$ [see (5.7)], hence the previous formula may be written in the form

$$R_n(f, x) = \omega(x)f(x_0, x_1, \ldots, x_n, x) \tag{6.3}$$

By theorem 1 of § 3, there exists a point ξ, $a < \xi < b$, such that

$$f(x_0, x_1, \ldots, x_n, x) = \frac{f^{(n+1)}(\xi)}{(n+1)!}.$$

Comparison of the last equation with (6.3) gives (6.2).

The representation (6.2) was obtained on the assumption that x was distinct from all of the knots. However, it obviously holds for $x = x_k$. By $W_{n+1}(M)$, we denote the class of functions defined on the interval [a, b] which have derivatives of the $(n+1)$st order which satisfy the inequality

$$|f^{(n+1)}(x)| \leqq M$$

on [a, b]. Suppose that the points of interpolation $x_0, x_1, \ldots, x_n$ belong to [a, b]. Then by the theorem on the representation of the remainder of interpolation in the form (6.2), one has for any function $f(x) \in W_{n+1}(M)$ that

$$|R_n(f, x)| \leqq \frac{M}{(n+1)!} |\omega(x)|. \tag{6.4}$$

On the other hand, the function

$$f_0(x) = \frac{M}{(n+1)!} \omega(x) \in W_{n+1}(M)$$

coincides exactly with its interpolating polynomial, and, consequently,

$$|R_n(f_0, x)| = \frac{M}{(n+1)!} |\omega(x)|.$$

It follows from this and (6.4) that

$$r(x, x_0, x_1, \ldots, x_n) = \sup |R_n(f, x)| = \frac{M}{(n+1)!} |\omega(x)| \tag{6.5}$$

where the exact upper bound is chosen for all functions in the class $W_{n+1}(M)$. Thus, the exact upper bound of the remainder of interpolation at x for functions in the class $W_{n+1}(M)$ depends upon the points of interpolation and upon x, which justifies the notation $r(x, x_0, x_1, \ldots, x_n)$.

It will be assumed that a table of the values of a function $f(x) \in W_{n+1}(M)$ is given on the interval [a, b] for the values of the argument

$$x_1', x_2', \ldots, x_m', \tag{6.6}$$

where $m > n+1$. The problem may be posed: Which of the arguments

(6.6) are to be taken as points of interpolation $x_0, x_1, \ldots, x_n$ so that at a given point x, the exact upper bound (6.5) will be the least? This is the problem of minimizing the remainder of interpolation at the point x for the class $W_{n+1}(M)$. It will be solved in the following way. As points of interpolation, one chooses from (6.6) the points for which the right-hand side of (6.5), or, what is the same, the quantity

$$|\omega(x)| = |x - x_0||x - x_1| \ldots |x - x_n|$$

assumes the smallest value.

Such a choice is easy to make. For x_0, the point of (6.6) is chosen for which the difference

$$|x - x'_k|, \qquad k = 1, 2, \ldots, m,$$

is the smallest. For x_1, out of the remaining $m - 1$ points of (6.6), the one is chosen for which $|x - x'_k|$ is the smallest, and so forth. Obviously, for this choice of knots, $|\omega(x)|$ has the smallest value.

Consideration of the problem of minimizing the remainder of interpolation at a point arises when one is given a function $f(x)$ defined by a table. Now, it will be assumed that the function $f(x) \in W_{n+1}(M)$ is defined analytically, and it is required to construct its interpolating polynomial on the interval [a, b] using $n + 1$ knots. In this case, one has complete freedom in the choice of the knots on the interval [a, b].

The problem may be posed: Distribute the knots $x_0, x_1, \ldots, x_n$ on [a, b] so that the maximum in x on [a, b] of the exact upper bound (6.5) will be a minimum. This is the problem of minimization of the remainder of interpolation on the interval [a, b] for the class of functions $W_{n+1}(M)$.

The maximum of the exact upper bound (6.5) on x on [a, b] is equal to

$$\frac{M}{(n+1)!} \max_{a \leq x \leq b} |\omega(x)|$$

and, consequently, the problem of minimizing the remainder of interpolation on the interval consists of choosing the knots $x_0, x_1, \ldots, x_n$ on [a, b] for which the quantity

$$\max_{a \leq x \leq b} |\omega(x)|$$

has the smallest value.

It may be shown that such a choice of knots is always possible. In the case that the interval [a, b] $\equiv$ [−1, 1], the desired knots are the zeros of the Čebyšev polynomial

$$T_{n+1}(t) = \cos [(n+1) \cos^{-1}t]. \tag{6.7}$$

These zeros are equal to

$$t_k = \cos \frac{(2k+1)\pi}{2(n+1)}, \qquad k = 0, 1, 2, \ldots, n.$$

They are real, distinct, and lie in the interior of the interval $(−1, 1)$. For the interval [a, b], the required knots x_k are

$$x_k = \frac{b-a}{2} t_k + \frac{a+b}{2}, \qquad k = 0, 1, 2, \ldots, n.$$

7. Interpolation with equally spaced points. Newton's formulas for interpolation at the beginning and end of tables

Various representations of the interpolating polynomial $P_n(x)$ were indicated above for arbitrary dispositions of the knots $x_0, x_1, \ldots, x_n$. In the theory and practice of approximate calculation, interpolation with equally spaced points plays a big rôle. We shall consider representations of the interpolating polynomial in this case.

To begin, Newton's formula for interpolation at the beginning of a table will be introduced. This formula gives the interpolating polynomial $P_n(x)$ for a function which, at the points

$$a+kh, \ k = 0, 1, 2, \ldots, N, \ N \geqq n, \tag{7.1}$$

takes on the values

$$y_k = f(a+kh). \tag{7.2}$$

The formula is applicable to the calculation of the value of the function $f(x)$ for an x which lies close to a, $a \leqq x \leqq a+h/2$. It will be assumed that for points to the left of a in the table, $f(x)$ is not defined, that is, the values $y_{-k} = f(a−kh)$ are unknown for $k = 1, 2, \ldots$. This means that the table begins at the point a, and this circumstance is noted in the name of the formula.

On the basis of what was said in § 6 about the minimum in the remainder of interpolation at a point, for interpolation one takes the $n+1$ knots

$$x_0 = a, \ x_1 = a+h, \ldots, \ x_n = a+nh, \tag{7.3}$$

which are closest to the point x. Newtons' interpolation formula (5.7) is used,

$$P_n(x) = f(x_0) + (x-x_0)f(x_0, x_1) +$$
$$+ (x-x_0)(x-x_1)f(x_0, x_1, x_2) \ldots$$
$$\ldots + (x-x_0)(x-x_1) \ldots (x-x_{n-1})f(x_0, x_1, \ldots, x_n). \tag{7.4}$$

If the knots of interpolation are permuted in order, for example, by taking

$$x_0 = a, \ x_1 = a+nh, \ x_2 = a+2h, \ldots$$
$$\ldots, x_{n-1} = a+(n-1)h, \ x_n = a+h,$$

then one obtains one and the same value of $P_n(x)$. However, for calculation, one essentially always chooses the knots in just the order (7.3). It will be assumed that in applying formula (7.4) for $n+1$ points, it is discovered that one would obtain sufficient accuracy if it were restricted to n points. Then, it is natural to reject the last term in formula (7.4) and use the interpolating polynomial

$$f(x_0) + (x-x_0)f(x_0, x_1) + \ldots + (x-x_0)(x-x_1) \ldots$$
$$\ldots (x-x_{n-2})f(x_0, x_1, \ldots, x_{n-1}).$$

For the numeration of the knots in the order indicated by (7.3), the previous polynomial has the minimal estimate of the remainder at the point x. This is true for all of the polynomials

$$f(x_0),$$
$$f(x_0) + (x-x_0)f(x_0, x_1),$$
$$f(x_0) + (x-x_0)f(x_0, x_1) + (x-x_0)(x-x_1)f(x_0, x_1, x_2),$$
$$\ldots \ldots \ldots \ldots \ldots \ldots \ldots \ldots \ldots \ldots \ldots \ldots \ldots \ldots$$
$$f(x_0) + (x-x_0)f(x_0, x_1) + \ldots + (x-x_0)(x-x_1) \ldots$$
$$\ldots (x-x_{n-1})f(x_0, x_1, \ldots, x_n).$$

This last assertion is verified in that the absolute value of the coefficient for the remainder with respect to formula (7.4) for the knots numbered in the order (7.3) is less than the absolute value of the corresponding coefficient of formula (7.4) for any other choice of the knots.

By transforming formula (7.4), one may replace the remainder of interpolation by a finite difference on the basis of (3.11):

$$f(x_0, x_1, \ldots, x_k) = \frac{\Delta^k f(x_0)}{k! h^k}, \qquad k = 1, 2, \ldots, n.$$

One obtains

$$P_n(x) = f(x_0) +$$

$$+ \frac{x - x_0}{1! h} \Delta f(x_0) + \frac{(x - x_0)(x - x_1)}{2! h^2} \Delta^2 f(x_0) + \cdots$$

$$\cdots + \frac{(x - x_0)(x - x_1) \cdots (x - x_{n-1})}{n! h^n} \Delta^n f(x_0). \tag{7.5}$$

We introduce the new independent variable

$$t = \frac{x - x_0}{h} = \frac{x - a}{h}. \tag{7.6}$$

Formula (7.5) may now be written as:

$$P_n(a + ht) = y_0 + \frac{t}{1!} \Delta y_0 + \frac{t(t-1)}{2!} \Delta^2 y_0 + \cdots$$

$$\cdots + \frac{t(t-1) \cdots (t-n+1)}{n!} \Delta^n y_0. \tag{7.7}$$

Formula (7.7) gives *Newton's interpolating polynomial for interpolation at the beginning of tables.* Note that on the right-hand side of formula (7.7), the finite differences enter which are located on the upper diagonal line of the table of finite differences.

By postulating the existence of $f^{(n+1)}(x)$ on $[a, a+nh]$, one may express the remainder term of interpolation by the polynomial (7.7) [see (6.2)]:

$$R_n(f, x) = f(x) - P_n(x) = h^{n+1} \frac{t(t-1) \cdots (t-n)}{(n+1)!} f^{(n+1)}(\xi), \tag{7.8}$$

where $a < \xi < a + nh$.

We now pass to the construction of Newton's interpolating polynomial for interpolation at the end of tables. Suppose that for equally spaced values of the argument,

$$a-kh, \quad k = 0, 1, 2, \ldots, N, \quad N \geq n, \tag{7.9}$$

there are given values of the function $f(x)$:

$$y_{-k} = f(a-kh). \tag{7.10}$$

It is required to construct the interpolating polynomial from $n+1$ values of the function $f(x)$ suitable for the calculation of $f(x)$ at a point x which lies close to a: $a-h/2 \leq x \leq a$. It will be supposed that the values of $f(x)$ are unknown for $x = a+kh$, $k = 1, 2, \ldots$, so that the table for $f(x)$ actually ends at the point a.

Newton's interpolation formula (7.4) will be used. Clearly, for interpolation, one needs to make into account the knots

$$x_0 = a, \ x_1 = a-h, \ldots, x_n = a-nh$$

in the order indicated here. One replaces the divided differences in formula (7.4) by the finite differences

$$f(x_0, x_1, \ldots, x_k) = f(a-kh, \ldots, a-h, a) =$$

$$= \frac{\Delta^k f(a-kh)}{k!h^k} = \frac{\Delta^k y_{-k}}{k!h^k}$$

One obtains that

$$P_n(x) = y_0 + \frac{x-x_0}{1!h}\Delta y_{-1} + \frac{(x-x_0)(x-x_1)}{2!h^2}\Delta^2 y_{-2} + \cdots$$

$$\cdots + \frac{(x-x_0)(x-x_1)\ldots(x-x_{n-1})}{n!h^n}\Delta^n y_{-n}. \tag{7.11}$$

Introducing the new independent variable

$$t = \frac{x-a}{h},$$

in formula (7.6), one gets *Newton's interpolating polynomial for interpolation at the end of tables*:

$$P_n(a+ht) = y_0 + \frac{t}{1!}\Delta y_{-1} + \frac{t(t+1)}{2!}\Delta^2 y_{-2} + \cdots$$

$$\cdots + \frac{t(t+1)\ldots(t+n-1)}{n!}\Delta^n y_{-n}. \tag{7.12}$$

The finite differences which lie on the lower diagonal line of the table of differences enter into the right-hand side of formula (7.12).

The remainder term for Newton's formula for interpolation at the end of tables is:

$$R_n(f, x) = h^{n+1} \frac{t(t+1)\ldots(t+n)}{(n+1)!} f^{(n+1)}(\xi), \qquad a-nh < \xi < a.$$
$$(7.13)$$

Introducing the notation

$$C_t^{(k)} = \frac{t(t-1)\ldots(t-k+1)}{k!}, \qquad k = 0, 1, 2, \ldots,$$

formula (7.7) may be rewritten in the following way:

$$P_n(a+ht) = y_0 + C_t^{(1)}\Delta y_0 + C_t^{(2)}\Delta y_0 + \ldots + C_t^{(n)}\Delta^n y_0. \qquad (7.14)$$

Obviously,

$$C_{-t}^{(k)} = (-1)^k \frac{t(t+1)\ldots(t+k-1)}{k!},$$

therefore, formula (7.12) may be written in the form

$$P_n(a+ht) = y_0 - C_{-t}^{(1)}\Delta y_{-1} + C_{-t}^{(2)}\Delta^2 y_{-2} - \ldots$$
$$\ldots + (-1)^n C_{-t}^{(n)}\Delta^n y_{-n}. \qquad (7.15)$$

There are tables of the coefficients $C_t^{(k)}$ which will be used in the following computations. We indicate first of all the tables [15], which contain tables of values of $C_t^{(k)}$. Small tables of $C_t^{(k)}$ are found in the books [16] and [25].

Example. A table of the function $y = \cos x$, $x = 0(0.1)0.6$ is given. The values of $\cos x$ are taken from [24]. We shall calculate the value of $\cos x$ for $x = 0.575$.

We construct a table of finite differences [see Table 20]. It is seen from the table that the fourth differences may be considered to be constant, and thus we shall interpolate $y = \cos x$ by the polynomial $P_4(x)$. To calculate $\cos 0.048$, we shall use formula (7.14), taking $a = 0$. The underlined differences in the upper diagonal line of the table enter into it. We obtain from formula (7.6) that

$$t = \frac{x-a}{h} = \frac{0.048-0}{0.1} = 0.48.$$

Table 20

x	y	Δy	$\Delta^2 y$	$\Delta^3 y$	$\Delta^4 y$
0	1				
		-500			
0.1	0.99500		-993		
		-1493		13	
0.2	0.98007		-980		12
		-2473		25	
0.3	0.95534		-955		10
		-3428		35	
0.4	0.92106		-920		9
		-4348		44	
0.5	0.87758		-876		
		-5224			
0.6	0.82534				

The numbers $C_t^{(k)}$ are taken from the tables [15] for $t = 0.48$ and $k = 1, 2, 3, 4$:

$$C_t^{(1)} = 0.48; \qquad C_t^{(2)} = -0.12480; \qquad C_t^{(3)} = 0.06323;$$
$$C_t^{(4)} = -0.03984.$$

As a result of the calculations, we obtain

$$\cos 0.048 = 0.99884.$$

The corresponding tabulated value is equal to 0.99885.
To calculate $\cos x$ for $x = 0.575$, we make use of formula (7.15) for $a = 0.6$. We obtain

$$t = \frac{x - a}{h} = \frac{0.575 - 0.6}{0.1} = -0.25.$$

We find $C_{-t}^{(k)}$ in the tables [15] for $-t = 0.25$ and $k = 1, 2, 3, 4$:

$$C_{-t}^{(1)} = 0.25; \qquad C_{-t}^{(2)} = -0.09375; \qquad C_{-t}^{(3)} = 0.05469;$$
$$C_{-t}^{(4)} = -0.03760.$$

The finite differences which figure in formula (7.15) lie along the lower diagonal line in the table. As a result of the calculations, we obtain

$$\cos 0.575 = 0.83919.$$

Here, all figures are correct.

8. Interpolation at equally spaced points. The formulas of Gauss, Stirling, and Bessel

Suppose that the values of the function $f(x)$ are known at the knots

$$\ldots, a-2h, \ a-h, \ a, \ a+h, \ a+2h, \ \ldots,$$

lying to the left and right of a, and suppose that the point of interpolation x satisfies the inequality

$$a < x < a+h. \tag{8.1}$$

Since the table of $f(x)$ extends in both directions from the point x, it is said in this case that the point of interpolation x is found in the middle of the table. The values of the function at the knots will be denoted by

$$y_k = f(a+kh), \ k = 0, \pm 1, \pm 2, \ldots$$

Newton's formula (5.7) will be used to construct the interpolating polynomial from $n+1$ values of the function, with the knots being chosen in the following way:

$$x_0 = a, x_1 = a+h, x_2 = a-h, x_3 = a+2h,$$

$$x_4 = a-2h, \ldots, x_n = a+(-1)^{n+1} \left[\frac{n+1}{2}\right] h. \tag{8.2}$$

Here, $[(n+1)/2]$ denotes the integral part of the number $(n+1)/2$. Of course, if the point x is found to be close to $a+h$, the smallest estimate for the remainder of interpolation would be obtained by choosing the knots in the order

$$x_0 = a+h, x_1 = a, x_2 = a+2h, x_3 = a-h, \ldots$$

Since the only information on the location of x is provided by inequality (8.1), we shall definitely settle on the choice of knots (8.2).

Divided differences will be expressed in terms of finite differences. One has from (3.11) that

$$f(x_0, x_1, \ldots, x_k) =$$

$$= f\left(a, a+h, \ldots, a+(-1)^k \left[\frac{k}{2}\right] h, a+(-1)^{k+1} \left[\frac{k+1}{2}\right] h\right) =$$

$$= \frac{\Delta^k f(a-[k/2]h)}{k!h^k} = \frac{\Delta^k y_{-[k/2]}}{k!h^k}, \qquad k = 1, 2, \ldots, n.$$

Substituting this expression for the divided difference in formula (5.7) gives:

$$P_n(x) = y_0 + \frac{x-x_0}{1!h} \Delta y_0 + \frac{(x-x_0)(x-x_1)}{2!h^2} \Delta^2 y_{-1} +$$

$$+ \frac{(x-x_0)(x-x_1)(x-x_2)}{3!h^3} \Delta^3 y_{-1} + \ldots$$

$$\ldots + \frac{(x-x_0)(x-x_1)\ldots(x-x_{n-1})}{n!h^n} \Delta^n y_{-[n/2]}. \qquad (8.3)$$

Make the change of variable

$$t = \frac{x-x_0}{h} = \frac{x-a}{h}.$$

In terms of the new variable, formula (8.3) may be rewritten as:

$$P_n(a+ht) = y_0 + \frac{t}{1!} \Delta y_0 + \frac{t(t-1)}{2!} \Delta^2 y_{-1} +$$

$$+ \frac{t(t-1)(t+1)}{3!} \Delta^3 y_{-1} + \frac{t(t-1)(t+1)(t-2)}{4!} \Delta^4 y_{-2} + \ldots$$

$$\ldots + \frac{t(t-1)(t+1)\ldots(t+(-1)^{n-1}[n/2])}{n!} \Delta^n y_{-[n/2]}. \qquad (8.4)$$

This formula gives Gauss' interpolating polynomial for interpolation in the middle of tables. In this formula, the finite differences which lie on two horizontal lines in the table of finite differences figure: The even differences on the same line as y_0, and the odd differences on the same line as Δy_0. In the table presented below, these differences are underlined:

x	y	Δy	$\Delta^2 y$	$\Delta^3 y$	$\Delta^4 y$
$a-2h$	y_{-2}				
		Δy_{-2}			
$a-h$	y_{-1}		$\Delta^2 y_{-2}$		
		Δy_{-1}		$\Delta^3 y_{-2}$	
a	$\underline{y_0}$		$\underline{\Delta^2 y_{-1}}$		$\underline{\Delta^4 y_{-2}}$
		$\underline{\Delta y_0}$		$\underline{\Delta^3 y_{-1}}$	
$a+h$	y_1		$\Delta^2 y_0$		
		Δy_1			
$a+2h$	y_2				

The remainder term in Gauss' formula is

$$R_n(f, x) = f(x) - P_n(x) =$$

$$= h^{n+1} \frac{t(t-1)(t+1)\ldots\left(t+(-1)^n\left[\frac{n+1}{2}\right]\right)}{(n+1)!} f^{(n+1)}(\xi), \qquad (8.5)$$

where

$$a+(-1)^n\left[\frac{n}{2}\right]h \lessgtr \xi \lessgtr a+(-1)^{n+1}\left[\frac{n+1}{2}\right]h.$$

We proceed to introduce Stirling's formula. Like Gauss' formula, it is intended for interpolation in the middle of tables. The point of interpolation x lies in the interval $(a, a+h)$, and is found closer to a than to the midpoint of the interval: $a < x < a+h/4$. In order to minimize the remainder at the point x (see § 6), it is appropriate in this case to choose an *odd* number $n+1$ knots, distributed in the following way:

$$x_0 = a, x_1 = a+h, x_2 = a-h, x_3 = a+2h,$$

$$x_4 = a-2h, \ldots, x_{n-1} = a+\frac{n}{2}h, x_n = a-\frac{n}{2}h.$$

The required interpolating polynomial coincides with (8.4), where n is an even number:

$$P_n(a+ht) = y_0 + \frac{t}{1!}\Delta y_0 + \frac{t(t-1)}{2!}\Delta^2 y_{-1} +$$

$$+ \frac{t(t-1)(t+1)}{3!}\Delta^3 y_{-1} + \ldots$$

$$\ldots + \frac{t(t-1)(t+1)(t-2)\ldots((t-n)/2)}{n!}\Delta^n y_{-n/2}. \qquad (8.6)$$

Formula (8.6) may be transformed into a more symmetric form, corresponding to the symmetry of the knots utilized. Namely, the sum of the last two terms in formula (8.6) may be written as

$$\frac{t(t-1)(t+1)\ldots(t-k)(t+k)}{(2k+1)!} \times$$

$$\times \left[\Delta^{2k+1}y_{-k} + \left(\frac{t}{2k+2} - \tfrac{1}{2}\right)\Delta^{2k+2}y_{-k-1}\right], \qquad (8.7)$$

$$k = 0, 1, 2, \ldots, \frac{n-2}{2},$$

and it may be noted that

$$\Delta^{2k+1}y_{-k} - \tfrac{1}{2}\Delta^{2k+2}y_{-k-1} =$$

$$= \frac{\Delta^{2k+1}y_{-k}}{2} + \frac{\Delta^{2k+1}y_{-k} - \Delta^{2k+2}y_{-k-1}}{2}$$

$$= \frac{\Delta^{2k+1}y_{-k} + \Delta^{2k+1}y_{-k-1}}{2}.$$

This expression for $\Delta^{2k+1}y_{-k} - \tfrac{1}{2}\Delta^{2k+2}y_{-k-1}$ may be substituted into (8.7). One obtains that

$$\frac{t(t^2-1)(t^2-2^2)\ldots(t^2-k^2)}{(2k+1)!} \times$$

$$\times \left[\frac{\Delta^{2k+1}y_{-k} + \Delta^{2k+1}y_{-k-1}}{2} + \frac{t}{2k+2}\Delta^{2k+2}y_{-k-1}\right].$$

As a result of these transformations, formula (8.6) takes the form

$$P_n(a+ht) = y_0 + \frac{t}{1!}\frac{\Delta y_0 + \Delta y_{-1}}{2} + \frac{t^2}{2!}\Delta^2 y_{-1} +$$

$$+ \frac{t(t^2-1)}{3!}\frac{\Delta^3 y_{-1} + \Delta^3 y_{-2}}{2} + \frac{t^2(t^2-1)}{4}\Delta^4 y_{-2} + \ldots$$

$$\ldots + \frac{t(t^2-1)(t^2-2^2)\ldots(t^2-((n-2)/2)^2)}{(n-1)!} \times$$

$$\times \frac{\Delta^{n-1}y_{-n-2/2} + \Delta^{n-1}y_{-n/2}}{2} +$$

$$+ \frac{t^2(t^2-1)(t^2-2^2)\ldots(t^2-((n-2)/2)^2)}{n!}\Delta^n y_{-n/2}. \qquad (8.8)$$

Formula (8.8) gives *Stirling's interpolating polynomial*. Into it enter the even differences on the line corresponding to $x = a$, and the averages of the odd difference located on the two neighboring horizontal line. These differences are underlined in the table.

x	y	Δy	$\Delta^2 y$	$\Delta^3 y$	$\Delta^4 y$
$a-h$	y_{-1}		$\Delta^2 y_{-2}$		$\Delta^4 y_{-2}$
		Δy_{-1}		$\Delta^3 y_{-2}$	
a	y_0		$\Delta^2 y_{-1}$		$\Delta^4 y_{-2}$
		Δy_0		$\Delta^3 y_{-1}$	
$a+h$	y_1		$\Delta^2 y_0$		$\Delta^4 y_{-1}$

The remainder term of Stirling's interpolation formula is

$$R_n(f, x) = f(x) - P_n(x) =$$

$$= h^{n+1} \frac{t(t^2-1)(t^2-2^2)\ldots(t^2-(n/2)^2)}{(n+1)!} f^{(n+1)}(\xi), \qquad (8.9)$$

$$a - \frac{n}{2}h < \xi < a + \frac{n}{2}h.$$

Suppose that the point of interpolation satisfies the inequality $a < x < a+h$ and is located close to the midpoint $a+h/2$ of the interval or coincides with it. From the considerations in § 6, it now follows that an *even* number $n+1$ of points are to be chosen, distributed in the following way:

$$x_0 = a, \ x_1 = a+h, \ x_2 = a-h, \ x_3 = a+2h, \ldots$$

$$\ldots, x_{n-1} = a - \frac{n-1}{2}h, \qquad x_n = a + \frac{n+1}{2}h.$$

The interpolating polynomial, constructed from the values at the knots indicated, is written by Gauss' formula (8.4) as:

$$P_n(a+ht) = y_0 + \frac{t}{1!}\Delta y_0 + \frac{t(t-1)}{2!}\Delta^2 y_{-1} +$$

$$+ \frac{t(t-1)(t+1)}{3!}\Delta^3 y_{-1} + \frac{t(t-1)(t+1)(t-2)}{4!}\Delta^4 y_{-2} + \ldots$$

$$\ldots + \frac{t(t-1)(t+1)\ldots(t+(n-)1/2)}{n!}\Delta^n y_{-((n-1)/2)}. \qquad (8.10)$$

Formula (8.10) will be transformed into a symmetric form. Rewriting the sum of the last two terms of this formula and using an obvious transformation gives:

$$\frac{t(t-1)(t+1)\dots(t+k-1)(t-k)}{(2k)!}\left[\varDelta^{2k}y_{-k}+\frac{t+k}{2k+1}\varDelta^{2k+1}y_{-k}\right]=$$

$$=\frac{t(t-1)(t+1)\dots(t+k-1)(t-k)}{(2k)!}\times$$

$$\times\left[\frac{\varDelta^{2k}y_{-k}+\varDelta^{2k}y_{-k+1}}{2}-\tfrac{1}{2}\varDelta^{2k+1}y_{-k}+\frac{t+k}{2k+1}\varDelta^{2k+1}y_{-k}\right]=$$

$$=\frac{t(t-1)(t+1)\dots(t+k-1)(t-k)}{(2k)!}\times$$

$$\times\left[\frac{\varDelta^{2k}y_{-k}+\varDelta^{2k}y_{-k+1}}{2}+\frac{t-\tfrac{1}{2}}{2k+1}\varDelta^{2k+1}y_{-k}\right],$$

$$k=0,1,2,\dots,\frac{n-1}{2}.$$

As a result of the transformation, formula (8.10) may be rewritten as:

$$P_n(a+ht)=\frac{y_0+y_1}{2}+\frac{t-\tfrac{1}{2}}{1!}\varDelta y_0+$$

$$+\frac{t(t-1)}{2!}\frac{\varDelta^2 y_{-1}+\varDelta^2 y_0}{2}+\frac{(t-\tfrac{1}{2})t(t-1)}{3!}\varDelta^3 y_{-1}+\dots$$

$$\dots+\frac{t(t-1)(t+1)\dots(t-(n-1)/2)}{(n-1)!}\times$$

$$\times\frac{\varDelta^{n-1}y_{-(n-1/2)}+\varDelta^{n-1}y((n-3)/2)}{2}+$$

$$+\frac{(t-\tfrac{1}{2})t(t-1)(t+1)\dots((t-n-1)/2)}{n!}\varDelta^n y_{-(n-1/2)}. \qquad (8.11)$$

This is *Bessel s interpolating polynomial*. The averages of the differences of even order which lie on the same horizontal lines as the arguments a and $a+h$, and the odd differences on the intermediate line enter into it. These differences are underlined in the table

x	y	Δy	$\Delta^2 y$	$\Delta^3 y$	$\Delta^4 y$	$\Delta^5 y$
$a-h$	y_{-1}		$\Delta^2 y_{-2}$		$\Delta^4 y_{-3}$	
		Δy_{-1}		$\Delta^3 y_{-2}$		$\Delta^5 y_{-3}$
a	y_0		$\Delta^2 y_{-1}$		$\Delta^4 y_{-2}$	
		Δy_0		$\Delta^3 y_{-1}$		$\Delta^5 y_{-2}$
$a+h$	y_1		$\Delta^2 y_0$		$\Delta^4 y_{-1}$	
		Δy_1		$\Delta^3 y_0$		$\Delta^5 y_{-1}$
$a+2h$	y_2		$\Delta^2 y_1$		$\Delta^4 y_0$	

The remainder term in Bessel's formula is

$$R_n(f, x) = f(x) - P_n(x) =$$

$$= h^{n+1} \frac{t(t-1)(t+1)\ldots(t+(n-1)/2)(t-(n+1)/2)}{(n+1)!} f^{(n+1)}(\xi),$$

$$a - \frac{n-1}{2} h < \xi < a + \frac{n+1}{2} h. \tag{8.12}$$

Note the frequent case for Bessel's formula (8.11) that $t = \frac{1}{2}$, that is, $x = a + h/2$:

$$P_n\left(a + \frac{h}{2}\right) = \frac{y_0 + y_1}{2} - \frac{1}{8} \frac{\Delta^2 y_{-1} + \Delta^2 y_0}{2} +$$

$$+ \frac{3}{128} \frac{\Delta^4 y_{-2} + \Delta^4 y_{-1}}{2} + \ldots$$

$$\ldots + (-1)^{n-1/2} \frac{(1 \cdot 3 \cdot 5 \ldots (n-2)]^2}{2^{n-1}(n-1)!} \frac{\Delta^{n-1} y_{-(n-1)/2} + \Delta^{n-1} y_{-(n-3)/2}}{2}. \tag{8.13}$$

The numerical values of the coefficients for the formulas of Gauss, Stirling, and Bessel are given in the tables [15]. A short table of the coefficients for Stirling's formula is given in [16].

Example. There is given a table of values of the function

$$y = \sinh x, \qquad x = 1(0.1)1.8.$$

We shall calculate the value of $\sinh x$ for $x = 1.41710$ and $x = 1.45224$.

Table 21

x	y	Δy	$\Delta^2 y$	$\Delta^3 y$	$\Delta^4 y$	$\Delta^5 y$
1	1.17520					
		16045				
1.1	1.33565		1336			
		17381		175		
1.2	1.50946		1511		14	
		18892		189		3
1.3	1.69838		1700		17	
		20592		206		2
1.4	1.90430		1906		19	
		22498		225		2
1.5	2.12928		2131		21	
		24629		246		4
1.6	2.37557		2377		25	
		27006		271		
1.7	2.64563		2648			
		29654				
1.8	2.94217					

We construct a table of finite differences (see Table 21). To calculate sinh 1.41710, we shall use Stirling's formula (8.8) with $a = 1.4$. The coefficients of Stirling's formula for $t = 0.1710$ are taken from the tables [15]:

$$\frac{t^2}{2!} = 0.01462, \quad \frac{t(t^2-1)}{3!} = -0.02767, \quad \frac{t^2(t^2-1)}{4!} = -0.00118.$$

We obtain

$$\sinh 1.41710 = 1.94136.$$

To calculate sinh 1.45224, Bessel's formula (8.11) will be applied, taking $a = 1.4$. We have that $t = 0.5224$. The coefficients of the polynomial (8.11) are found from [15]:

$$\frac{t(t-1)}{2!} = -0.12475, \quad \frac{(t-\frac{1}{2})t(t-1)}{3!} = 0.00093,$$

$$\frac{t(t^2-1)(t-2)}{4!} = 0.02339.$$

As a result of the calculation, we obtain that

sinh 1.45224 = 2.01931.

The values found for the hyperbolic sine are correct in all figures.

9. Inverse Interpolation.   Interpolation without differences

The solution of the following problem has been discussed: Given a table of values of a function $y = f(x)$, one is required to find an approximation to the value $f(x^*)$ of the function for a given value x^* of the argument. The problem of inverse interpolation may be stated thusly: Given a table of the function $y = f(x)$, for a known value y^* of the function, find an argument x^* for which

$$f(x^*) = y^*.$$

It will be assumed that the portion of the table of the function f considered is monotone, and, consequently, it has a single-valued inverse $f^{-1}: x = f^{-1}(y)$. In this case, inverse interpolation coincides with ordinary interpolation for the function $f^{-1}(y)$. In order to obtain an approximate value of x^*, one may use Lagrange's or Newton's formula (for unequally spaced knots).

Suppose that the function $y = f(x)$ does not have a single-valued inverse. One constructs the polynomial $P_n(x)$ which interpolates the function $f(x)$. The problem of determining x^* reduces to solving the equation

$$P_n(x) = y^*. \tag{9.1}$$

We shall consider in the most detail the case that the table of the function is given for equally spaced values of the argument

$$x_0, x_0+h, x_0+2h, \ldots,$$

$$f(x_0+kh) = y_k, \qquad k = 0, 1, 2, \ldots$$

For definiteness, it will be decided that $\Delta y_0 > 0$, and y^* satisfies the inequality $y_0, < y^* < y_1$, so that the desired x^* is to be found between x_0 and x_0+h. Under these conditions, one would replace $f(x)$ by Newton's interpolating polynomial (7.7). Equation (9.1) takes the form

$$y_0 + \frac{t}{1!} \Delta y_0 + \frac{t(t-1)}{2!} \Delta^2 y_0 + \ldots$$

$$\ldots + \frac{t(t-1) \ldots (t-n+1)}{n!} \Delta^n y_0 = y^*.$$

Solving this equation for t:

$$t = \frac{1}{\Delta y_0} \left[y^* - y_0 - \frac{t(t-1)}{2!} \Delta^2 y_0 - \ldots \right.$$

$$\left. \ldots - \frac{t(t-1) \ldots (t-n+1)}{n!} \Delta^n y_0 \right]. \tag{9.2}$$

To determine the solution of equation (9.2), one may apply the method of iterations, choosing $t_0 = 0$ as the initial approximation. The method of iterations will converge, for example, if $|\Delta^2 y_0| \leq |\Delta y_0|$, and the differences $\Delta^3 y_0, \Delta^4 y_0, \ldots$ of higher order are small in comparison to Δy_0 and $\Delta^2 y_0$. One may convince oneself of this by considering the derivative of the right-hand side of equation (9.2) with respect to t. Suppose that $t^* = \lim t_k$ is the desired solution. Then, the value of the argument sought is equal to

$$x^* = x_0 + ht^*.$$

The method of inverse interpolation based on the solution of equation (9.1) is, of course, suitable also in the case that $y = f(x)$ has a single-valued inverse.

Example. There is given a table of the function $y = e^x$, $x = 0.65(0.1)1.15$. We shall find ln 2. It is clear that the problem is one of inverse interpolation: Find the x for which $e^x = 2$. We construct a table of finite differences (see Table 22). The desired value of x is to be found at the beginning of the table, so the corresponding formula of Newton will be used. Equation (9.2) may be written as:

$$t = \tfrac{1}{20146} [8446 - 2119 C_t^{(2)} - 222 C_t^{(3)} - 25 C_t^{(4)}]$$

To solve this equation by the method of iterations the coefficients $C_t^{(k)}$ are obtained from [15]. We find that

$$t_0 = 0, \quad t_1 = 0.41924, \quad t_2 = 0.43138, \quad t_3 = t_4 = 0.43148.$$

Table 22

x	$y = e^x$	Δy	$\Delta^2 y$	$\Delta^3 y$	$\Delta^4 y$
0.65	1.91554				
		20146			
0.75	2.11700		2119		
		22265		222	
0.85	2.33965		2341		25
		24606		247	
0.95	2.58571		2588		25
		27194		272	
1.05	2.85765		2860		
		30054			
1.15	3.15819				

Thus, $t^* = 0.43148$, and $\ln 2 = 0.65 + 0.1\ t^* = 0.693148$.

In this example, the function $y = e^x$ is monotone increasing, and we could have constructed, for instance, Newton's interpolating polynomial for the function $x = \ln y$ and calculated its value for $y = 2$.

We conclude with a method of interpolation which is called Aitken's method [25] in the literature *). Suppose that at the knots

$$x_0, x_1, x_2, \ldots, x_N,$$

which are pairwise distinct, and, generally speaking, are unequally spaced, there are given the values

$$y_0, y_1, y_2, \ldots, y_N.$$

We shall denote by

$$P_{i_1, i_2, \ldots, i_k}(x) \tag{9.3}$$

the interpolating polynomial constructed from the conditions

$$P_{i_1, i_2, \ldots, i_k}(x_{i_j}) = y_{i_j}, \qquad j = 1, 2, \ldots, k.$$

Here, $i_1, i_2, \ldots, i_k$ are non-negative, pairwise distinct integers. It is clear that they may be written in any order in the notation (9.3).

For example, $P_{0, 1}(x)$ is a linear function, the graph of which passes

---

* In so far as finite differences are not required in the application of Aitken's method, it and methods similar to it are said to be difference-free.

through the points (x_0, y_0), (x_1, y_1). Obviously, one has that

$$P_{0,1}(x) = \frac{1}{x_1-x_0} \begin{vmatrix} y_0 & x_0-x \\ y_1 & x_1-x \end{vmatrix}.$$

Consider the determinant

$$P(x) = \frac{1}{x_2-x_0} \begin{vmatrix} P_{0,1}(x) & x_0-x \\ P_{1,2}(x) & x_2-x \end{vmatrix}.$$

It is a polynomial in x of degree not higher than two. Obviously,

$$P(x_0) = P_{0,1}(x_0) = y_0,$$

$$P(x_2) = P_{1,2}(x_2) = y_2,$$

$$P(x_1) = \frac{1}{x_2-x_0} \begin{vmatrix} y_1 & x_0-x_1 \\ y_1 & x_2-x_1 \end{vmatrix} = y_1,$$

hence

$$P(x) = P_{0,1,2}(x).$$

Thus, one may calculate the value of the interpolating polynomial of second degree by applying linear interpolation to $P_{0,1}(x)$ and $P_{1,2}(x)$. Of course, linear interpolation may be applied to two other possible pairs: $P_{0,1}(x)$ and $P_{0,2}(x)$; $P_{0,2}(x)$ and $P_{1,2}(x)$.

In general, the value of the interpolating polynomial constructed from $n+1$ knots may be obtained by linear interpolation of the values of two distinct interpolating polynomials constructed from n knots selected from the $n+1$ knots considered. For example,

$$P_{0,1,2,\ldots,n}(x) = \frac{1}{x_n-x_0} \begin{vmatrix} P_{0,1,2,\ldots,n-1}(x) & x_0-x \\ P_{1,2,3,\ldots,n}(x) & x_n-x \end{vmatrix}.$$

Aitken's method consists of calculating the value of the interpolating polynomial, say $P_{0,1,2,3}(x)$, by performing a series of linear interpolations according to the scheme

x_0	y_0				x_0-x
x_1	y_1	$P_{0,1}(x)$			x_1-x
x_2	y_2	$P_{0,2}(x)$	$P_{0,1,2}(x)$		x_2-x
x_3	y_3	$P_{0,3}(x)$	$P_{0,1,3}(x)$	$P_{0,1,2,3}(x)$	$x_3-x.$

Here, every value computed may be accumulated on a calculating machine, and later divided. For example, $P_{0,1,3}(x)$ is calculated as the determinant with the elements underlined in the diagram, which is then divided by $x_3 - x_1$.

Example. Find sin 0.57891, given the table of the function

$$y = \sin x, \quad x = 0.4(0.1)0.8.$$

The results of the calculation by Aitken's method are presented in Table 23. The calculations are carried out to one extra place. The

Table 23

x_i	$\sin x_i$	$P_{0,i}$	$P_{0,1,i}$	$P_{0,1,2,i}$	$x_i - 0.57891$
0.4	0.38942				−0.17891
0.5	0.47943	0.550457			−0.07891
0.6	0.56464	0.546163	0.547069		0.02109
0.7	0.64422	0.541374	0.546873	0.547110	0.12109
0.8	0.71735	0.536099	0.546680	0.547110	0.22109

agreement of all digits in the values written in the last column indicate that the calculation has to be terminated. The final result is rounded to five places:

$$\sin 0.57891 = 0.54711.$$

10. Hermite Interpolation

Suppose that there are given m distinct real numbers

$$x_1, x_2, \ldots, x_m$$

as knots of interpolation. It will be assumed that at the knot x_j, the value of the function $f(x)$ and all of its derivatives up to order $p_j - 1$ are given,

$$f(x_j), f'(x_j), \ldots, f^{(p_j - 1)}(x_j), \qquad j = 1, 2, \ldots, m. \tag{10.1}$$

Thus,

$$p_1 + p_2 + \ldots + p_m = n + 1$$

pieces of information are given concerning the function $f(x)$.

We shall consider the problem of constructing a polynomial of degree not higher than n

$$P_n(x) = a_0 x^n + a_1 x^{n-1} + \ldots + a_{n-1} x + a_n, \tag{10.2}$$

which satisfies the conditions

$$P_n^{(s)}(x_j) = f^{(s)}(x_j), \quad j = 1, 2, \ldots, m,$$

$$s = 0, 1, 2, \ldots, p_j - 1. \tag{10.3}$$

Here, $f^{(0)}(x)$ is understood to be $f(x)$. Clearly, this is a frequent case of the general problem of interpolation as considered in § 4. Conditions (10.3) represent a system of linear algebraic equations for the coefficients $a_0, a_1, \ldots, a_n$ of the polynomial (10.2). A similarly expressed system of this form was presented in § 4, as formula (4.4).

Construction of the polynomial (10.2) from the conditions (10.3) is called *Hermite interpolation* or *interpolation with multiple knots*. The number p_j is called the *multiplicity of the knot* $x_j (j = 1, 2, \ldots, m)$.

We remark that there can exist only one polynomial (10.2) which satisfies the conditions (10.3). Indeed, suppose that there turned out to be two such polynomials: $P_1(x)$ and $P_2(x)$. Then, their difference $P_1(x) - P_2(x)$ would be a polynomial of degree not greater than n, moreover, at x_j this polynomial would have a zero of multiplicity $p_j (j = 1, 2, \ldots, m)$, that is, it would have $p_1 + p_2 + \ldots + p_m = n + 1$ zeros. This is possible only in the case that $P_1(x) - P_2(x) \equiv 0$.

The existence of the Hermite interpolation polynomial follows from its uniqueness. Indeed, the Hermite polynomial constructed from the function $f(x) \equiv 0$ is unique also, which is equivalent to the homogeneous system corresponding to the linear algebraic system (10.3) having only zero solutions. It follows from this that the determinant of the system (10.3) is different from zero, and, consequently, it has solutions for any right-hand sides.

We shall show how it is possible to construct the Hermite interpolation polynomial. The coefficients $a_0, a_1, \ldots, a_n$ of the polynomial (10.2) are defined by the system (10.3), hence they are linear combinations of the values of $f^{(s)}(x_j)$ with coefficients which depend only on $x_j (j = 1, 2, \ldots, m)$:

$$a_k = \sum_{j=1}^{m} \sum_{s=0}^{p_j-1} a_{kjs} f^{(s)}(x_j), \quad k = 0, 1, 2, \ldots, n.$$

Substituting this expression for the coefficients into (10.2), one obtains that

$$P_n(x) = \sum_{j=1}^{m} \sum_{s=0}^{p_j-1} l_{js}(x) f^{(s)}(x_j),$$ (10.4)

where $l_{js}(x)$ is a polynomial of degree n. In view of this, one may look for the polynomial (10.2) in the form (10.4).

Instead of the polynomial $P_n(x)$, we shall first construct the remainder of interpolation

$$R_n(f, x) = f(x) - P_n(x) = f(x) - \sum_{j=1}^{m} \sum_{s=0}^{p_j-1} l_{js}(x) f^{(s)}(x_j).$$ (10.5)

It will be assumed from the start that $f(z)$ is an analytic function of the complex variable z, which is single-valued and regular in some singly-connected region D which contains the point of interpolation x and the knots $x_1, x_2, \ldots, x_m$ in its interior. The point x and the knots $x_1, x_2, \ldots, x_m$ are considered to be real (but they could just as well be taken to be complex). It is also assumed that x is different from all of the $x_j, j = 1, 2, \ldots, m$.

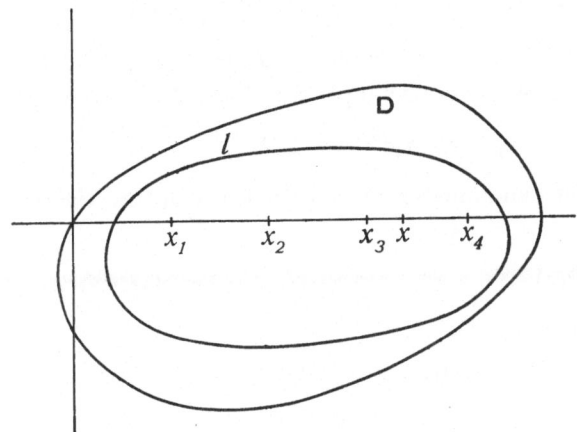

Fig. 13

We draw a closed contour l which lies entirely in D and encloses the knots $x_1, x_2, \ldots, x_m$ and the point x (see Figure 13). By Cauchy's formula,

$$f(x) = \frac{1}{2\pi i} \int_{l} \frac{f(z)}{z-x} \, dz.$$ (10.6)

Here, the integration along the contour l is performed in the counter-clockwise direction. One also has that

$$f^{(s)}(x_j) = \frac{s!}{2\pi i} \int_l \frac{f(z)}{(z-x_j)^{s+1}} \, dz. \tag{10.7}$$

On the right-hand side of (10.5) one may substitute for $f(x)$ and $f^{(s)}(x)$ their representations by the Cauchy integrals (10.6) and (10.7). One obtains that

$$R_n(f, x) = \frac{1}{2\pi i} \int_l f(z) \left[\frac{1}{z-x} - \sum_{j=1}^{m} \sum_{s=0}^{p_j-1} l_{js}(x) \frac{s!}{(z-x_j)^{s+1}} \right] dz. \tag{10.8}$$

The remainder of Hermite interpolation of the function

$$\frac{1}{z-x},$$

considered as a function of x is in the square brackets:

$$R_n \left(\frac{1}{z-x}, x \right) = \frac{1}{z-x} - \sum_{j=1}^{m} \sum_{s=0}^{p_j-1} l_{js}(x) \frac{s!}{(z-x_j)^{s+1}}. \tag{10.9}$$

The right-hand side of formula (10.9) is a rotational function of z, represented in the form of simple fractions. We introduce the notation

$$\Omega(z) = (z-x_1)^{p_1}(z-x_2)^{p_2} \ldots (z-x_m)^{p_m}. \tag{10.10}$$

The common denominator of the fractions on the right-hand side of (10.9) is equal to

$$(z-x)\Omega(z)$$

and one has that

$$R_n \left(\frac{1}{z-x}, x \right) = \frac{Q(z)}{(z-x)\Omega(z)}, \tag{10.11}$$

where the degree of the numerator is less than the degree of the denominator, since the degree of $(z-x)\Omega(z)$ is equal to $n+2$ and the degree of $Q(z)$ does not exceed $n+1$.

We shall show that $Q(z)$ does not depend on z. To do this, we expand

$$R_n \left(\frac{1}{z-x}, x \right)$$

in a series of powers of $1/z$, assuming that

$$|z| > |x|, |z| > |x_j|, \qquad j = 1, 2, \ldots, m.$$

One has that

$$\frac{1}{z-x} = \frac{1}{z}\frac{1}{1-x/z} = \sum_{v=0}^{\infty}\frac{x^v}{z^{v+1}},$$

$$\frac{s!}{(z-x_j)^{s+1}} = \sum_{v=0}^{\infty}\frac{v(v-1)\ldots(v-s+1)}{z^{v+1}}x_j^{v-s}, \qquad j = 1, 2, \ldots, m.$$

Substituting this expression into formula (10.9) one obtains that

$$R_n\left(\frac{1}{z-x}, x\right) =$$

$$= \sum_{v=0}^{\infty}\frac{x^v}{z^{v+1}} - \sum_{j=1}^{m}\sum_{s=0}^{p_j-1}l_{js}(x)\sum_{v=0}^{\infty}\frac{v(v-1)\ldots(v-s+1)}{z^{v+1}}x_j^{v-s} =$$

$$= \sum_{v=0}^{\infty}\frac{1}{z^{v+1}}\left(x^v - \sum_{j=1}^{m}\sum_{s=0}^{p_j-1}l_{js}(x)v(v-1)\ldots(v-s+1)x_j^{v-s}\right).$$

The expression in parentheses is the remainder term in Hermite interpolation for x^v. Thus, one obtains finally that

$$R_n\left(\frac{1}{z-x}, x\right) = \sum_{v=0}^{\infty}\frac{1}{z^{v+1}}R_n(x^v, x).$$

Now, obviously, $R_n(x^v, x) = 0$ for $v \leq n$ (since x is taken as the interpolating polynomial), hence, in the previous formula, one may begin the summation with $v = n+1$:

$$R_n\left(\frac{1}{z-x}, x\right) = \sum_{v=n+1}^{\infty}\frac{1}{z^{v+1}}R_n(x^v, x).$$

It is clear from this that as $z \to \infty$,

$$R_n\left(\frac{1}{z-x}, x\right) \to 0$$

at least as fast as $1/z^{n+2}$. Now, it follows from relationship (10.11) that the degree of $Q(z)$ in z is equal to zero, and $Q(z)$ does not depend on z:

$$Q(z) = A.$$

It follows from the representation (10.9) that

$$\lim_{z \to x} (z-x)R_n \left(\frac{1}{z-x}, x \right) = 1, \tag{10.12}$$

since it was assumed that $x \neq x_j, j = 1, 2, \ldots, m$. By virtue of (10.11),

$$\lim_{z \to x} (z-x)R_n \left(\frac{1}{z-x}, x \right) = \frac{A}{\Omega(x)}. \tag{10.13}$$

Comparing (10.12) and (10.13), one obtains that $A = \Omega(x)$. Therefore,

$$R_n \left(\frac{1}{z-x}, x \right) = \frac{\Omega(x)}{(z-x)\Omega(z)}$$

and, consequently, one has on the basis of (10.8) that

$$R_n(f, x) = \frac{\Omega(x)}{2\pi i} \int_l \frac{f(z)}{(z-x)\Omega(z)} \, dz. \tag{10.14}$$

This is the required expression for the remainder of Hermite interpolation. One finds with the aid of (10.14) that

$$P_n(x) = f(x) - R_n(f, x) =$$

$$= f(x) - \frac{\Omega(x)}{2\pi i} \int_l \frac{f(z)}{(z-x)\Omega(z)} \, dz. \tag{10.15}$$

We have obtained a representation for the Hermite interpolating polynomial in terms of a Cauchy integral. In order to write this polynomial in an explicit form, we shall evaluate the contour integral on the right-hand side of (10.15). To calculate the contour integral, the following theorem from the theory of functions of a complex variable will be used. If a function is regular in a closed region, except at a finite number of points of the region which are poles or essential singular points lying in the interior of the region, then the integral of a function around the contour bounding the region is equal to the product of $2\pi i$ and the sum of the residues at the singular points indicated.

We recall that the residue of a function $f(z)$ at a singular point a (a pole or essential singular point) is the coefficient of $(z-a)^{-1}$ in the expansion of $f(z)$ in a Laurent series.

The integrand in (10.15),

$$\frac{f(z)}{(z-x)\Omega(z)} \tag{10.16}$$

has as singular points the poles

$$z = x, \ z = x_j, \qquad j = 1, 2, \ldots, m.$$

The residue of the function (10.16) at the pole $z = x$ is equal to

$$R_0 = \frac{f(x)}{\Omega(x)}. \tag{10.17}$$

We shall calculate the residues of the function (10.16) at the poles $z = x_j$. In a neighbourhood of x_j, there hold the power series expansions

$$f(z) = \sum_{q=0}^{\infty} \frac{f^{(q)}(x_j)}{q!}(z - x_j)^q, \tag{10.18}$$

$$\frac{1}{z - x} = \frac{1}{z - x_j - (x - x_j)} = -\frac{1}{x - x_j} \frac{1}{1 - \dfrac{z - x_j}{x - x_j}} =$$

$$= -\sum_{q=0}^{\infty} \frac{1}{(x - x_j)^{q+1}} (z - x_j)^q, \tag{10.19}$$

$$\frac{(z - x_j)^{p_j}}{\Omega(z)} = \sum_{q=0}^{\infty} c_q^{(j)}(z - x_j)^q. \tag{10.20}$$

In order to obtain the residue of the function (10.16),

$$\frac{f(z)}{(z - x)\Omega(z)} = \frac{1}{(z - x_j)^{p_j}} \cdot f(z) \cdot \frac{1}{z - x} \cdot \frac{(z - x_j)^{p_j}}{\Omega(z)}$$

at the pole $z = x_j$, one must find the coefficient of $(z - x_j)^{p_j - 1}$ in the product of the series (10.18), (10.19), (10.20). The coefficient required is equal to

$$\sum_{q=0}^{p_j - 1} \frac{f^{(q)}(x_j)}{q!} A_{p_j - 1 - q},$$

where $A_{p_j - 1 - q}$ is the coefficient of $(z - x_j)^{p_j - 1 - q}$ in the product of the series (10.19) and (10.20). It is clear that

$$A_{p_j - 1 - q} = -\sum_{r=0}^{p_j - 1 - q} c_r^{(j)} \frac{1}{(x - x_j)^{p_j - q - r}}.$$

Hence, the residue at the pole $z = x_j$ is equal to

$$R_j = -\sum_{q=0}^{p_j - 1} \frac{f^{(q)}(x_j)}{q!} \sum_{r=0}^{p_j - 1 - q} c_r^{(j)}(x - x_j)^{-p_j + q + r}. \tag{10.21}$$

By the theorem from the theory of complex variables, one has that

$$P_n(x) = f(x) - 2\pi i \cdot \frac{\Omega(x)}{2\pi i} \sum_{j=0}^{m} R_j.$$

Replacing R_0 by formula (10.17) and R_j, j 1, 2, ..., m, by formula (10.21) here, we obtain the required representation of the Hermite interpolating polynomial $P_n(x)$:

$$P_n(x) = \sum_{j=1}^{m} \frac{\Omega(x)}{(x-x_j)^{p_j}} \sum_{q=0}^{p_j-1} \frac{f^{(q)}(x_j)}{q!} \sum_{r=0}^{p_j-1-q} c_r^{(j)}(x-x_j)^{q+r}. \qquad (10.22)$$

Obviously, the representation (10.22) for the interpolating polynomial remains valid in the case that $f(x)$ is not an analytic function. Further, (10.22) gives the polynomial which satisfies conditions (10.3) for arbitrary numbers (10.1), whether or not it is known beforehand that they are derivatives of some function.

We conclude with some frequent cases of Hermite's formula. If the knots $x_1, x_2, ..., x_m$ are simple, that is,

$$p_1 = p_2 = \ldots = p_m = 1, \ m = n+1,$$
$$\Omega(z) = \omega(z) = (z-x_1)(z-x_2) \ldots (z-x_{n+1}),$$

formula (10.22) reduces to Lagrange's formula.

Suppose that $m = 1$, that is, there is a single knox x, at which is given

$$f(x_1), f'(x_1), \ldots, f^{(p_1-1)}(x_1).$$

In this case,

$$p_1 = n+1,$$

and the right-hand side of formula (10.22) becomes a partial sum of the Taylor series for the function $f(x)$:

$$P_n(x) = \sum_{q=0}^{n} \frac{f^{(q)}(x_1)}{q!} (x-x_1)^q.$$

Now we consider Hermite interpolation with double knots: $f(x_j)$ and $f'(x_j)$ are given at the knot x_j, $j = 1, 2, ..., m$. In this case,

$$p_1 = p_2 = \ldots = p_m = 2, \ 2m = n+1,$$
$$\Omega(z) = (z-x_1)^2(z-x_2)^2 \ldots (z-x_m)^2 = \omega^2(z).$$

In order to write the Hermite polynomial, one needs to find the first two terms in the expansion (10.20),

$$\frac{(z-x_j)^2}{\omega(z)^2} = c_0^{(j)} + c_1^{(j)}(z-x_j) + \ldots$$

One has that

$$c_0^{(j)} = \lim_{z \to x_j} \frac{(z-x_j)^2}{\omega(z)^2} = \frac{1}{\omega'(x_j)^2},$$

$$c_1^{(j)} = \lim_{z \to x_j} \frac{d}{dz} \left[\frac{z-x_j}{\omega(z)} \right]^2 =$$

$$= \lim_{z \to x_j} 2 \left[\frac{z-x_j}{\omega(z)} \right] \frac{\omega(z) - (z-x_j)\omega'(z)}{\omega^2(z)} =$$

$$= 2 \frac{1}{\omega'(x_j)} \lim_{z \to x_j} \frac{\omega(z) - (z-x_j)\omega'(z)}{\omega^2(z)}.$$

Calculating the last limit by L'Hospital's rule, one finds that

$$c_1^{(j)} = - \frac{\omega''(x_j)}{\omega'(x_j)^3}.$$

The coefficients $c_0^{(j)}$ and $c_1^{(j)}$ found thus are substituted into formula (10.22). One obtains that

$$P_n(x) = \sum_{j=1}^{m} \frac{\omega^2(x)}{\omega'(x_j)^2 (x-x_j)^2} \left\{ f(x_j) \left[1 - \frac{\omega''(x_j)}{\omega'(x_j)} (x-x_j) \right] + \right.$$

$$\left. + f'(x_j)(x-x_j) \right\}. \tag{10.23}$$

In the process of constructing the Hermite polynomial, we obtained the representation (10.14) for the remainder term of interpolation, supposing that $f(x)$ was a regular analytic function. We shall establish a theorem on the representation of the remainder term of Hermite interpolation for real functions which are differentiable $n+1$ times.

Theorem. If $f(x)$ is differentiable $n+1$ times on the interval $[a, b]$ which contains the knots of interpolation $x_1, x_2, \ldots, x_n$, then for any point $x \in [a, b]$,

$$R_n(f, x) = \frac{\Omega(x)}{(n+1)!} f^{(n+1)}(\xi), \tag{10.24}$$

where $a < \xi < b$,

$$\Omega(x) = (x-x_1)^{p_1}(x-x_2)^{p_2} \ldots (x-x_m)^{p_m}$$

and

$$n+1 = p_1+p_2+ \ldots +p_m.$$

Proof. Choose a point $x \in [a, b]$ which is different from all of the knots $x_1, x_2, \ldots, x_m$, and introduce the auxiliary function

$$F(z) = f(z)-P_n(z)- \frac{\Omega(z)}{\Omega(x)} [f(x)-P_n(x)].$$

The function $F(z)$ is differentiable $n+1$ times on $[a, b]$. Obviously, x is a zero of $F(z)$. The knot x_j is a zero of multiplicity p_j of the function $F(z)$, so that $F(z)$ has at least $p_1+p_2+ \ldots +p_m+1 = n+2$ zeros (counting multiplicities). On the basis of the generalized Rolle's theorem, one may assert that $F'(z)$ has at least $n+1$ zeros, $F''(z)$, n zeros, and so forth, $F^{(n+1)}(z)$ having at least one zero ξ. This gives

$$F^{(n+1)}(z) = f^{(n+1)}(z)- \frac{(n+1)!}{\Omega(x)} [f(x)-P_n(x)].$$

One obtains (10.24) by setting $z = \xi$ in this.

Example. We shall construct the polynomial which satisfies the conditions

x	$P(x)$	$P'(x)$	$P''(x)$
0	1	1	-2
1	2	1	
3	3	-1	

One has that

$$\Omega(x) = x^3(x-1)^2(x-3)^2.$$

Writing the first three terms of the expansion (10.20) for the triple knot $x_1 = 0$:

$$\begin{aligned}
\frac{z^3}{\Omega(z)} &= \frac{1}{(z-1)^2(z-3)^2} = \frac{1}{9} \frac{1}{(1-z)^2} \frac{1}{(1-z/3)^2} = \\
&= \tfrac{1}{9}(1+2z+3z^2+ \ldots)(1+\tfrac{2}{3}z+\tfrac{1}{3}z^2+ \ldots) = \\
&= \tfrac{1}{9}(1+\tfrac{8}{3}z+\tfrac{14}{3}z^2+ \ldots).
\end{aligned}$$

One finds the first two terms of the expansion (10.20) for the knots $x_2 = 1$, $x_3 = 3$ in an analogous way:

$$\frac{(z-1)^2}{\Omega(z)} = \frac{1}{z^3(z-3)^2} = \tfrac{1}{4}[1-2(z-1)+ \ldots],$$

$$\frac{(z-3)^2}{\Omega(z)} = \frac{1}{z^3(z-1)^2} = \tfrac{1}{108}[1-2(z-3)+ \ldots].$$

Rewriting formula (10.22), one has

$$P(x) = (x-1)^2(x-3)^2 \left\{ P(0)\tfrac{1}{9}[1+\tfrac{8}{3}x+\tfrac{14}{3}x^2]+ \right.$$
$$\left. + \frac{P'(0)}{1!} \tfrac{1}{9}[x+\tfrac{8}{3}x^2]+ \frac{P''(0)}{2!} \tfrac{1}{9}x^2 \right\} +$$
$$+ x^3(x-3)^2 \left\{ P(1)\tfrac{1}{4}[1-2(x-1)]+ \frac{P'(1)}{1!} \tfrac{1}{4}(x-1) \right\} +$$
$$+ x^3(x-1)^2 \left\{ P(3)\tfrac{1}{108}[1-2(x-3)]+ \frac{P'(3)}{1!} \tfrac{1}{108}(x-3) \right\}.$$

Substituting the given values of the polynomial and its derivatives at the knots, we obtain the required polynomial

$$P(x) = \frac{(x-1)^2(x-3)^2}{9} (1+\tfrac{11}{3}x+\tfrac{19}{3}x^2)+$$
$$+ \frac{x^3(x-3)^2}{4} (5-3x)+ \frac{x^3(x-1)^2}{108} (24-7x).$$

11. Numerical Differentiation

We conclude with a brief discussion of the problem of calculating the derivative of a function $f(x)$, given a table of its values at the knots $x_0, x_1, \ldots, x_n$ (which are not required to be equally spaced),

$$f(x_0), f(x_1), \ldots, f(x_N).$$

This problem has the following solution. By constructing the interpolating polynomial $P_n(x)$ with the values of $f(x)$ at $n+1$ knots $(n \le N)$,

$$f(x) \cong P_n(x) \tag{11.1}$$

The value of the mth derivative of the interpolating polynomial $P_n(x)$

gives an approximate value of the mth derivative of the function $f(x)$,

$$f^{(m)}(x) \cong P_n^{(m)}(x), \qquad m \leq n. \tag{11.2}$$

We shall indicate an expression for the derivatives of $P_n(x)$ in the case that $f(x)$ is given for equally spaced values of the argument with spacing h, and the interpolating polynomial $P_n(x)$ is given by Newton's formula (7.7) for interpolation at the beginning of tables,

$$P_n(x_0 + ht) = \sum_{k=0}^{n} C_t^{(k)} \Delta^k f(x_0). \tag{11.3}$$

Differentiating both sides of equation (11.3) $m(\leq n)$ times with respect to t, one finds that

$$\frac{d^m P_n(x_0 + ht)}{dt^m} = h^m \frac{d^m P_n(x)}{dx^m} = \sum_{k=m}^{n} \frac{d^m C_t^{(k)}}{dt^m} \Delta^k f(x_0), \tag{11.4}$$

from which

$$P_n^{(m)}(x) = \frac{1}{h^m} \sum_{k=m}^{n} \frac{d^m C_t^{(k)}}{dt^m} \Delta^k f(x_0). \tag{11.5}$$

In order to give an explicit expression for the polynomial $C_t^{(k)}$, the polynomial $k! \, C_t^{(k)}$ is written in the form

$$t(t-1) \ldots (t-k+1) = S_k^{(k)} t^k + S_k^{(k-1)} t^{k-1} + \ldots + S_k^{(1)} t. \tag{11.6}$$

The numbers $S_k^{(j)}$, $j \leq k = 1, 2, 3, \ldots$ are obviously integers. They are known as *Stirling numbers of the first kind*. The numbers $S_k^{(j)}$ may be determined otherwise by means of the equations

$$\frac{[\ln (1+x)]^j}{j!} = \sum_{k=j}^{\infty} \frac{S_k^{(j)}}{k!} x^k, \qquad j = 1, 2, 3, \ldots \tag{11.7}$$

We shall not pause to prove this assertion.
One finds from the relationship (11.6) that

$$k! \frac{d^m C_t^{(k)}}{dt^m} = \sum_{j=m}^{k} S_k^{(j)} j(j-1) \ldots (j-m+1) t^{j-m}.$$

Substituting the expression obtained for the mth derivative of $C_t^{(k)}$ into (11.5), one obtains that

$$P_n^{(m)}(x) = \frac{1}{h^m} \sum_{k=m}^{n} \sum_{j=m}^{k} \frac{S_k^{(j)}}{k!} j(j-1) \ldots$$

$$\ldots (j-m+1) t^{j-m} \Delta^k f(x_0). \tag{11.8}$$

This is the expression required for the mth derivative of the interpolating polynomial (11.3).

Tables of the numbers

$$c_{jk} = \frac{S_k^{(j)}}{k!}$$

are given in the book by V. N. Faddeeva [6] to all figures for $1 \leq j \leq k \leq 20$.

Set $t = 0$ $(x = x_0)$ in (11.8). One obtains that

$$P_n^{(m)}(x_0) = \frac{m!}{h^m} \sum_{k=m}^{n} \frac{S_k^{(m)}}{k!} \Delta^k f(x_0).$$

If (11.7) is taken into account, then the previous formula may be written in the symbolic form:

$$P_n^{(m)}(x_0) = \frac{1}{h^m} [\ln (1+\Delta)]^m f(x_0). \tag{11.9}$$

It is assumed that $\Delta^v f(x_0) = 0$ for $v > n$. Finally, formula (11.9) may be established independently, without deriving equation (11.7).

In an analogous way, one may write an expression for the mth derivative of Newton's interpolating polynomial (7.12), which is used for interpolation at the end of tables:

$$P_n(x_n+ht) = \sum_{k=0}^{n} (-1)^k C_{-t}^{(k)} \Delta^k f(x_{n-k}). \tag{11.10}$$

One has that

$$k! (-1)^k C^{(k)} = \sum_{j=1}^{k} (-1)^{k-j} S_k^{(j)} t^j$$

and, consequently,

$$(-1)^k k! \frac{d_m}{dt^m} C_{-t}^{(k)} =$$

$$= \sum_{j=m}^{k} (-1)^{k-j} S_k^{(j)} j(j-1) \ldots (j-m+1) t^{j-m}.$$

Calculating the derivative of order m with respect to t of the left and right sides of (11.10), and taking into account the last equation, one obtains the required expression for the derivative

$$P_n^{(m)}(x) = \frac{1}{h^m} \sum_{k=m}^{n} \sum_{j=m}^{k} (-1)^{k-j} \frac{S_k^{(j)}}{k!} j(j-1) \cdots$$

$$\cdots (j-m+1)t^{j-m}\Delta^k f(x_{n-k}). \tag{11.11}$$

Set $t = 0$ $(x = x_n)$ in (11.11). One obtains that

$$P_n^{(m)}(x_n) = \frac{1}{h^m} \sum_{k=m}^{n} (-1)^{k-m} \frac{S_k^{(m)}}{k!} m! \Delta^k f(x_{n-k}). \tag{11.12}$$

Formula (11.12) may be written in symbolic form if one uses another notation for finite differences.
We introduce the notation

$$\nabla f(x_q) = f(x_q) - f(x_{q-1}).$$

As arguments of the function under the ∇ sign, one writes the index decreasing from that in $f(x_q)$, in contrast to the Δ notation, in which the index of the argument increases. Analogously, from the differences of order p, $\nabla^p f(x_q)$ and $\nabla^p f(x_{q-1})$, one defines the difference of order $p+1$,

$$\nabla^{p+1} f(x_q) = \nabla^p f(x_q) - \nabla^p f(x_{q-1}).$$

The difference $\Delta^p f(x_q)$ is called a *forward* difference, and $\nabla^p f(x_q)$ is called a *backward* difference.
A section of a table of backward finite differences is recorded thusly:

x	$f(x)$	$\nabla f(x)$	$\nabla^2 f(x)$	$\nabla^3 f(x)$	$\nabla^4 f(x)$
x_{n-4}	$f(x_{n-4})$				
		$\nabla f(x_{n-3})$			
x_{n-3}	$f(x_{n-3})$		$\nabla^2 f(x_{n-2})$		
		$\nabla f(x_{n-2})$		$\nabla^3 f(x_{n-1})$	
x_{n-2}	$f(x_{n-2})$		$\nabla^2 f(x_{n-1})$		$\nabla^4 f(x_n)$
		$\nabla f(x_{n-1})$		$\nabla^3 f(x_n)$	
x_{n-1}	$f(x_{n-1})$		$\nabla^2 f(x_n)$		
		$\nabla f(x_n)$			
x_n	$f(x_n)$				

The differences of the function with the same argument, for example

$$f(x_n), \ \nabla f(x_n), \ \nabla^2 f(x_n), \ \nabla^3 f(x_n), \ \nabla^4 f(x_n)$$

are located on a rising diagonal line. The designation "backward difference" is due to this. It is seen from the table that

$$\Delta^k f(x_{n-k}) = \nabla^k f(x_n).$$

Using this equation, (11.12) may be rewritten as:

$$P_n^{(m)}(x_n) = \frac{m!}{h^m} \sum_{k=m}^{n} (-1)^{k-m} \frac{S_k^{(m)}}{k!} \nabla^k f(x_n). \tag{11.13}$$

In (11.7), set $j = m$, replace x by $-x$, and multiply both sides by $(-1)^m$. One obtains that

$$(-1)^m \frac{[\ln (1-x)]^m}{m!} = \sum_{k=m}^{\infty} (-1)^{k-m} \frac{S_k^{(m)}}{k!} x^k.$$

By virtue of the last equation, formula (11.13) may be written in symbolic form:

$$P_n^{(m)}(x_n) = \frac{(-1)^m}{h^m} [\ln (1-\nabla)]^m f(x_n). \tag{11.14}$$

It is assumed that $\nabla^v f(x_n) = 0$ for $v > n$.

We shall occupy ourselves with the topic of representation of the remainder term for formulas for numerical differentiation. By the definition of the remainder in interpolation [see (6.1)], one has that

$$R_n(f, x) = f(x) - P_n(x).$$

Take the mth derivative $(m \leq n)$ of both sides of this equation. One obtains that

$$R_n^{(m)}(f, x) = f^{(m)}(x) - P_n^{(m)}(x). \tag{11.15}$$

Therefore, the remainder term of the formula for calculating the mth derivative is equal to the mth derivative with respect to x of the remainder term of interpolation. A theorem will be proved on the representation of the remainder term (11.15) for a function $f(x)$ which is differentiable $n+1$ times.

Theorem. Suppose that x is not contained in the interior of $[\alpha, \beta]$, the smallest interval which contains the points of interpolation $x_0, x_1, \ldots, x_n$. Assume that the function $f(x)$ is differentiable $n+1$ times on the smallest interval $[a, b]$ which contains the points of interpolation and the point x. Then, there exists a point ξ, $a < \xi < b$, such that

$$R_n^{(m)}(f, x) = \frac{\omega^{(m)}(x)}{(n+1)!} f^{(n+1)}(\xi). \tag{11.16}$$

where $\omega^{(m)}(x)$ is the mth derivative of the polynomial

$$\omega(x) = (x-x_0)(x-x_1)\ldots(x-x_n).$$

Proof. Consider the auxiliary function

$$F(z) = f(z)-P_n(z)-K\omega(z),\tag{11.17}$$

where K is a constant which will be determined later. The function $F(z)$ is differentiable $n+1$ times on $[a, b]$. The points of interpolation $x_0, x_1, \ldots, x_n$ are zeros of the function $F(z)$, hence one may assert on the basis of Rolle's theorem that $F'(z)$ has at least n distinct zeros interior to (α, β), $F''(z)$ has $n-1$ zeros interior to (α, β), and so forth, $F^{(m)}(z)$ having at least $n+1-m$ distinct zeros interior to (α, β).
The constant K is chosen so that $F^{(m)}(z) = 0$ for $z = x$, explicitly,

$$F^{(m)}(x) = f^{(m)}(x)-P_n^{(m)}(x)-K\omega^{(m)}(x) = 0.\tag{11.18}$$

Such a choice of K is possible, since $\omega^{(m)}(z)$ has $n+1-m$ zeros which are located interior to (α, β), and x does not belong to (α, β), so, consequently, $\omega^{(m)}(x) \neq 0$. One finds from (11.18) that

$$K = \frac{f^{(m)}(x)-P_n^{(m)}(x)}{\omega^{(m)}(x)}.$$

For the given choice of K, the derivative $F^{(m)}(z)$ has at least $n+2-m$ distinct zeros in $[a, b]$. By Rolle's theorem, $F^{(m+1)}(z)$ has at least $n+1-m$ zeros, $F^{(m+2)}(z)$ has at least $n-m$ zeros, and so forth, $F^{(n+1)}(z)$ having at least one zero ξ interior to $[a, b]$: $F^{(n+1)}(\xi) = 0$. By using the definition (11.17) of the function $F(z)$, one obtains that

$$F^{(n+1)}(\xi) = f^{(n+1)}(\xi)-(n+1)!K = 0,$$

from which one finds that

$$K = \frac{f^{(n+1)}(\xi)}{(n+1)!}.$$

This value of K may be substituted into (11.18). One obtains that

$$f^{(m)}(x)-P_n^{(m)}(x)-\frac{f^{(n+1)}(\xi)}{(n+1)!}\omega^{(m)}(x) = 0,$$

which coincides with (11.16).
The representation for the remainder term (11.16) is easy to remember.

It suffices to write the mth derivative with respect to x of the remainder term of interpolation (6.2),

$$R_n(f, x) = \frac{\omega(x)}{(n+1)!} f^{(n+1)}(\xi),$$

considering the quantity $f^{(n+1)}(\xi)$ to be constant. In fact, $f^{(n+1)}(\xi)$ indeed depends on x, and the ξ, in the representations of $R_n(f, x)$ and $R_n^{(m)}(f, x)$ is not the same. We wish to emphasize further that the representation (11.16) only holds for x which do not belong to (α, β), where $[\alpha, \beta]$ is the smallest interval which contains the points of interpolation $x_0, x_1, \ldots, x_n$.

EXERCISES FOR CHAPTER II

1.

1. Find the polynomial of degree two or less which assumes the same values as the function $\phi(x)$ at

$$x = a, \frac{a+b}{2}, \ b.$$

Integrate $P(x)$ from a to b to obtain Simpson's formula.

2. Construct a polynomial of degree two or less such that

$$P(x_0) = \phi(x_0), \ P'(x_0) = \phi'(x_0), \ P''(x_0) = \phi''(x_0)$$

for a given twice-differentiable function $\phi(x)$ at $x = x_0$. Use this result to approximate a solution of the equation $\phi(x) = 0$ which is close to $x = x_0$, assuming that $\phi''(x_0) \neq 0$. Under what condition will this approximation fail to be real?

3. Show that Simpson's formula gives the exact value for the integral from a to b of the cubic polynomial

$$\phi(x) = Ax^3 + Bx^2 + Cx + D.$$

2.

1. Construct a table of values of the polynomial

$$P(x) = x^4 - 2.000x^3 + 1.3611x^2 - 0.3611x + 0.0309$$

for $x = 0.0(0.1)1.0$. Calculate the finite differences ΔP, $\Delta^2 P$, $\Delta^3 P$, $\Delta^4 P$ to check the validity of the computation.

2. Prove the validity of the formulas

$$\Delta^k y_l = y_{k+l} - \frac{k}{1!} y_{k+l-1} + \frac{k(k-1)}{2!} y_{k+l-2} - \ldots + (-1)^k y_l,$$

and

$$y_{k+l} = y_l + \frac{k}{1!} \Delta y_l + \frac{k(k-1)}{2!} \Delta^2 y_l + \ldots + \Delta^k y_l$$

by mathematical induction.

3. The following table of values was calculated for a polynomial $P_3(x)$ of degree three: ·

x	$P_3(x)$
−0.1035	0.4051
−0.1040	0.4020
−0.1045	0.4090
−0.1050	0.4105
−0.1055	0.4150
−0.1060	0.4238
−0.1065	0.4374
−0.1070	0.4560
−0.1075	0.4804
−0.1080	0.5111

By using finite differences, check the validity of the calculation, and find . and correct any errors.

3.

1. Construct a table of divided differences of orders one, two, and three for the function
$$f(x) = \log x(= \log_{10} x)$$
for the system of arguments
$$x = 1.1,\ 1.3,\ 1.6,\ 1.8,\ 2.2.$$
Use five-place tables. Would approximation of $\log x$ by a polynomial of degree three give accurate results for these values of x?

2. For
$$P(x) = x^4 + 2.5x^3 - 3x^2 + 2x + 1.8,$$
and the divided differences given in Table 18, Theorem 1 states that points ξ_1, ξ_2, ξ_3 exist,
$$1 < \xi_i < 1.8,\ i = 1, 2, 3,$$
such that
$$\frac{P'(\xi_1)}{1!} = 9.86800,$$
$$\frac{P''(\xi_2)}{2!} = 16.93000,$$
$$\frac{P'''(\xi_3)}{3!} = 8.20000.$$

Calculate ξ_1, ξ_2, ξ_3 to five decimal places.
Hint: By Theorem 1, $1 < \xi_1 < 1.2$.

3. Prove formula (3.11) by mathematical induction.

4.

1. Determine the interpolating polynomial of the form

$$S_2(x) = a + bx + cx^2$$

for the function $f(x)$ such that

$$f(0) = 0, \qquad f'(0) = 1, \qquad f(1) = 0.$$

2. Determine the trigonometric interpolating polynomial of the form

$$S_2(x) = a + b \sin \pi x + c \cos \pi x$$

for the function $f(x)$ of Problem 1.

3. Determine the generalized interpolating polynomial of the form

$$S_2(x) = ae^{-x} + be^x + ce^{2x}$$

for the function $f(x)$ such that

$$f(0) = 0, \qquad f'(0) = 1, \qquad f(\ln 2) = 0.$$

5.

1. Construct the interpolating polynomial $P_2(x)$ by formula (5.4) for

$$x = 1.2, \ 1.4, \ 2.0,$$
$$f(1.2) = 0.3849, \qquad f(1.4) = 0.4192, \qquad f(2.0) = 0.4773.$$

2. Construct the interpolating polynomial $P_2(x)$ for x and $f(x)$ as given in Problem 1 by Lagrange's formula (5.6).

3. Construct the interpolating polynomial $P_2(x)$ for x and $f(x)$ as given in Problem 1 by Newton's interpolation formula (5.7). Given the additional information

$$x = 1.7, \ f(1.7) = 0.4554,$$

construct the interpolating polynomial $P_3(x)$.

6.

1. For $f(x) = \log x (= \log_{10} x)$,

$$x_0 = 2.5, \ x_1 = 3.0, \ x_2 = 3.6,$$

find ξ, $2.5 < \xi < 3.6$, to five decimal places such that

$$R_2(f, 3.2) = \frac{\omega(3.2)}{3!} f'''(\xi),$$

that is, for which formula (6.2) holds. Use five-place tables.

2. For $f(x) = \log x \ (= \log_{10} x)$, determine the smallest values of M to five decimal places for which $f(x)$ belongs to $W_2(M)$ and $W_3(M)$ for x in the interval $[2.5, 3.6]$. For

$$x_0 = 2.5, \qquad x_1 = 3.0, \qquad x_2 = 3.6,$$

calculate upper bounds for

$$\sup|R_1(f, x)|, \qquad \sup|R_2(f, x)|, \qquad x \in [2.5, 3.6],$$

by inequality (6.4). Also by inequality (6.4), calculate an upper bound for

$$|R_2(f, 3.2)|.$$

3. Write the Čebyšev polynomials

$$T_1(t), \ T_2(t), \ T_3(t), \ T_4(t)$$

defined by formula (6.7) explicitly in terms of t.

7.

1. Prove that

$$C_t^{(k+1)} = \frac{t-k}{k+1} C_t^{(k)},$$

and

$$C_{-t}^{(k+1)} = - \frac{t-k}{k+1} C_{-t}^{(k)},$$

$k = 1, 2, 3, \ldots$, with

$$C_t^{(1)} = t, \qquad C_{-t}^{(1)} = -t.$$

2. Given the values for $\log x(= \log_{10} x)$:

x	$\log x$
3.00	0.47712
3.10	0.49136
3.20	0.50515
3.30	0.51851
3.40	0.53148
3.50	0.54407
3.60	0.55630

Calculate $\log 3.048$ by (7.14) and $\log 3.575$ by (7.15) to five decimal places. Compare with the values obtained from a five-place table.

3. Prove the validity of formulas (7.8) and (7.13) on the basis of (6.2).

8.

An important transcendental function is the complete elliptic integral

$$K(\theta) = \int_0^{\pi/2} \frac{d\phi}{\sqrt{1 - \sin^2 \theta \sin^2 \phi}}.$$

Given the table of values

σ	$K(\theta)$
5°	1.5738
10°	1.5828
15°	1.5981
20°	1.6200
25°	1.6490
30°	1.6858
35°	1.7312

construct a table of finite differences, and perform the interpolations indicated in the following exercises.

1. Calculate $K(18°)$ to four decimal places by Gauss' interpolation formula (8.4). Compare the result with the one given in a four-place table.

2. Calculate $K(16°)$ by Stirling's interpolation formula (8.8). Compare the result with the one given in a four-place table.

3. Calculate $K(17.5°)$ by Bessel's formula (8.13).

9.

1. Using the values of sinh x given in Table 21, calculate $x^* = \sinh^{-1} 1.6$ by inverse interpolation, retaining five decimal places.

2. From the table of values of $K(\theta)$ given in the Exercises for § 8, calculate $K(18°)$ to four decimal places by Aitken's method of interpolation.

3. Using the values of e^x given in Table 22, use Aitken's method of interpolation to find ln 2. Retain five decimal places in the calculations.

10.

1. For the example given in the text, construct the interpolating polynomial

$$P_6(x) = a_0 x^6 + a_1 x^5 + \ldots + a_5 x + a_6$$

directly by setting up and solving the corresponding system (4.4).

2. Prove that the formula (10.22) for the Hermite interpolation polynomial reduces to Lagrange's formula (5.6) in the case that all of the knots are simple, that is,

$$p_1 = p_2 = \ldots = p_m = 1, \ m = n+1.$$

3. Using formula (10.23), construct the polynomial which satisfies the conditions

x	-1	0	1	2
$P(x)$	1	2	1	0
$P'(x)$	-1	-2	-1	1

11.

1. From the values given in Table 21, use formula (11.14) to calculate

$$\cosh x = \frac{d}{dx}(\sinh x) \qquad \text{for} \quad x = 1.7,$$

and

$$\sinh x = \frac{d^2}{dx^2}(\sinh x) \qquad \text{for} \quad x = 1.7.$$

Retain five decimal places in the calculations. Compare the first result with

$$\cosh 1.7 = 2.82832.$$

2. From the values given in Table 20, use formula (11.9) to obtain

$$\frac{d}{dx} \cos x = -\sin x \quad \text{at} \quad x = 0.1,$$

and

$$\frac{d^2}{dx^2} \cos x = -\cos x \quad \text{at } x = 0.1.$$

Compare the first result with

$$\sin 0.1 = 0.09983.$$

3. Obtain an upper bound for the remainder of interpolation (11.6) for each of the values calculated in Problems 1−2. Compare these bounds with the actual errors.

Chapter III

APPROXIMATE CALCULATION OF INTEGRALS

1. Interpolation quadrature formulas

Calculation of definite integrals by the fundamental formula of integral calculus,

$$\int_a^b f(x)dx = F(b) - F(a),$$

where $f(x)$, let us say, is a continuous function on $[a, b]$ and $F(x)$ is its primitive function, is made difficult by the fact that the actual determination of $F(x)$ is possible only in rare cases. For this reason, formulas for the approximate calculation of integrals are of great significance. In this chapter, we shall become acquainted with the most important of them. Many formulas for the approximate calculation of definite integrals have the form

$$\int_a^b p(x)f(x)dx \cong \sum_{k=1}^n A_k^{(n)} f(x_k^{(n)}) \tag{1.1}$$

and are called *mechanical quadrature formulas*. The sum on the right hand side of (1.1) is called the *quadrature sum*. The numbers $x_k^{(n)}$ belong to the interval $[a, b]$, and are called the *knots* of the quadrature formula, and the numbers $A_k^{(n)}$ are the *coefficients* of the quadrature formula. We shall always consider the knots of the quadrature formula to be numbered in increasing order:

$$x_1^{(n)} < x_2^{(n)} < \ldots < x_n^{(n)}.$$

The interval of integration $[a, b]$ can also be infinite. The integrand is written in the form of the product of the two functions $p(x)$ and $f(x)$. The first of these, $p(x)$, is assumed to be fixed for the given formula (1.1) and is called the *weight function*. The function $f(x)$ belongs to some sufficiently wide class of functions, for example, continuous and such that the integral on the left hand side of (1.1) exists.

The equation (1.1) is approximate. The difference between the integral and the quadrature sum

$$R_n(f) = \int_a^b p(x)f(x)dx - \sum_{k=1}^{n} A_k^{(n)}f(x_k^{(n)})$$

is called the *remainder term of the quadrature formula* (1.1). The remainder term represents the error arising from replacing the integral by the quadrature sum.

When calculating the integral with the help of formula (1.1), the quadrature sum itself is calculated only approximately. Instead of $f(x_k^{(n)})$ we get some $\tilde{f}(x_k^{(n)})$, so that

$$\tilde{f}(x_k^{(n)}) = f(x_k^{(n)}) + \varepsilon_k^{(n)},$$

where $\varepsilon_k^{(n)}$ is the round off error. We shall assume that $|\varepsilon_k^{(n)}| \leq \varepsilon$ for $k = 1, 2, \ldots, n$. If one does not take into account the round off errors arising in the calculation of the sum of products

$$\sum_{k=1}^{n} A_k^{(n)}f(x_k^{(n)}),$$

then we can assert that the round off error in the calculation of the quadrature sum will not exceed the number

$$\varepsilon \sum_{k=1}^{n} |A_k^{(n)}|$$

and can be equal to it. From this it follows that the larger $\sum_{k=1}^{n} |A_k^{(n)}|$, is, the greater the round off error of the quadrature sum can be.

We shall assume that in (1.1), $p(x) \geq 0$ on $[a, b]$, and the formula is exact for $f(x) \equiv 1$. The latter condition is equivalent to the equation

$$\sum_{k=1}^{n} A_k^{(n)} = \int_a^b p(x)dx,$$

therefore, it is obvious that $\sum_{k=1}^{n} |A_k^{(n)}|$ will have the smallest possible value, if $A_k^{(n)} > 0$ for $k = 1, 2, \ldots, n$. This circumstance partially explains the importance for applications of the quadrature formulas with positive coefficients.

A method of constructing quadrature formulas, based on algebraic interpolation, is often applied. Let us assume that we are considering the calculation of

$$\int_a^b p(x)f(x)dx. \tag{1.2}$$

The interval of integration we consider to be arbitrary, the weight function $p(x)$ is assumed to be different from zero on a set of positive measure and such that the integrals

$$\mu_k = \int_a^b p(x)x^k dx, \qquad k = 0, 1, 2, \ldots$$

exist. The numbers μ_k are called the *moments* of the weight function $p(x)$. The function $f(x)$ is assumed to be continuous and such that the integral (1.2) exists. Of course, this latter restriction is only for the case that the interval $[a, b]$ is infinite.

We shall choose a finite number n of points $x_1^{(n)}, x_2^{(n)}, \ldots, x_n^{(n)}$ from the interval $[a, b]$ and carry out interpolation of the function $f(x)$ using its values at these points:

$$f(x) = \sum_{k=1}^n \frac{\omega(x)}{(x-x_k^{(n)})\omega'(x_k^{(n)})} f(x_k^{(n)}) + R_n(f, x). \tag{1.3}$$

Here $R_n(f, x)$ is the remainder of the interpolation and

$$\omega(x) = (x-x_1^{(n)})(x-x_2^{(n)}) \ldots (x-x_n^{(n)}).$$

In place of $f(x)$, we substitute the right hand side of (1.3) under the integral sign in (1.2). We obtain

$$\int_a^b p(x)f(x)dx = \sum_{k=1}^n A_k^{(n)}f(x_k^{(n)}) + \int_a^b p(x)R_n(f, x)dx, \tag{1.4}$$

where

$$A_k^{(n)} = \int_a^b p(x) \frac{\omega(x)}{(x-x_k^{(n)})\omega'(x_k^{(n)})} dx. \tag{1.5}$$

If $f(x)$ is approximated on $[a, b]$ with sufficient accuracy by the interpolation polynomial, in other words, if the remainder $R_n(f, x)$ is sufficiently small on the whole interval $[a, b]$, then in formula (1.4), which represents an exact equation, one can discard

$$\int_a^b p(x)R_n(f, x)dx,$$

and obtain approximately

$$\int_a^b p(x)f(x)dx \cong \sum_{k=1}^n A_k^{(n)} f(x_k^{(n)}). \tag{1.6}$$

In the quadrature formula (1.6), the coefficients are determined by formula (1.5). The arbitrariness in the construction of formula (1.6) is reduced to the arbitrariness in the selection of knots in the interval $[a, b]$. The quadrature formula (1.6), whose coefficients are determined by (1.5), we shall call an *interpolation quadrature formula*.

Theorem. In order that the quadrature formula (1.1) with n knots be an interpolation quadrature formula, it is necessary and sufficient, that it be exact, when $f(x)$ is a polynomial of degree not higher than $n-1$.

Proof. Necessity. Let formula (1.1) be an interpolation quadrature formula. If $f(x)$ is a polynomial of a degree not higher than $n-1$, then in equation (1.3), $R_n(f, x) \equiv 0$ and, consequently, (1.6) becomes an exact equation. But the quadrature sum in (1.6) coincides with the quadrature sum in (1.1), since, by hypothesis, the coefficients of formula (1.1) are determined with the help of (1.5).

Sufficiency. If formula (1.1) is exact for any polynomial of degree not higher than $n-1$, then, in particular, it must be accurate for the polynomial of the degree $n-1$

$$\omega_m(x) = \frac{\omega(x)}{(x-x_m^{(n)})\omega'(x_m^{(n)})}, \qquad m = 1, 2, \ldots, n.$$

Substituting $\omega_m(x)$ in (1.1), we get

$$\int_a^b p(x)\omega_m(x)dx = \sum_{k=1}^n A_k^{(n)} \omega_m(x_k^{(n)}) = A_m^{(n)},$$

and this means, that formula (1.1) is an interpolation quadrature formula. The sufficiency is proved.

We say that the quadrature formula has algebraic accuracy of degree m, if it is exact for $f(x) = x^j, j = 0, 1, 2, \ldots, m$ (or, for any polynomial of a degree not higher than m), and does not give an exact result for $f(x) = x^{m+1}$. From the theorem proved, it follows that the interpolation quadrature formulas with n knots have algebraic accuracy of degree smaller than $n-1$.

2. The simplest interpolation quadrature formulas

Let us examine the interpolation quadrature formulas which correspond to the case that the knots are chosen to be equidistant. We assume that the interval of integration $[a, b]$ is finite and the weight function $p(x)$ is identically equal to one.

We will begin with the rectangular formula with a single knot of interpolation. We take as the knot a certain point c of the interval $[a, b]$. The interpolation polynomial in the case of one knot is the constant $f(c)$, and the quadrature formula, consequently, has the form

$$\int_a^b f(x)dx \cong (b-a)f(c). \tag{2.1}$$

In this way, the area under the graph of the curve $y = f(x)$ is approximated by the area of the rectangle with height equal to the value of the function $f(x)$ at $x = c$ (see figure 14). Therefore, formula (2.1) is called the rectangular formula.

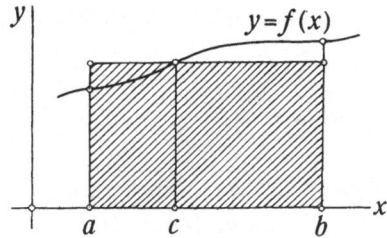

Figure 14

We note important special cases of formula (2.1). If the knot c coincides with the left end of the interval of integration: $c = a$, the formula (2.1) is called the left rectangular formula. If $c = b$, then formula (2.1) is called the right rectangular formula. Finally, if the knot c coincides with the middle of the interval of integration: $c = (a+b)/2$, then formula (2.1) is called the midpoint rectangular formula.

We shall find the representation of the remainder term of formula (2.1),

$$R(f, c) = \int_a^b f(x)dx - (b-a)f(c) \tag{2.2}$$

for

$$c = a, \quad c = b \quad \text{and} \quad c = \frac{a+b}{2}.$$

In the case of the left and right rectangular formulas, we assume that $f(x)$ has a continuous first derivative. According to Taylor's formula, we have

$$f(x) - f(a) = (x-a)f'(\xi),$$

where $a < \xi < b$. Integrating both sides of this equality with respect to x from a to b, we get

$$R(f, a) = \int_a^b (x-a)f'(\xi)dx. \tag{2.3}$$

We set

$$m = \min_{[a, b]} f'(x), \qquad M = \max_{[a, b]} f'(x). \tag{2.4}$$

The multiplier $x - a$ does not change sign in the interval $[a, b]$, therefore, to the integral of the right side of (2.3) one may apply the generalized mean value theorem,

$$R(f, a) = L \int_a^b (x-a)dx = L \frac{(b-a)^2}{2}, \tag{2.5}$$

where $m \le L \le M$. Since $f'(x)$ is continuous, then by virtue of the previous inequality an η, $a \le \eta \le b$, exists such that

$$L = f'(\eta).$$

One may now write down equality (2.5) thusly:

$$R(f, a) = \frac{(b-a)^2}{2} f'(\eta), \qquad a \le \eta \le b. \tag{2.6}$$

This is the representation of $R(f, a)$ which was to be found.
In this way one may get a representation for the remainder term of the right rectangular formula,

$$R(f, b) = -\frac{(b-a)^2}{2} f'(\eta), \qquad a \le \eta \le b. \tag{2.7}$$

In the case of the midpoint rectangular formula, we assume that the function $f(x)$ has a continuous second derivative. According to Taylor's formula, we have

$$f(x) - f\left(\frac{a+b}{2}\right) = \left(x - \frac{a+b}{2}\right) f'\left(\frac{a+b}{2}\right) + \frac{1}{2}\left(x - \frac{a+b}{2}\right)^2 f''(\xi),$$

where ξ is between x and $(a+b)/2$, and depends on x. Integrating both sides of the last equality with respect to x from a to b, we get

$$R\left(f, \frac{a+b}{2}\right) = \frac{1}{2}\int_a^b \left(x - \frac{a+b}{2}\right)^2 f''(\xi)dx.$$

We may apply the mean value theorem to the integral in the right side, since $(x-(a+b)/2)^2$ does not change sign on $[a, b]$. Carrying through the same reasoning as in the case of the left rectangular formula, we get:

$$R\left(f, \frac{a+b}{2}\right) = \frac{(b-a)^3}{24} f''(\eta), \qquad a \leqq \eta \leqq b. \tag{2.8}$$

If the interval $[a, b]$ is large, then the rectangular formulas are inaccurate. We divide the interval $[a, b]$ into n subintervals of length $h = (b-a)/n$ by the division points

$$x_k = a + kh, \qquad k = 0, 1, 2, \ldots, n.$$

We apply the rectangular formula (2.1) to the integral over each subinterval $[x_k, x_{k+1}]$

$$\int_{x_k}^{x_{k+1}} f(x)dx \cong hf(\alpha + kh), \qquad k = 0, 1, 2, \ldots, n-1, \tag{2.9}$$

where α is some point in the interval $[x_0, x_1] \equiv [a, a+h]$.

Summing both sides of formula (2.9) on k from 0 to $n-1$, we get

$$\int_a^b f(x)dx \cong h\sum_{k=0}^{n-1} f(\alpha + kh). \tag{2.10}$$

Formula (2.10) is called the *"large" rectangular formula* as distinct from the "small" formula (2.1). In the special cases $\alpha = a$, $\alpha = a+h$ and $\alpha = a+h/2$ formula (2.10) is called, respectively, the *large left, right,* and *midpoint* rectangular formula.

The remainder term $R_n(f, \alpha)$ of the rectangular formula (2.10) is obviously equal to the sum of the remainder terms

$$R(f, \alpha, k) = \int_{x_k}^{x_{k+1}} f(x)dx - hf(\alpha + kh)$$

of the small rectangular formulas (2.9):

$$R_n(f, \alpha) = \sum_{k=0}^{n-1} R(f, \alpha, k). \tag{2.11}$$

In particular, if $\alpha = a$, on the basis of (2.6),

$$R(f, a, k) = \frac{(b-a)^2}{2n^2} f'(\xi_k), \qquad x_k \leq \xi_k \leq x_{k+1},$$

and from formula (2.11) we get

$$R_n(f, a) = \frac{(b-a)^2}{2n^2} \sum_{k=0}^{n-1} f'(\xi_k). \tag{2.12}$$

We shall indicate a more compact representation for $R_n(f, a)$. We have

$$m \leq \frac{f'(\xi_0)+f'(\xi_1)+ \cdots +f'(\xi_{n-1})}{n} \leq M,$$

where m and M are defined by the relations (2.4). By virtue of the continuity of $f'(x)$ there exists an η, $a \leq \eta \leq b$, such that

$$\frac{f'(\xi_0)+f'(\xi_1)+ \cdots +f'(\xi_{n-1})}{n} = f'(\eta),$$

and from (2.12) we get a representation of the remainder term of the large left rectangular formula:

$$R_n(f, a) = \frac{(b-a)^2}{2n} f'(\eta), \qquad a \leq \eta \leq b. \tag{2.13}$$

By analogy, one can indicate the representations of the remainder terms of the large right rectangular formula

$$R_n(f, a+h) = - \frac{(b-a)^2}{2n} f'(\eta), \qquad a \leq \eta \leq b, \tag{2.14}$$

and the large midpoint rectangular formula

$$R_n\left(f, a+\frac{h}{2}\right) = \frac{(b-a)^3}{24n^2} f''(\eta), \qquad a \leq \eta \leq b. \tag{2.15}$$

We go on now to the derivation of the Newton-Cotes quadrature formulas. On the interval $[a, b]$ as $n+1$ equidistant knots, we take

$$x_k^{(n)} = a+kh, \qquad k = 0, 1, 2, \ldots, n, \ n \geq 1. \tag{2.16}$$

where $h = (b-a)/n$. Interpolation quadrature formulas having the knots (2.16) are called *Newton-Cotes formulas*. These formulas have the form

$$\int_a^b f(x)dx \cong \sum_{k=0}^n A_k^{(n)} f(a+kh). \tag{2.17}$$

Our problem consists of determining the coefficients. By virtue of (1.5),

$$A_k^{(n)} = \int_a^b \frac{\omega(x)}{(x-x_k^{(n)})\omega'(x_k^{(n)})} dx, \tag{2.18}$$

where

$$\omega(x) = (x-a)(x-a-h)\ldots(x-a-nh).$$

In the integral (2.18) we make a change of the variable $x = a+ht$. Obviously,

$$\omega(x) = h^{n+1}t(t-1)(t-2)\ldots(t-n),$$
$$x-x_k^{(n)} = h(t-k),$$
$$\omega'(x_k^{(n)}) = (x_k^{(n)}-x_0^{(n)})\ldots(x_k^{(n)}-x_{k-1}^{(n)})(x_k^{(n)}-x_{k+1}^{(n)})\ldots$$
$$\ldots(x_k^{(n)}-x_n^{(n)}) = (-1)^{n-k}h^n k!(n-k)!,$$

and we get

$$A_k^{(n)} = h\frac{(-1)^{n-k}}{k!(n-k)!}\int_0^n t(t-1)\ldots(t-k+1)(t-k-1)\ldots(t-n)dt.$$

We have

$$A_k^{(n)} = (b-a)B_k^{(n)}, \tag{2.19}$$

where

$$B_k^{(n)} = \frac{(-1)^{n-k}}{nk!(n-k)!}\int_0^n t(t-1)\ldots(t-k+1)(t-k-1)\ldots(t-n)dt. \tag{2.20}$$

Obviously, the coefficients $B_k^{(n)}$ do not depend on the interval $[a, b]$.

We write the $B_k^{(n)}$ for n from 1 to 5.

$$n = 1 \quad B_0^{(1)} = B_1^{(1)} = \tfrac{1}{2}$$

$$n = 2 \quad B_0^{(2)} = B_2^{(2)} = \tfrac{1}{6}, \qquad B_1^{(2)} = \tfrac{4}{6}$$

$$n = 3 \quad B_0^{(3)} = B_3^{(3)} = \tfrac{1}{8}, \qquad B_1^{(3)} = B_2^{(3)} = \tfrac{3}{8}$$

$$n = 4 \quad B_0^{(4)} = B_4^{(4)} = \tfrac{7}{90}, \qquad B_1^{(4)} = B_3^{(4)} = \tfrac{32}{90}, \qquad B_2^{(4)} = \tfrac{12}{90}$$

$$n = 5 \quad B_0^{(5)} = B_5^{(5)} = \tfrac{19}{288}, \qquad B_1^{(5)} = B_4^{(5)} = \tfrac{75}{288},$$

$$B_2^{(5)} = B_3^{(5)} = \tfrac{50}{288}.$$

For $n = 8$ and $n = 10$, negative numbers occur among the coefficients $B_k^{(n)}$. R. O. Kuz'min established asymptotic formulas for the coefficients $B_k^{(n)}$, from which, in particular, it follows that $\sum_{k=0}^{n} |B_k^{(n)}|$ increases without bound as $n \to \infty$. In view of the equality

$$\sum_{k=1}^{n} B_k^{(n)} = \frac{1}{b-a} \int_a^b 1 \, dx = 1$$

it follows that for large n there are positive as well as negative numbers among the coefficients.

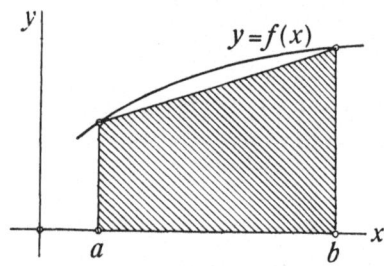

Fig. 15

We examine the simplest of the Newton-Cotes formulas. For $n = 1$, formula (2.17) will be written in the following way:

$$\int_a^b f(x)dx \cong \frac{b-a}{2} \left[f(a) + f(b) \right]. \tag{2.21}$$

This is the trapezoidal rule (see figure 15).

If one assumes that $f''(x)$ is continuous on $[a, b]$, then it is not difficult to indicate the representation of the remainder term of the trapezoidal rule,

$$R^{(1)}(f) = -\frac{(b-a)^3}{12}f''(\xi), \qquad a \leq \xi \leq b. \tag{2.22}$$

For this, it is sufficient to write the representation of the remainder term of interpolation for $f(x)$ according to its values at the knots a and b (see (6.2) from Chap. II)

$$f(x) - P_1(x) = \tfrac{1}{2}(x-a)(x-b)f''(\eta)$$

and to integrate both sides with respect to x from a to b.
We get

$$R^{(1)}(f) = \frac{1}{2}\int_a^b (x-a)(x-b)f''(\eta)dx.$$

It remains only to use the general mean-value theorem, which is possible since $(x-a)(x-b) \leq 0$ on $[a, b]$.
For $n = 2$ we get Simpson's rule from (2.17),

$$\int_a^b f(x)dx \cong \frac{b-a}{6}\left[f(a) + 4f\left(\frac{a+b}{2}\right) + f(b)\right]. \tag{2.23}$$

It is impossible to obtain the representation of the remainder in the form (2.22) in the same way as in the case of trapezoidal rule, since

$$\omega(x) = (x-a)\left(x-\frac{a+b}{2}\right)(x-b)$$

changes sign on $[a, b]$.
It is not hard to verify that if one constructs the Hermite interpolating polynomial according to the conditions

$$P(a) = f(a), \qquad P\left(\frac{a+b}{2}\right) = f\left(\frac{a+b}{2}\right),$$

$$P'\left(\frac{a+b}{2}\right) = f'\left(\frac{a+b}{2}\right), \qquad P(b) = f(b),$$

then

$$\int_a^b P(x)dx = \frac{b-a}{6}\left[f(a) + 4f\left(\frac{a+b}{2}\right) + f(b)\right]. \tag{2.24}$$

We write the representation of the remainder term of interpolation $f(x) - P(x)$ according to formula (10.24) Chap. II:

$$f(x) - P(x) = \frac{1}{4!} \Omega(x) f^{(IV)}(\xi), \qquad a < \xi < b,$$

where

$$\Omega(x) = (x-a) \left(x - \frac{a+b}{2}\right)^2 (x-b)$$

does not change sign on $[a, b]$.

By comparing (2.23) and (2.24), we see that the remainder term of formula (2.23) is equal to

$$R^{(2)}(f) = \frac{1}{4!} \int_a^b \Omega(x) f^{(IV)}(\xi) dx.$$

If one supposes that $f^{(IV)}(x)$ is continuous on $[a, b]$, then we get

$$R^{(2)}(f) = -\frac{(b-a)^5}{2880} f^{(IV)}(\eta), \qquad a \leq \eta \leq b. \tag{2.25}$$

It is not expedient to apply the quadrature formulas of Newton-Cotes for large n. As has already been noted, there are negative numbers among the coefficients of these formulas for large n. It is much more convenient to obtain an increase of accuracy by subdividing the interval $[a, b]$ and applying quadrature formulas with rather small n on the subintervals. This we did previously in the case of the rectangular quadrature formulas.

Without proof, we indicate the large forms of the trapezoidal and Simpson's rule, and their remainder terms. The trapezoidal rule has the form:

$$\int_a^b f(x) dx \cong \frac{b-a}{2n} [f(x_0^{(n)}) + 2(f(x_1^{(n)}) + f(x_2^{(n)}) + \ldots$$

$$\ldots + f(x_{n-1}^{(n)})) + f(x_n^{(n)})], \tag{2.26}$$

here, the $x_k^{(n)}$ are defined by (2.16). The remainder term of formula (2.26), assuming that $f''(x)$ is continuous on $[a, b]$, is

$$R_n^{(1)}(f) = -\frac{(b-a)^3}{12n^2} f''(\xi), \qquad a \leq \xi \leq b. \tag{2.27}$$

Simpson's rule is

$$\int_a^b f(x)dx \cong \frac{b-a}{3n} [f(x_0^{(n)})+4(f(x_1^{(n)})+f(x_3^{(n)})+ \cdots$$

$$\cdots +f(x_{n-1}^{(n)}))+2(f(x_2^{(n)})+f(x_4^{(n)})+ \cdots +f(x_{n-2}^{(n)}))+f(x_n^{(n)})],$$

$$(2.28)$$

here the $x_k^{(n)}$ are defined by (2.16) and n is assumed to be even. The remainder term of formula (2.28) is:

$$R_n^{(2)}(f) = -\frac{(b-a)^5}{180n^4} f^{(IV)}(\xi), \qquad a \leq \xi \leq b. \qquad (2.29)$$

This formula is valid on the assumption that $f^{(IV)}(x)$ exists and is continuous on $[a, b]$.

We will note that the interpolation formulas are only small formulas (rectangular, trapezoidal, and Simpson's).

Example. We compute the value of the sine integral

$$Si(x) = \int_0^x \frac{\sin t}{t} dt$$

for $x = 1$. We shall compute this integral with the help of the large rectangular, trapezoidal and Simpson formulas and compare the results obtained.

The question of the selection of the knots in the application of mechanical quadrature formulas is an important problem. The number of knots must be such that the required accuracy of the result of the computation is obtained. A precise way of solving the problem of the number of knots is by the use of the remainder term and an estimate for it. However, there are difficulties with this method: 1) derivates of the integrand functions which are often difficult to calculate and estimate enter into the remainder term. 2) the estimate obtained in this way for the remainder term turns out, as a rule, to be overly pessimistic. The correctness of the result may be verified by calculating the integral according to another formula (with the same knots), or according to the same formula, but with a different number of knots.

In our example, the derivatives of the integrand function are easily estimated. We have

$$\frac{\sin x}{x} = \int_0^1 \cos \xi x \, d\xi.$$

We calculate the derivative of order k with respect to x:

$$\frac{d^k}{dx^k} \left(\frac{\sin x}{x} \right) = \int_0^1 \xi^k \cos \left(x\xi + k \frac{\pi}{2} \right) d\xi.$$

Since the absolute value of the integral on the right-hand side is less than $1/(k+1)$, we have

$$\left| \frac{d^k}{dx^k} \left(\frac{\sin x}{x} \right) \right| \le \frac{1}{k+1}. \tag{2.30}$$

The estimate (2.30) of the derivative allows one to solve the problem of the choice of the number of knots. We shall determine, for example, how many knots one should take in the large trapezoidal formula (2.26) so that the approximate value of the integral obtained differs from its exact value by less than three units in the fourth decimal place. By virtue of formula (2.27) and the estimate (2.30), we find

$$\left| R_n^{(1)} \left(\frac{\sin x}{x} \right) \right| \le \frac{1}{12n^2} \cdot \frac{1}{3}.$$

We shall answer the question of the number of knots by solving the inequality

$$\frac{1}{36n^2} < 3 \cdot 10^{-4}.$$

It is easy to calculate that the inequality will be satisfied for $n \ge 10$, therefore, we choose $n = 10$.
We now estimate the remainder term of Simpson's rule (2.29) for $n = 10$. We have

$$\left| R_n^{(2)} \left(\frac{\sin x}{x} \right) \right| \le \frac{1}{180 \cdot 10^4} \cdot \tfrac{1}{5} = \tfrac{1}{9} 10^{-6} < 1.2 \cdot 10^{-7}.$$

Thus, one may obtain an approximate value of Si(1) which differs from the exact value by not more than 1.2 units of the seventh decimal place by Simpson's rule (2.28). We shall carry out the calculation with six decimal places, and the calculated value of the integral will differ from the exact value only on account of round off errors.

Table 24

x_k	$\sin x_k$	$f(x_k) = \dfrac{\sin x_k}{x_k}$	$10\,A_k$	$10\,A_k$	$20\,A_k$	$30\,A_k$
0	0	1	1		1	1
0.1	0.099833	0.998330	1	1	2	4
0.2	0.198669	0.993345	1	1	2	2
0.3	0.295520	0.985067	1	1	2	4
0.4	0.389418	0.973545	1	1	2	2
0.5	0.479426	0.958852	1	1	2	4
0.6	0.564642	0.941070	1	1	2	2
0.7	0.644218	0.920311	1	1	2	4
0.8	0.717356	0.896695	1	1	2	2
0.9	0.783327	0.870363	1	1	2	4
1	0.841471	0.841471		1	1	1
			9.537578	9.379049	18.916627	28.382473
		$\Sigma\,A_k f(x_k)$	0.953758	0.937905	0.945831	0.946082

The calculations according to the left and right rectangular formulas, the trapezoidal and Simpson's rule are shown in table 24. The calculations by the midpoint rectangular rule are given in table 25. The value of the calculated integral to six places is:

$$\text{Si}(1) = 0.946083.$$

The value calculated by Simpson's rule differs from the exact value by only one unit in the sixth decimal place. According to the trapezoidal formula the value obtained differs from the exact value by 2.5 units in the fourth decimal place, which is very close to the estimate of the remainder term found above: $3 \cdot 10^{-4}$.

The left and right rectangular formulas have little accuracy: the difference from the exact value is about $8 \cdot 10^{-3}$. In this connection, the formula of the left rectangles gives a value which is too large and the formula of the right rectangles one which is too small. This is explained by the fact that $\sin x/x$ is decreasing on $[0, 1]$. The midpoint rectangular formula gives a good result.

We note that, in our example, the second derivative of the integrand function is not positive on $[0, 1]$. In comparing the expressions of the

Table 26

x_k	$\sin x_k$	$f(x_k) = \dfrac{\sin x_k}{x_k}$	$10\,A_k$
0.05	0.049979	0.999580	1
0.15	0.149438	0.996253	1
0.25	0.247404	0.989616	1
0.35	0.342898	0.979708	1
0.45	0.434966	0.966591	1
0.55	0.522687	0.950340	1
0.65	0.605186	0.931055	1
0.75	0.681639	0.908852	1
0.85	0.751280	0.883859	1
0.95	0.813416	0.856227	1

| | | 9.462081 | |
| $\Sigma\,A_k f(x_k)$ | | 0.946208 | |

remainder terms of the midpoint rectangular formula (2.15) and of the trapezoidal rule (2.27), we conclude that the first of them gives a value which is too large, the second a value which is too small. Therefore, we can write the inequality

$$0.945831 \leq \text{Si}(1) \leq 0.946208,$$

from which we conclude that each of the values obtained differs from the exact value Si(1) by not more than 4 units in the fourth decimal place, and their average 0.946020 by not more than 2 units in the fourth decimal place.

3. Calculation of integrals of periodic functions and the rectangular quadrature formula

We consider the problem of the calculation of integral

$$\int_0^{2\pi} f(x)\,dx,$$

where $f(x)$ is a periodic function with period 2π. We shall construct a quadrature formula with n knots, belonging to the interval of integration,

$$\int_0^{2\pi} f(x)\,dx \cong \sum_{k=1}^{n} A_k f(x_k). \tag{3.1}$$

Since the integrand function $f(x)$ is 2π periodic, when constructing the formula (3.1) it is natural to approximate the function $f(x)$ by trigonometric (and not algebraic) polynomials,

$$T_m(x) = a_0 + \sum_{k=1}^{m} (a_k \cos kx + b_k \sin kx). \tag{3.2}$$

We say, that formula (3.1) has trigonometric degree of accuracy m, if it is exact for any trigonometric polynomial (3.2) of degree m and is not exact for polynomials of degree $m+1$.

We shall prove that formula (3.1) with n knots cannot be exact for all trigonometric polynomials of order n, no matter how the knots x_k and the coefficients A_k are chosen.

We examine the function

$$f(x) = \prod_{k=1}^{n} \sin^2 \frac{x - x_k}{2}, \tag{3.3}$$

where the x_k are knots of formula (3.1). Each of the n factors is a trigonometric polynomial of first degree

$$\sin^2 \frac{x - x_k}{2} = \tfrac{1}{2}[1 - \cos (x - x_k)] = \tfrac{1}{2}[1 - \cos x_k \cos x - \sin x_k \sin x].$$

It is known, that the product of trigonometric polynomials with real coefficients is a trigonometric polynomial, whose order is equal to the sum of orders of the polynomial factors. From this it follows that the function $f(x)$ defined by formula (3.3) is a trigonometric polynomial of order n. For this polynomial, formula (3.1) is not exact, since

$$\int_0^{2\pi} f(x)dx = \int_0^{2\pi} \prod_{k=1}^{n} \sin^2 \frac{x - x_k}{2} \, dx > 0$$

and at the same time

$$\sum_{k=1}^{n} A_k f(x_k) = 0.$$

We have established that the trigonometric degree of accuracy of formula (3.1) is not higher than $n-1$. We shall prove, that if, as formula (3.1), one takes the large rectangular formula (2.10), which in our case will be written in the following form

$$\int_0^{2\pi} f(x)dx \cong \frac{2\pi}{n} \sum_{k=1}^{n} f\left(\alpha+(k-1)\frac{2\pi}{n}\right),   \tag{3.4}$$

where α is any number from the interval $[0, 2\pi/n]$, then it will be exact for all trigonometric polynomials of order $n-1$.
One must prove that formula (3.4) is exact if

$$f(x) = \cos mx, \quad \sin mx$$

for $m = 0, 1, 2, \ldots, n-1$. It is sufficient to verify that it is exact if

$$f(x) = e^{imx}, \quad m = 0, 1, 2, \ldots, n-1,$$

where $i = \sqrt{-1}$. For $m = 0$, $f(x) = 1$ and formula (3.4), obviously, is exact. Let $0 < m \leq n-1$. We have

$$\int_0^{2\pi} f(x)dx = \int_0^{2\pi} e^{imx} dx = \frac{1}{im} e^{imx} \Big|_0^{2\pi} = 0.$$

The quadrature sum is also equal to zero:

$$\sum_{k=1}^{n} f(x_k) = \sum_{k=1}^{n} e^{im(\alpha+(k-1)2\pi/n)} = e^{im\alpha} \sum_{k=1}^{n} e^{im(k-1)2\pi/n} =$$

$$= e^{im\alpha} \frac{e^{im(2\pi/n)n}-1}{e^{im(2\pi/n)}-1} = 0,$$

since the numerator of the fraction is equal to zero, and the denominator is different from zero.
Thus, we have established that the trigonometric degree of accuracy of the quadrature formula (3.4) is equal to $n-1$.
It is obvious that the quadrature formula

$$\int_0^{T} f(x)dx \cong \frac{T}{n} \sum_{k=1}^{n} f\left(\alpha+(k-1)\frac{T}{n}\right), \quad *   \tag{3.5}$$

where α is any number from the interval $[0, T/n]$, becomes an exact equality when $f(x)$ is any trigonometrical polynomial of order $n-1$:

$$T_{n-1}(x) = a_0 + \sum_{k=1}^{n-1} \left(a_k \cos\frac{2\pi}{T} kx + b_k \sin\frac{2\pi}{T} kx\right).$$

---

* If $f(x)$ is a T-periodic function, then the integral in (3.5) may be replaced by an integral over any interval of length T.

Example. Calculate the integral

$$I = \int_0^{\frac{1}{2}\pi} \frac{d\varphi}{\sqrt{1-0.5\sin^2\varphi}}.$$

The integrand function is even and π-periodic, therefore

$$I = \frac{1}{2}\int_{-\frac{1}{2}\pi}^{\frac{1}{2}\pi} \frac{d\varphi}{\sqrt{1-0.5\sin^2\varphi}}$$

and we can use formula (3.5) for $T = \pi$. We will take $n = 6$ and we will situate the knots symmetrically with respect to the point $\varphi = 0$:

$$\varphi_{\pm 1} = \pm \tfrac{1}{12}\pi, \qquad \varphi_{\pm 2} = \pm \tfrac{1}{4}\pi, \qquad \varphi_{\pm 3} = \tfrac{5}{12}\pi.$$

Obviously, we have

$$2I \cong \frac{\pi}{3}\sum_{k=1}^{3} f(\varphi_k).$$

We will carry out the calculations to six decimal places. Here it is convenient to make use of a table of $\sin\varphi$, in which the argument is given in degrees. The calculations are shown in table 26. We obtain $I = 1.854007$. The value of integral I to six correct decimal places is equal to 1.854075, so that the error consists of 7 units in the fifth decimal place.

Table 26

φ_k	$\sin\varphi_k$	$\sin^2\varphi_k$	$0.5\sin^2\varphi_k$	$1-0.5\sin^2\varphi_k$	$(1-0.5\sin^2\varphi_k)^{\frac{1}{2}}$	$f(\varphi_k) =$ $= (-10.5\sin^2\varphi_k)^{-\frac{1}{2}}$
15°	0.258819	0.066987	0.033494	0.966506	0.983110	1.017180
45°	0.707107	0.5	0.25	0.75	0.866025	1.154701
75°	0.965926	0.933013	0.466506	0.533494	0.730407	1.369010
						3.540891
						1.854007

4. Gaussian type quadrature formulas

In § 1 it was established that the interpolation quadrature formula

$$\int_a^b p(x)f(x)dx \cong \sum_{k=1}^n A_k f(x_k) \tag{4.1}$$

is exact, if $f(x)$ is a polynomial of degree not greater than $n - 1$. This is true for any choice of the knots from $[a, b]$. To simplify the notation in the following presentation, as in (4. 1) we will not indicate in the notation the dependence on n of the knots and coefficients of the quadrature formula.

Gauss considered the following problem in the case of finite intervals $[a, b]$ and $p(x) = 1$: choose the knots $x_1, x_2, \ldots, x_n$ so that the quadrature formula (4.1) is accurate for all polynomials of a possibly higher degree. Since one has at his disposal the choice of n parameters (the knots), then it is natural to expect an increase in the degree of accuracy of formula (4.1) by n units. Gauss proved that it is actually possible to choose the knots $x_1, x_2, \ldots, x_n$, so that formula (4.1) will be accurate if $f(x)$ is a polynomial of degree not higher than $2n-1$. Later Gauss's result was generalized to the case of any interval and weight function $p(x)$ satisfying the conditions formulated below.

We pass to the presentation of these general results. We assume that $p(x) \geqq 0$ on $[a, b]$ and that it is integrable, with

$$\int_a^b p(x)dx > 0. * \tag{4.2}$$

We assume the existence of the moments of the function $p(x)$,

$$\mu_k = \int_a^b p(x)x^k dx, \qquad k = 0, 1, 2, \ldots . \tag{4.3}$$

In the following presentation, we assume that the weight function $p(x)$ satisfies the conditions above, unless the contrary is stated.

We say that the functions $\varphi(x)$ and $\psi(x)$ are *orthogonal with respect to the weight function* $p(x)$ on the interval $[a, b]$, if

---

* For the applications which we consider, it would be sufficient to make a more general assumption: $p(x)$ is non-negative and integrable on $[a, b]$, and the equality $p(x) = 0$ can take place only at a finite number of points.

$$\int_a^b p(x)\varphi(x)\psi(x)dx = 0.$$

In particular, if $p(x) \equiv 1$, then we simply say that $\varphi(x)$ and $\psi(x)$ are *orthogonal* on $[a, b]$.

Theorem 1. In order for the quadrature formula (4.1) to be exact for any polynomial of degree $2n-1$, it is necessary and sufficient that it be an interpolation formula and its knots $x_1, x_2, \ldots, x_n$ be roots of the polynomial of degree n

$$\omega(x) = (x-x_1)(x-x_2)\ldots(x-x_n), \tag{4.4}$$

which is orthogonal with respect to the weight function $p(x)$ to any polynomial $Q(x)$ of degree lower than n:

$$\int_a^b p(x)\omega(x)Q(x)dx = 0. \tag{4.5}$$

Proof. Necessity. Let the formula (4.1) be exact for all polynomials of degree $2n-1$. From this, by virtue of the theorem in § 1, it follows that formula (4.1) is an interpolation formula. We choose any polynomial $Q(x)$ of degree lower than n. We assume

$$f(x) = \omega(x)Q(x),$$

where $\omega(x)$ is the polynomial of degree n defined by formula (4.4). Obviously, the polynomial $f(x)$ has degree less than or equal to $2n-1$, therefore, formula (4.1) is exact for it:

$$\int_a^b p(x)\omega(x)Q(x)dx = \sum_{k=1}^n A_k\omega(x_k)Q(x_k) = 0.$$

The necessity of condition (4.5) has been proved.

Sufficiency. We assume that formula (4.1) is an interpolation formula and the polynomial $\omega(x)$ is orthogonal with respect to the weight function $p(x)$ to any polynomial of degree less than n. It is necessary to prove that if $f(x)$ is a polynomial of a degree less or equal to $2n-1$, then formula (4.1) is exact for it. Performing the division of $f(x)$ by $\omega(x)$, we obtain

$$f(x) = \omega(x)Q(x)+r(x), \tag{4.6}$$

where the degrees of $Q(x)$ and $r(x)$ are less than n

We multiply both sides of equation (4.6) by $p(x)$ and integrate from a to b:

$$\int_a^b p(x)f(x)dx = \int_a^b p(x)\omega(x)Q(x)dx + \int_a^b p(x)r(x)dx.$$

The first integral on the right hand side is equal to zero by hypothesis, and the second integral is exactly equal to the quadrature sum, since the degree of $r(x)$ is less than n and formula (4.1) is an interpolation formula. Thus, we get

$$\int_a^b p(x)f(x)dx = \sum_{k=1}^n A_k r(x_k),$$

but by virtue of (4.6), $r(x_k) = f(x_k)$, and we finally get

$$\int_a^b p(x)f(x)dx = \sum_{k=1}^n A_k f(x_k).$$

This means that formula (4.1) is exact for $f(x)$.

The question may be posed concerning the existence of formula (4.1), exact for all polynomials of degree $2n-1$ or, as we say, of Gaussian type formulas; the theorem reduces this to the question of the existence of a polynomial $\omega(x)$ of degree n, orthogonal to all polynomials of degree $n-1$, and such that its roots are real, distinct and lie in the interval $[a, b]$. With the assumptions made above concerning the weight function $p(x)$, the required polynomial $\omega(x)$ exists and its roots possess the properties required. We shall deduce the proofs of these facts.

We form the determinant

$$\Delta_j = \begin{vmatrix} \mu_0 & \mu_1 & \cdots & \mu_j \\ \mu_1 & \mu_2 & \cdots & \mu_{j+1} \\ \cdots & \cdots & \cdots & \cdots \\ \mu_j & \mu_{j+1} & \cdots & \mu_{2j} \end{vmatrix},$$

where the μ_k are defined by formula (4.3), and we shall prove that $\Delta_j \neq 0$ for $j = 0, 1, 2, \ldots$.

We shall prove that the linear homogeneous system for the unknowns $a_0, a_1, \ldots, a_j$

$$\begin{cases} \mu_0 a_0 + \mu_1 a_1 + \ldots + \mu_j a_j = 0, \\ \mu_1 a_0 + \mu_2 a_1 + \ldots + \mu_{j+1} a_j = 0, \\ \cdots \cdots \cdots \cdots \cdots \cdots \cdots \\ \mu_j a_0 + \mu_{j+1} a_1 + \ldots + \mu_{2j} a_j = 0 \end{cases}$$

has only the zero solution. From this it follows that $\Delta_j \neq 0$. Using (4.3), we rewrite this system in the form

$$\int_a^b p(x)[a_0 + a_1 x + \ldots + a_j x^j]dx = 0,$$

$$\int_a^b p(x)[a_0 + a_1 x + \ldots + a_j x^j]x\,dx = 0,$$

$$\cdots\cdots\cdots\cdots\cdots\cdots\cdots\cdots$$

$$\int_a^b p(x)[a_0 + a_1 x + \ldots + a_j x^j]x^j\,dx = 0.$$

We sum all of these equations after multiplying the first by a_0, the second by $a_1, \ldots$, the jth by a_j. We get

$$\int_a^b p(x)[a_0 + a_1 x + \ldots + a_j x^j]^2\,dx = 0.$$

Since $p(x) \geq 0$ and $\int_a^b p(x)dx > 0$,
then it follows from this that $a_0 + a_1 x + \ldots + a_j x^j \equiv 0$,
which is possible only in the case that $a_0 = a_1 = \ldots = a_j = 0$.
The polynomial

$$\omega(x) = \frac{1}{\Delta_{n-1}} \begin{vmatrix} \mu_0 & \mu_1 & \cdots & \mu_{n-1} & 1 \\ \mu_1 & \mu_2 & \cdots & \mu_n & x \\ \cdots & \cdots & \cdots & \cdots & \cdots \\ \mu_{n-1} & \mu_n & \cdots & \mu_{2n-2} & x^{n-1} \\ \mu_n & \mu_{n+1} & \cdots & \mu_{2n-1} & x^n \end{vmatrix} \qquad (4.7)$$

is of degree n (since the determinant $\Delta_{n-1} \neq 0$), and is orthogonal with respect to the weight function $p(x)$ to any polynomial of degree $n-1$. In fact,

$$\int_a^b p(x)\omega(x)x^k\,dx = \frac{1}{\Delta_{n-1}} \begin{vmatrix} \mu_0 & \mu_1 & \cdots & \mu_{n-1} & \mu_k \\ \mu_1 & \mu_2 & \cdots & \mu_n & \mu_{k+1} \\ \cdots & \cdots & \cdots & \cdots & \cdots \\ \mu_{n-1} & \mu_n & \cdots & \mu_{2n-2} & \mu_{k+n-1} \\ \mu_n & \mu_{n+1} & \cdots & \mu_{2n-1} & \mu_{k+n} \end{vmatrix} = 0$$

for $k = 0, 1, 2, \ldots, n-1$, which is equivalent to the orthogonality of $\omega(x)$ with respect to the weight function $p(x)$ to any polynomial of degree $n-1$. We note that in the polynomial (4.7), the coefficient of x^n is equal to one so that denoting it by $\omega(x)$ is justified.

We prove now that the roots of $\omega(x)$ are real, distinct and lie in the interval (a, b).

First of all, there are roots of odd multiplicity of $\omega(x)$ in (a, b), since

$$\int_a^b p(x)\omega(x) \cdot 1\, dx = 0$$

by virtue of the orthogonality of $\omega(x)$ with respect to the weight function $p(x)$ to 1. Let m be the number of roots of $\omega(x)$ of odd multiplicity, situated in (a, b). We denote them by $x_1', x_2', \ldots, x_m'$. We form the auxiliary polynomial

$$Q(x) = (x-x_1')(x-x_2') \ldots (x-x_m').$$

The product $\omega(x)Q(x)$ is a polynomial, which has all roots of even multiplicity in (a, b), and this means that it does not change sign on $[a, b]$. But then,

$$\int_a^b p(x)\omega(x)Q(x)dx \neq 0 \qquad (4.8)$$

by virtue of (4.2) and the fact that $p(x) \geq 0$. Since $\omega(x)$ is orthogonal to all polynomials of degree $n-1$, then it follows from (4.8) that the degree of $Q(x)$ is equal to n or, what is the same, that $m = n$. The assertion has been proved.

Thus, with the assumptions made about the weight function $p(x)$, a Gaussian type formula always exists. Let us convince ourselves that for a given n, such a formula is unique. To do this, we prove that the polynomial $\omega(x)$ which satisfies the condition (4.5) is unique. If there were two such polynomials: $\omega(x)$ and $\omega^*(x)$, then their difference

$$Q(x) = \omega(x)-\omega^*(x)$$

(we recall that the coefficient of x^n for $\omega(x)$ and $\omega^*(x)$ is equal to 1) would be a polynomial of degree not greater than $n-1$, and according to property (4.5), we would obtain

$$\int_a^b p(x)[\omega(x)-\omega^*(x)]Q(x)dx = \int_a^b p(x)Q^2(x)dx = 0,$$

which is possible only if

$$Q(x) = \omega(x)-\omega^*(x) \equiv 0.$$

We show now, that no quadrature formula of the form (4.1) can be exact for polynomials of degree $2n$ whatever the choice of the knots and coefficients. We call the knots of the quadrature formula $x_1, x_2, \ldots, x_n$ and we consider the polynomial of degree $2n$

$$f(x) = \omega^2(x) = (x-x_1)^2(x-x_2)^2 \ldots (x-x_n)^2.$$

Obviously formula (4.1) is not accurate for it, since

$$\int_a^b p(x)\omega^2(x)dx > 0,$$

and the quadrature sum

$$\sum_{k=1}^n A_k \omega^2(x_k) = 0$$

for any choice of the coefficients A_k.

Thus, the algebraic degree of accuracy of the Gaussian type formulas is equal to $2n-1$. The Gaussian type quadrature formulas are also called the quadrature formulas of the highest algebraic degree of accuracy, as we have proved, no quadrature formula of the form (4.1) can have algebraic degree of accuracy greater than $2n-1$ (under the assumptions made about the weight function $p(x)$).

We note another highly important characteristic of the Gaussian type formulas: the coefficients A_k of such formulas are positive.

In fact, let us take the polynomial of the degree $2n-2$

$$f(x) = \left[\frac{\omega(x)}{x-x_k}\right]^2,$$

for which, obviously, we have

$$f(x_j) = \begin{cases} 0 & \text{for } j \neq k, \\ [\omega'(x_k)]^2 & \text{for } j = k. \end{cases}$$

For this polynomial the Gaussian type formula is exact:

$$\int_a^b p(x)f(x)dx = A_k[\omega'(x_k)]^2.$$

From this equation it is obvious that $A_k > 0$, since

$$\int_a^b p(x)f(x)dx > 0$$

by virtue of the assumptions about $p(x)$ and the definition of $f(x)$. We now prove a theorem on the representation of the remainder term for a Gaussian type formula.

Theorem 2. If the function $f(x)$ has a continuous derivative of order $2n$, on the interval $[a, b]$, then there exists a point $\xi \in [a, b]$, such that the remainder term of the Gaussian type quadrature formula with n knots $x_1, x_2, \ldots, x_n$ has the form

$$R_n(f) = \frac{f^{(2n)}(\xi)}{(2n)!} \int_a^b p(x)\omega^2(x)dx, \tag{4.9}$$

where

$$\omega(x) = (x - x_1)(x - x_2) \ldots (x - x_n).$$

Proof. We construct the Hermitian interpolation polynomial $P(x)$ corresponding to the following conditions

$$P(x_i) = f(x_i), \qquad P'(x_i) = f'(x_i), \qquad i = 1, 2, \ldots, n. \tag{4.10}$$

According to formula (10.24) of Chapter II, which gives the representation of the remainder term of Hermitian interpolation, we have

$$f(x) = P(x) + \frac{\omega^2(x)}{(2n)!} f^{(2n)}(\eta), \tag{4.11}$$

where n depends on x and is situated in the interval containing x and the knots $x_1, x_2, \ldots, x_n$ of interpolation. We assume that x does not fall outside the limits of the interval $[a, b]$, therefore $\eta \in (a, b)$.

We multiply both sides of (4.11) by $p(x)$ and integrate with respect to x from a to b:

$$\int_a^b p(x)f(x)dx = \int_a^b p(x)P(x)dx +$$

$$+ \frac{1}{(2n)!} \int_a^b p(x)f^{(2n)}(\eta)\omega^2(x)dx. \tag{4.12}$$

We assume, of course, that the integral on the left hand side exists. As the polynomial $P(x)$ has degree not greater than $2n - 1$, the first integral in right hand side of (4.12) can be replaced by the quadrature sum

$$\sum_{k=1}^n A_k P(x_k).$$

In the quadrature sum, by virtue of (4.10), one can replace $P(x_k)$ by $f(x_k)$, so that (4.12) may be written in the form

$$\int_a^b p(x)f(x)dx = \sum_{k=1}^n A_k f(x_k) + \frac{1}{(2n)!} \int_a^b p(x)f^{(2n)}(\eta)\omega^2(x)dx.$$

From this it is obvious that the integral on the right hand side represents the remainder of the quadrature formula,

$$R_n(f) = \frac{1}{(2n)!} \int_a^b p(x)f^{(2n)}(\eta)\omega^2(x)dx.$$

Since $p(x)\omega^2(x) \geq 0$, then one can use the generalized mean-value theorem, and obtain formula (4.9).

We now prove a theorem on the convergence of Gaussian type quadrature formulas.

Theorem 3. If the interval of integration $[a, b]$ is finite and $f(x)$ is continuous on it, then the convergence of the quadrature sum of the Gaussian type formulas to the integral takes place as $n \to \infty$:

$$\lim_{n \to \infty} \sum_{k=1}^n A_k^{(n)} f(x_k^{(n)}) = \int_a^b p(x)f(x)dx. \qquad (4.13)$$

Proof: It is necessary to prove that

$$R_n(f) = \int_a^b p(x)f(x)dx - \sum_{k=1}^n A_k^{(n)} f(x_k^{(n)}) \to 0$$

as $n \to \infty$. Let $\varepsilon > 0$. Since $[a, b]$ is finite and $f(x)$ is continuous on it, then according to the Weierstrass theorem a polynomial $P(x)$ exists such that

$$|f(x) - P(x)| \leq \varepsilon \qquad (4.14)$$

for all $x \in [a, b]$. We have

$$R_n(f) = \int_a^b p(x)(f(x) - P(x))dx$$
$$+ \left[\int_a^b p(x)P(x)dx - \sum_{k=1}^n A_k^{(n)} P(x_k^{(n)}) \right] +$$
$$+ \sum_{k=1}^n A_k^{(n)} (P(x_k^{(n)}) - f(x_k^{(n)})). \qquad (4.15)$$

The expression in the square brackets represents $R_n(P)$ which is the remainder of the quadrature formula for the polynomial $P(x)$. If we denote the degree of this polynomial by N, then for $2n-1 > N$ the equality $R_n(P) = 0$ takes place. The two remaining expressions on the right hand side of (4.15) can be estimated easily on the basis of inequality (4.14):

$$\left| \int_a^b p(x)(f(x) - P(x))dx \right| \leq \varepsilon \int_a^b p(x)dx,$$

$$\left| \sum_{k=1}^n A_k^{(n)}(P(x_k^{(n)}) - f(x_k^{(n)})) \right| \leq \varepsilon \sum_{k=1}^n A_k^{(n)} = \varepsilon \int_a^b p(x)dx.$$

Thus for $2n-1 > N$

$$|R_n(f)| \leq 2\varepsilon \int_a^b p(x)dx$$

the theorem is proved.

We consider the sequence of the quadrature formulas (not necessarily the Gauss type) of the general form

$$\int_a^b p(x)f(x)dx \cong \sum_{k=1}^n A_k^{(n)}f(x_k^{(n)}). \tag{4.16}$$

We assume that the interval $[a, b]$ is finite, and the weight function $p(x)$ is any integrable function on $[a, b]$. We cite without proof the theorem on the convergence of the quadrature formulas (4.16).

Theorem 4. In order that

$$\lim_{n \to \infty} \sum_{k=1}^n A_k^{(n)}f(x_k^{(n)}) = \int_a^b p(x)f(x)dx \tag{4.17}$$

for any continuous function $f(x)$ on $[a, b]$, the fulfilment of two conditions is necessary and sufficient:
1) the limit relation (4.17) holds for any polynomial $f(x)$;
2) there is a number K such that

$$\sum_{k=1}^n |A_k^{(n)}| \leq K, \qquad n = 1, 2, 3, \ldots .$$

If the quadrature formulas (4.16) are interpolation formulas and have positive coefficients ($A_k^{(n)} > 0$, $k = 1, 2, \ldots, n$, $n = 1, 2, 3, \ldots$), then the conditions of the theorem are fulfilled. Thus, theorem 3 is a particular case of theorem 4.

5. Legendre polynomials and the Gauss formula

The Gauss quadrature formula represents a particular case of Gaussian type formulas for the weight function $p(x) \equiv 1$ and the finite interval of integration $[a, b]$. The hypothesis on the finiteness of $[a, b]$ provides the existence of the moments of the function $p(x) = 1$. We shall consider the interval to be $[-1, 1]$.

The Gauss quadrature formula

$$\int_{-1}^{1} f(x)dx \cong \sum_{k=1}^{n} A_k f(x_k) \tag{5.1}$$

has as its knots x_k the roots of the polynomial (4.7), $\omega(x)$, of degree n, which is orthogonal to all polynomials of degree less than n. The polynomials which possess the indicated property are called the Legendre polynomials. Of course, by the conditions of orthogonality, the Legendre polynomial is determined only up to a constant multiplier. For the following presentation (4.7) is not very suitable. We will use the representation of the Legendre polynomial according to the Rodrigues formula:

$$P_n(x) = \frac{1}{2^n n!} \frac{d^n (x^2 - 1)^n}{dx^n}. \tag{5.2}$$

We shall show that the polynomial $P_n(x)$ possesses the orthogonality property on the interval $[-1, 1]$. We introduce the notation

$$\varphi(x) = (x^2 - 1)^n. \tag{5.3}$$

Obviously, we have

$$P_n(x) = \frac{1}{2^n n!} \varphi^{(n)}(x), \quad \varphi^{(k)}(\pm 1) = 0, \qquad k = 0, 1, 2, \ldots, n-1. \tag{5.4}$$

Let $Q(x)$ be any n times continuously differentiable function on $[-1, 1]$. Integrating by parts and taking into account relationship (5.4), we get

$$\int_{-1}^{1} P_n(x)Q(x)dx = \frac{1}{2^n n!} \int_{-1}^{1} Q(x)\varphi^{(n)}(x)dx =$$

$$= \frac{1}{2^n n!} [Q(x)\varphi^{(n-1)}(x)]_{-1}^{1} - \frac{1}{2^n n!} \int_{-1}^{1} Q'(x)\varphi^{(n-1)}(x)dx =$$

$$= -\frac{1}{2^n n!} \int_{-1}^{1} Q'(x)\varphi^{(n-1)}(x)dx = \ldots$$

$$\ldots = (-1)^n \frac{1}{2^n n!} \int_{-1}^{1} Q^{(n)}(x)\varphi(x)dx.$$

Thus, we have the equality

$$\int_{-1}^{1} P_n(x)Q(x)dx = (-1)^n \frac{1}{2^n n!} \int_{-1}^{1} Q^{(n)}(x)\varphi(x)dx. \qquad (5.5)$$

If $Q(x)$ is any polynomial of a degree less than n, then $Q^{(n)}(x) \equiv 0$, and from (5.5) we get

$$\int_{-1}^{1} P_n(x)Q(x)dx = 0.$$

The polynomial $P_n(x)$ differs from the polynomial $\omega(x)$, defined by formula (4.7) by a constant multiplier which is easy to determine. From formula (5.2), we get

$$P_n(x) = \frac{(2n)!}{2^n(n!)^2} x^n - \dots . \qquad (5.6)$$

Thus,

$$\omega(x) = \frac{2^n(n!)^2}{(2n)!} P_n(x). \qquad (5.7)$$

We compute now the integral of the square of the polynomial $P_n(x)$. We set $Q(x) = P_n(x)$ in formula (5 5). In addition, we take into account that, by virtue of (5.6),

$$Q^{(n)}(x) = P_n^{(n)}(x) = \frac{(2n)!}{2^n \cdot n!}.$$

We get

$$\int_{-1}^{1} P_n^2(x)dx = (-1)^n \frac{(2n)!}{2^{2n}(n!)^2} \int_{-1}^{1} (x^2-1)^n dx =$$

$$= 2 \frac{(2n)!}{2^{2n}(n!)^2} \int_{0}^{1} (1-x^2)^n dx, \qquad (5.8)$$

and remaining is the calculation of the integral

$$I_n = \int_{0}^{1} (1-x^2)^n dx.$$

Supposing that $x = \sin \varphi$, we get

$$I_n = \int_{0}^{\frac{1}{2}\pi} \cos^{2n+1} \varphi \, d\varphi = \frac{(2n)!!}{(2n+1)!!} = \frac{2^{2n}(n!)^2}{(2n+1)!}.$$

From (5.8), we find that

$$\int_{-1}^{1} P_n^2(x)dx = \frac{2}{2n+1}. \tag{5.9}$$

According to Leibnitz' rule for the derivative of the product of two functions, we find

$$P_n(x) = \frac{1}{2^n n!} \frac{d^n}{dx^n} [(x+1)^n(x-1)^n] =$$

$$= \frac{1}{2^n n!} \sum_{k=0}^{n} C_n^{(k)} \frac{d^{(n-k)}}{dx^{n-k}}(x+1)^n \cdot \frac{d^k}{dx^k}(x-1)^n =$$

$$= \frac{1}{2^n n!} \sum_{k=0}^{n} C_n^{(k)} \frac{n!}{k!}(x+1)^k \frac{n!}{(n-k)!}(x-1)^{n-k} =$$

$$= \frac{1}{2^n} \sum_{k=0}^{n} [C_n^{(k)}]^2(x+1)^k(x-1)^{n-k}.$$

In particular, for $x = 1$ and -1, we get

$$P_n(1) = 1, \qquad P_n(-1) = (-1)^n. \tag{5.10}$$

We proceed now to the determination of the knots and coefficients of the Gauss formula. In order to find the knots of the formula, it is necessary to find all the roots of the algebraic equation

$$P_n(x) = 0.$$

After the knots are determined, the coefficients may be found by formula (1.5)

$$A_k = \int_{-1}^{1} \frac{P_n(x)dx}{(x-x_k)P_n'(x_k)}. \tag{5.11}$$

We note, that the polynomial $\omega(x)$ appears in formula (1.5) with the coefficient of x^n equal to one. However, from the form of formula (1.5), it is clear that it will be exact if this condition is not satisfied by $\omega(x)$. Formula (5.11) is not convenient for calculation, and we shall indicate a simpler formula. We consider the integral

$$S_k = \int_{-1}^{1} \frac{P_n(x)}{x-x_k} P_n'(x)dx. \tag{5.12}$$

Since the integrand represents a polynomial of degree $2n-2$, then according to the Gauss formula,

$$S_k = A_k[P'_n(x_k)]^2. \tag{5.13}$$

On the other hand, applying the formula of integration by parts to the integral (5.12) and setting

$$U = \frac{P_n(x)}{x-x_k}, \qquad dV = P'_n(x)dx,$$

we get

$$S_k = \frac{P_n^2(x)}{x-x_k}\bigg|_{-1}^{1} - \int_{-1}^{1} P_n(x)U'(x)dx.$$

The integral on the right side equals zero by orthogonality. The first term on the right-hand side is easy to calculate if one considers (5.10):

$$S_k = \frac{P_n^2(x)}{x-x_k}\bigg|_{-1}^{1} = \frac{1}{1-x_k} + \frac{1}{1+x_k} = \frac{2}{1-x_k^2}.$$

By comparing this expression for S_k with the expression (5.13), we get

$$A_k = \frac{2}{(1-x_k^2)[P'_n(x_k)]^2}. \tag{5.14}$$

This is the formula we wanted to obtain. In calculating the A_k by formula (5.14) it is not necessary to compute the integrals. We note that from (5.14), the positiveness of the coefficients again occurs in the Gauss formula.

We indicate the representation of the remainder term of the Gauss formula. We suppose that function $f(x)$ has a continuous derivative of the order $2n$ on the interval $[-1, 1]$. Then, on the basis of the general representation of the remainder term of the Gaussian type formula (4.9), we get

$$R_n(f) = \frac{f^{(2n)}(\xi)}{(2n)!} \int_{-1}^{1} \omega^2(x)dx.$$

By using formulas (5.7) and (5.9), we find that

$$R_n(f) = \frac{f^{(2n)}(\xi)}{(2n)!} \int_{-1}^{1} \frac{2^{2n}(n!)^4}{[(2n)!]^2} P_n^2(x)dx =$$

$$= \frac{f^{(2n)}(\xi)}{(2n)!} \frac{2^{2n}(n!)^4}{[(2n)!]^2} \frac{2}{2n+1}.$$

We have obtained the representation of the remainder term of the Gauss quadrature formula (5.1)

$$R_n(f) = \frac{2^{2n+1}}{2n+1} \frac{(n!)^4}{[(2n)!]^3} f^{(2n)}(\xi), \qquad -1 \leq \xi \leq 1. \qquad (5.15)$$

We cite the nodes and coefficients of the Gauss formula for $n = 1(1)8$:

$$n = 1$$
$$x_1 = 0 \qquad\qquad A_1 = 2$$

$$n = 2$$
$$-x_1 = x_2 = 0.5773502692 \qquad A_1 = A_2 = 1$$

$$n = 3$$
$$-x_1 = x_3 = 0.7745966692 \qquad A_1 = A_3 = 0.5555555556$$
$$x_2 = 0 \qquad\qquad A_2 = 0.8888888889$$

$$n = 4$$
$$-x_1 = x_4 = 0.8611363116 \qquad A_1 = A_4 = 0.3478548451$$
$$-x_2 = x_3 = 0.3399810436 \qquad A_1 = A_3 = 0.6521451549$$

$$n = 5$$
$$-x_1 = x_5 = 0.9061798459 \qquad A_1 = A_5 = 0.2369268851$$
$$-x_2 = x_4 = 0.5384693101 \qquad A_2 = A_4 = 0.4786286705$$
$$x_3 = 0 \qquad\qquad A_3 = 0.5688888889$$

$$n = 6$$
$$-x_1 = x_6 = 0.9324695142 \qquad A_1 = A_6 = 0.1713244924$$
$$-x_2 = x_5 = 0.6612093865 \qquad A_2 = A_5 = 0.3607615730$$
$$-x_3 = x_4 = 0.2386191861 \qquad A_3 = A_4 = 0.4679139346$$

$$n = 7$$
$$-x_1 = x_7 = 0.9491079123 \qquad A_1 = A_7 = 0.1294849662$$
$$-x_2 = x_6 = 0.7415311856 \qquad A_2 = A_6 = 0.2797053915$$
$$-x_3 = x_5 = 0.4058451514 \qquad A_3 = A_5 = 0.3818300505$$
$$x_4 = 0 \qquad\qquad A_4 = 0.4179591837$$

$$n = 8$$
$$-x_1 = x_8 = 0.9602898565 \qquad A_1 = A_8 = 0.1012285363$$
$$-x_2 = x_7 = 0.7966664774 \qquad A_2 = A_7 = 0.2223810345$$
$$-x_3 = x_6 = 0.5255324099 \qquad A_3 = A_6 = 0.3137066459$$
$$-x_4 = x_5 = 0.1834346425 \qquad A_4 = A_5 = 0.3626837834$$

In V. I. Krylov's book [18] the values of the knots and coefficients of the Gauss quadrature formula (5.1) are given to fifteen decimal places for $n = 1(1)16$.

We give the Gauss quadrature formula for an integral over an arbitrary interval $[a, b]$,

$$\int_a^b f(t)dt.$$

In the integral we make the change of the variable of integration,

$$t = \frac{b-a}{2} x + \frac{a+b}{2}.$$

We get

$$\int_a^b f(t)dt = \frac{b-a}{2} \int_{-1}^1 f\left(\frac{b-a}{2} x + \frac{a+b}{2}\right) dx.$$

By applying formula (5.1) to the integral on the right-hand side, we get

$$\int_a^b f(t)dt \cong \frac{b-a}{2} \sum_{k=1}^n A_k f(t_k), \tag{5.16}$$

where

$$t_k = \frac{b-a}{2} x_k + \frac{a+b}{2},$$

x_k are the knots of the Gauss quadrature formula for the interval $[-1, 1]$, and A_k are the coefficients corresponding to them. It follows that formula (5.16) must be used in calculating, moreover, the knots x_k and coefficients A_k for the interval $[-1, 1]$ are to be taken from the table cited above. Example. Calculate the integral

$$I = \int_0^1 \frac{dt}{\sqrt{(t^2+1)(3t^2+4)}}.$$

We use formula (5.16) with $n = 4$. The calculations are cited in table 27. We calculate, by choice, the quadrature sum

$$\sum_{k=1}^4 A_k f(t_k)$$

Table 27

x_k	$t_k = \frac{1}{2}(x_k+1)$	t_k^2	$3t_k^2+4$	$(t_k^2+1)(3t_k^2+4)$	$\sqrt{(t_k^2+1)(3t_k^2+4)}$	$f(t_k)$	A_k
—0.861136	0.069432	0.004821	4.014463	4.033817	2.008437	0.497897	0.347855
—0.339981	0.330009	0.108906	4.326718	4.797924	2.190417	0.456534	0.652145
0.339981	0.669991	0.448888	5.346664	7.746717	2.783293	0.359287	0.652145
0.861136	0.930568	0.865957	6.597871	12.311337	3.508752	0.285002	0.347855

we get

$$I = 0.402184.$$

The integral considered is an elliptic integral of the first kind

$$I = \frac{1}{2}\int_0^{\frac{1}{2}\pi} \frac{d\varphi}{\sqrt{1-0.25\sin^2\varphi}} = \frac{1}{2}F\left(0.5; \frac{\pi}{4}\right).$$

Its tabular value is equal to 0.402184.

6. Other special cases of quadratur formula of the Gaussian type

We shall study the quadrature formula of Gaussian type for calculation of integrals over a finite interval, which we can assume to coincide with $[-1, 1]$, with the weight function

$$p(x) = (1-x)^\alpha(1+x)^\beta. \tag{6.1}$$

Here, α and β are any real numbers which satisfy the condition

$$\alpha > -1, \qquad \beta > -1.$$

These inequalities insure the existence of the integral of $p(x)$ over the interval $[-1, 1]$. Of course, all the other conditions which the function $p(x)$ must satisfy are fulfilled.

The quadrature formula has the form

$$\int_{-1}^1 (1-x)^\alpha(1+x)^\beta f(x)dx \cong \sum_{k=1}^n A_k f(x_k). \tag{6.2}$$

The knots x_k are the roots of the polynomial of the degree n, orthogonal with respect to the weight function (6.1) to all polynomials of degree $n-1$. Such a polynomial is called a Jacobi polynomial. We shall not pause to study the properties of Jacobi polynomials, but cite the basic formulas without proof.

A representation is valid which is analogous to the Rodrigues formula for Legendre polynomials:

$$P_n^{(\alpha, \beta)}(x) = \frac{(-1)^n}{2^n n!} (1-x)^{-\alpha}(1+x)^{-\beta} \times$$

$$\times \frac{d^n}{dx^n} [(1-x)^{\alpha+n}(1+x)^{\beta+n}]. \tag{6.3}$$

The following representation is correct for the coefficients of formula (6.2)

$$A_k = 2^{\alpha+\beta+1} \frac{\Gamma(\alpha+n+1)\Gamma(\beta+n+1)}{n!\Gamma(\alpha+\beta+n+1)} \cdot \frac{1}{(1-x_k^2)[P_n^{(\alpha, \beta)\prime}(x_k)]^2} \cdot \tag{6.4}$$

Here, the function $\Gamma(z)$ is determined by an Euler integral of the second kind

$$\Gamma(z) = \int_0^\infty e^{-t} t^{z-1} dt,$$

where $z > 0$.

We now write the remainder term of the formula (6.2)

$$R_n(f) = \frac{2^{\alpha+\beta+2n+1}}{\alpha+\beta+2n+1} \times$$

$$\times \frac{\Gamma(\alpha+n+1)\Gamma(\beta+n+1)\Gamma(\alpha+\beta+n+1)}{\Gamma^2(\alpha+\beta+2n+1)} \frac{n!}{(2n)!} f^{(2n)}(\xi). \tag{6.5}$$

For $\alpha = \beta = 0$, Jacobi polynomials coincide with Legendre polynomials, moreover, formulas (6.2), (6.3), (6.4), (6.5) go over to formulas (5.1), (5.2), (5.14), (5.15), respectively.

We consider the particular case of formula (6.2), for $\alpha = \beta = -\frac{1}{2}$. In this case, the weight function (6.1) becomes the function

$$p(x) = \frac{1}{\sqrt{1-x^2}} \cdot \tag{6.6}$$

It is well known that the orthogonal polynomials with the weight function (6.6) on $[-1, 1]$ are the Čebyšev polynomials

$$T_n(x) = \cos n \arccos x. \tag{6.7}$$

This statement is not hard to verify. We calculate the integral

$$I = \int_{-1}^{1} \frac{1}{\sqrt{1-x^2}} x^m \cos n \arccos x \, dx.$$

We make the change of the variable of integration $x = \cos \theta$. We get

$$I = \int_{0}^{\pi} \cos^m \theta \cos n\theta \, d\theta = 0$$

for $m < n$, because, as one can show,

$$\cos^m \theta = \sum_{k=0}^{m} \alpha_k \cos k\theta$$

and

$$\int_{0}^{\pi} \cos k\theta \cdot \cos n\theta \, d\theta = 0$$

for $k = 0, 1, 2, \ldots, n-1$.

Thus, the knots of the quadrature formula of the highest degree of accuracy for the weight function (6.6) are the roots of the Čebyšev polynomial (6.7):

$$x_{n+1-k} = \cos \frac{2k-1}{2n} \pi, \qquad k = 1, 2, \ldots, n. \tag{6.8}$$

The coefficients of the formula may be written in the form

$$A_k = \frac{1}{T_n'(x_k)} \int_{-1}^{1} \frac{1}{\sqrt{1-x^2}} \frac{T_n(x)}{x-x_k} \, dx.$$

We calculate this integral. We will make the change of the variable of integration $x = \cos \theta$:

$$A_k = \frac{1}{T_n'(x_k)} \int_{0}^{\pi} \frac{\cos n\theta}{\cos \theta - x_k} \, d\theta.$$

In view of the evenness of the integrand,

$$A_k = \frac{1}{2T_n'(x_k)} \int_{-\pi}^{\pi} \frac{\cos n\theta}{\cos \theta - x_k} \, d\theta. \tag{6.9}$$

In § 3, it was proved that the rectangular quadrature formula with n

knots gives an exact value of the integral over an interval of length 2π of a trigonometrical polynomial of order $n-1$. A trigonometrical polynomial of order $n-1$ is under the integral sign in (6.9), as

$$\frac{T_n(x)}{x-x_k}$$

is a polynomial of the degree $n-1$ in x. Consequently, if one applies the rectangular formula for the computation of the integral (6.9), having taken the $2n$ knots

$$\theta_j = \frac{2j-1}{2n}\,\pi, \qquad j = -n+1, -n+2, \ldots, 0, 1, 2, \ldots, n,$$

then one obtains the exact value of A_k.
The value of the integrand

$$f(\theta) = \frac{\cos n\theta}{\cos\theta - x_k} \tag{6.10}$$

for $\theta = \theta_j, j = 1, 2, \ldots, n$, or, what is the same, the value of

$$\frac{T_n(x_j)}{x_j - x_k}$$

is equal to zero for $j \neq k$ and is equal to $T_n'(x_k)$ for $j = k$. Since the integrand (6.10) is even and $\theta_{-j+1} = -\theta_j$, then $f(\theta_{-j+1}) = f(\theta_j)$ for $j = 1, 2, \ldots, n$. Thus, we get

$$A_k = \frac{1}{2T_n'(x_k)} \frac{2\pi}{2n} [T_n'(x_k) + T_n'(x_k)] = \frac{\pi}{n}.$$

In this way, the quadrature formula of the highest algebraic degree of accuracy with the weight (6.6) has the form

$$\int_{-1}^{1} \frac{1}{\sqrt{1-x^2}} f(x)dx \cong \frac{\pi}{n} \sum_{k=1}^{n} f(x_k), \tag{6.11}$$

where the knots x_k are determined by formula (6.8). The quadrature formula (6.11) is called Mellor's formula.
We indicate the remainder term of formula (6.11). We use the relationship

$$\cos(j+1)\theta = 2\cos\theta \cdot \cos j\theta - \cos(j-1)\theta$$

$\theta = \arccos x$. We get the recurrence relation for the Čebyšev polynomials

$$T_{j+1}(x) = 2xT_j(x) - T_{j-1}(x). \tag{6.12}$$

Knowing that $T_0(x) = 1$ and $T_1(x) = x$, we consequently can find the Čebyšev polynomials. From relationship (6.12) it follows that the coefficient of x^n for the polynomial $T_n(x)$ is equal to 2^{n-1} for $n = 1, 2, \ldots$

$$T_n(x) = 2^{n-1}x^n + \alpha x^{n-1} + \cdots$$

On the basis of the general formula (4.9) for the remainder term of the Gaussian type formula, we have

$$R_n(f) = \frac{f^{(2n)}(\xi)}{(2n)!} \int_{-1}^{1} \frac{1}{\sqrt{1-x^2}} \frac{1}{2^{2n-2}} T_n^2(x)dx.$$

It is easy to verify that

$$\int_{-1}^{1} \frac{1}{\sqrt{1-x^2}} T_n^2(x)dx = \frac{\pi}{2}$$

and we get the representation of the remainder term of formula (6.11),

$$R_n(f) = \frac{\pi}{(2n)!2^{2n-1}} f^{(2n)}(\xi), \qquad -1 \leq \xi \leq 1. \tag{6.13}$$

We assume, of course, that $f^{(2n)}(x)$ exists and is continuous on $[-1, 1]$. The function

$$p(x) = e^{-x^2} \tag{6.14}$$

on the interval $(-\infty, +\infty)$ satisfies all the conditions which were imposed on the weight function $p(x)$ in the construction of the general quadrature formula of Gaussian type in § 4. Therefore, one can construct the quadrature formula

$$\int_{-\infty}^{\infty} e^{-x^2}f(x)dx \cong \sum_{k=1}^{n} A_k f(x_k), \tag{6.15}$$

which is exact, when $f(x)$ is any polynomial of degree $2n-1$. The knots of the quadrature formula are the roots of the polynomial of degree n, orthogonal with respect to the weight function e^{-x^2} on the interval $(-\infty, +\infty)$ to all polynomials of the degree $n-1$. Such polynomials are called the Čebyšev-Hermite orthogonal polynomials.

A valid representation for the Čebyšev-Hermite polynomial is

$$H_n(x) = (-1)^n e^{x^2} \frac{d^n}{dx^n} e^{-x^2}. \tag{6.16}$$

The coefficients of the quadrature formula (6.15) are determined by the formula

$$A_k = \frac{2^{n+1} n! \sqrt{\pi}}{[H'_n(x_k)]^2}, \qquad k = 1, 2, \ldots, n. \tag{6.17}$$

The remainder term of formula (6.15) is

$$R_n(f) = \frac{n! \sqrt{\pi}}{2^n (2n)!} f^{(2n)}(\xi). \tag{6.18}$$

In V. I. Krylov's book, the values of the knots and coefficients of formula (6.15) for $n = 1(1)10$ are given: moreover, the fractional part of the values contains twelve significant figures for $n = 1(1)5$ and nine significant figures for $n = 6(1)10$. In [31] the values of the knots and coefficients (6.15) are given for $n = 1(1)20$, moreover, the fractional part of the coefficients is given with thirteen significant figures, and the fractional part of the knots is given with fifteen significant figures for $n = 1(1)13$, for $n = 14(1)16$ with fourteen significant figures, and for $n = 17(1)20$ with thirteen significant figures.

The values of the knots and coefficients of formula (6.15) for $n = 1(1)8$ are given below. These values are taken from [31].

$$n = 1$$
$$x_1 = 0 \qquad\qquad\qquad A_1 = 1.7724538509$$

$$n = 2$$
$$-x_1 = x_2 = 0.7071067812 \qquad A_1 = A_2 = 0.8862269255$$

$$n = 3$$
$$-x_1 = x_3 = 1.2247448714 \qquad A_1 = A_3 = 0.2954089752$$
$$x_2 = 0 \qquad\qquad\qquad A_2 = 1.1816359006$$

$$n = 4$$
$$-x_1 = x_4 = 1.6506801239 \qquad A_1 = A_4 = 0.08131283545$$
$$-x_2 = x_3 = 0.5246476233 \qquad A_2 = A_3 = 0.8049140900$$

$$n = 5$$

$$-x_1 = x_5 = 2.0201828705 \qquad A_1 = A_5 = 0.01995324206$$
$$-x_2 = x_4 = 0.9585724646 \qquad A_2 = A_4 = 0.3936193232$$
$$x_3 = 0 \qquad\qquad\quad A_3 = 0.9453087205$$

$$n = 6$$

$$-x_1 = x_6 = 2.3506049737 \qquad A_1 = A_6 = 0.004530009906$$
$$-x_2 = x_5 = 1.3358490740 \qquad A_2 = A_5 = 0.1570673203$$
$$-x_3 = x_4 = 0.4360774119 \qquad A_3 = A_4 = 0.7246295952$$

$$n = 7$$

$$-x_1 = x_7 = 2.6519613568 \qquad A_1 = A_7 = 0.0009717812451$$
$$-x_2 = x_6 = 1.6735516288 \qquad A_2 = A_6 = 0.05451558282$$
$$-x_3 = x_5 = 0.8162878829 \qquad A_3 = A_5 = 0.4256072526$$
$$x_4 = 0 \qquad\qquad\quad A_4 = 0.8102646176$$

$$n = 8$$

$$-x_1 = x_8 = 2.9306374203 \qquad A_1 = A_8 = 0.0001996040722$$
$$-x_2 = x_7 = 1.9816567567 \qquad A_2 = A_7 = 0.01707798301$$
$$-x_3 = x_6 = 1.1571937124 \qquad A_3 = A_6 = 0.2078023258$$
$$-x_4 = x_5 = 0.3811869902 \qquad A_4 = A_5 = 0.6611470126$$

We now consider the weight function

$$p(x) = x^3 e^{-x} \tag{6.19}$$

on the interval $[0, \infty)$, where $s > -1$. The function (6.19) satisfies the conditions of §4. The corresponding quadrature formula of the highest degree of accuracy has the form

$$\int_0^\infty x^3 e^{-x} f(x) dx \cong \sum_{k=1}^n A_k f(x_k). \tag{6.20}$$

The knots of this formula are the roots of the polynomial of degree n, orthogonal with respect to the weight function (6.19) on the interval $[0, \infty)$ to any polynomial of degree $n-1$. The polynomials $L_n^{(s)}(x)$ which possess the indicated property of orthogonality are called the Čebyšev-Laguerre polynomials.
The following representation is correct for $L_n^{(s)}(x)$

$$L_n^{(s)}(x) = (-1)^n x^{-s} e^x \frac{d^n}{dx^n} (x^{s+n} e^{-x}). \tag{6.21}$$

The coefficients of the quadrature formula (6.20) are

$$A_k = \frac{n!\Gamma(s+n+1)}{x_k[L_n^{(s)'}(x_k)]^2}.$$ (6.22)

The remainder term of formula (6.20) is

$$R_n(f) = \frac{n!\Gamma(s+n+1)}{(2n)!}f^{(2n)}(\xi).$$ (6.23)

The numerical values of the nodes and coefficients of the quadrature formula (6.20) with $s = 0$ are given in V. I. Krylov's book [18] for $n = 1(1)15$. The fractional part of the knots and coefficients is given with twelve significant figures. Below we present the values of the knots and coefficients of formula (6.20) for $n = 1(1)6$.

$$n = 1$$

$x_1 = 1$ $A_1 = 1$

$$n = 2$$

$x_1 = $ 0.5857864376 $A_1 = 0.8535533906$
$x_2 = $ 3.4142135624 $A_2 = 0.1464466094$

$$n = 3$$

$x_1 = $ 0.4157745568 $A_1 = 0.7110930099$
$x_2 = $ 2.2942803603 $A_2 = 0.2785177336$
$x_3 = $ 6.2899450829 $A_3 = 0.(1)1038925650*$

$$n = 4$$

$x_1 = $ 0.3225476896 $A_1 = 0.6031541043$
$x_2 = $ 1.7457611012 $A_2 = 0.3574186924$
$x_3 = $ 4.5366202969 $A_3 = 0.(1)3888790852$
$x_4 = $ 9.3950709123 $A_4 = 0.(3)5392947056$

$$n = 5$$

$x_1 = $ 0.2635603197 $A_1 = 0.5217556106$
$x_2 = $ 1.4134030591 $A_2 = 0.3986668111$
$x_3 = $ 3.5964257710 $A_3 = 0.(1)7594244968$
$x_4 = $ 7.0858100059 $A_4 = 0.(2)3611758680$
$x_5 = $ 12.6408008443 $A_5 = 0.(4)2336997239$

---

* The figure in parentheses signifies the number of zeros which should be written after the decimal point before the significant figure, for example, $0.(1)10389\ldots = 0.010389\ldots$.

$$n = 6$$

$$
\begin{aligned}
x_1 &= 0.2228466042 & A_1 &= 0.4589646740 \\
x_2 &= 1.1889321017 & A_2 &= 0.4170008308 \\
x_3 &= 2.9927363261 & A_3 &= 0.1133733821 \\
x_4 &= 5.7751435691 & A_4 &= 0.(1)1039919745 \\
x_5 &= 9.8374674184 & A_5 &= 0.(3)2610172028 \\
x_6 &= 15.9828739806 & A_6 &= 0.(6)8985479064
\end{aligned}
$$

We cite numerical examples.

Example 1. Calculate the integral

$$\int_{-1}^{1} \frac{e^{2x}}{\sqrt{1-x^2}}\, dx.$$

We will use Mellor's formula (6.11). The remainder term (6.13) does not exceed two units in the fifth decimal place for $n = 5$. We will make calculations to six decimal places and take $n = 5$. The calculations are displayed in Table 28. We find that the approximate value of the integral is equal to 7.161529, which differs from the exact value of the integral by one unit in the sixth decimal place.

Table 28

x_k	$2x_k$	e^{2x_k}
—0.951057	—1.902114	0.149253
—0.587785	—1.175570	0.308643
0	0	1
0.587785	1.175570	3.239989
0.951057	1.902114	6.700043
		11.397928

Example 2. Calculate the integral

$$\int_{-\infty}^{\infty} e^{-x^2} \cos x\, dx.$$

We apply the quadrature formula (6.15) with $n = 5$.

Since the knots of the quadrature formula are located symmetrically with respect to $x = 0$, here, equal coefficients correspond to the symmetrical knots, and we get

$$\int_{-\infty}^{\infty} e^{-x^2} \cos x\, dx \cong A_3 \cos x_3 + 2 \sum_{k=4}^{5} A_k \cos x_k =$$

$$= 0.945309 + 2(0.393619 \cdot 0.574689 -$$

$$-0.019953 \cdot 0.434413) = 1.380390.$$

The remainder term for (6.18) with $n = 5$ does not exceed three units in the sixth decimal place. The exact value of the integral is equal to

$$\sqrt{\pi} e^{-\frac{1}{4}} = 1.3803885\ldots,$$

so that the actual error is equal to two units in the sixth decimal place.

7. A. A. Markov's quadrature formulas

We shall study the problem of constructing quadrature formulas of the form

$$\int_a^b p(x) f(x)\, dx \cong \sum_{j=1}^{m} B_j f(a_j) + \sum_{k=1}^{n} A_k f(x_k), \tag{7.1}$$

where the knots a_j and x_k lie in the interval $[a, b]$, here, the a_j are given previously, and the x_k are selected so that they do not coincide with any a_j, and so that formula (7.1) is exact for all polynomials of a possibly higher degree.

We assume that the weight function $p(x)$ satisfies the conditions enumerated in §4: $p(x) \geq 0$, the moments of $p(x)$ and $\mu_0 > 0$ exist. With proper choice of the coefficients, one can make formula (7.1) exact for polynomials of degree $m + n - 1$. One can hope to select the n knots $x_1, x_2, \ldots, x_n$, so that the formula is exact if $f(x)$ is a polynomial of the degree $m + 2n - 1$.

We shall use the notation

$$\sigma(x) = (x - a_1)(x - a_2) \ldots (x - a_m), \tag{7.2}$$

$$\omega(x) = (x - x_1)(x - x_2) \ldots (x - x_n). \tag{7.3}$$

Theorem. In order for the quadrature formula (7.1) to be exact for any polynomial $f(x)$ of degree $m + 2n - 1$, it is necessary and sufficient that it be an interpolation formula and the polynomial (7.3)

$$\omega(x) = (x - x_1)(x - x_2) \ldots (x - x_n)$$

be orthogonal with respect to the weight function

$$p(x)\sigma(x) = p(x)(x-a_1)(x-a_2)\ldots(x-a_m) \tag{7.4}$$

on the interval $[a, b]$ to any polynomial $Q(x)$ of degree less than n:

$$\int_a^b p(x)\sigma(x)\omega(x)Q(x)dx = 0. \tag{7.5}$$

Proof. Necessity.
Suppose formula (7.1) is exact, if $f(x)$ is any polynomial of degree $m+2n-1$. According to the theorem in §1, formula (7.1) is an interpolation formula. We take any polynomial $Q(x)$ of degree less than n. Then,

$$f(x) = \sigma(x)\omega(x)Q(x)$$

is a polynomial, with degree not higher than $m+2n-1$, and, by hypothesis, formula (7.1) is exact for it

$$\int_a^b p(x)f(x)dx = \int_a^b p(x)\sigma(x)\omega(x)Q(x)dx =$$

$$= \sum_{j=1}^m B_j f(a_j) + \sum_{k=1}^n A_k f(x_k) = 0.$$

The necessity of condition (7.5) is proved.
Sufficiency. We assume that formula (7.1) is an interpolation formula and condition (7.5) is satisfied. It is necessary to prove that formula (7.1) is exact, if $f(x)$ is any polynomial of degree $m+2n-1$. We divide $f(x)$ by the polynomial $\sigma(x)\omega(x)$ of degree $m+n$:

$$f(x) = \sigma(x)\omega(x)Q(x)+r(x), \tag{7.6}$$

where the degree of $Q(x)$ is less than n and the degree of $r(x)$ is less than $m+n$. We multiply both sides of equation (7.6) by $p(x)$ and integrate with respect to x from a to b. We get

$$\int_a^b p(x)f(x)dx = \int_a^b p(x)\sigma(x)\omega(x)Q(x)dx + \int_a^b p(x)r(x)dx.$$

According to condition (7.5), the first integral on the right hand side is equal to zero, and the second integral is equal to the quadrature sum

$$\sum_{j=1}^m B_j r(a_j) + \sum_{k=1}^n A_k r(x_k).$$

Since the degree of $r(x)$ is less than $m+n$ and formula (7.1) is an interpolation formula, by virtue of (7.6)

$$r(a_j) = f(a_j), \qquad r(x_k) = f(x_k),$$

and we finally get

$$\int_a^b p(x)f(x)dx = \sum_{j=1}^m B_j f(a_j) + \sum_{k=1}^n A_k f(x_k).$$

The sufficiency is established.

The theorem proved reduces the question of the existence of formula (7.1), exact for all polynomials of degree $m+2n-1$, to the question of the existence of a polynomial $\omega(x)$ of degree n, orthogonal with respect to the weight function (7.4) on $[a, b]$ to any polynomial of degree $n-1$. To apply this theorem, we must be convinced that the roots of $\omega(x)$ are real, different, lie in the interval $[a, b]$ and satisfy the inequalities

$$x_k \neq a_j, \qquad k = 1, 2, \ldots, n, \qquad j = 1, 2, \ldots, m. \qquad (7.7)$$

A. A. Markov pointed out three cases in which formula (7.1) can be constructed. In order for the orthogonal polynomials with weight $p(x)\sigma(x)$ to exist, it is sufficient to require that the polynomial $\sigma(x)$ does not change sign on $[a, b]$. This condition will obviously be violated if only one of the knots $a_1, a_2, \ldots, a_m$ lies inside of the interval (a, b). On the other hand, the knots must belong to $[a, b]$, therefore, $\sigma(x)$ will not change sign on $[a, b]$ in three cases:

1. one fixed knot is taken and this knot coincides with the left end of the interval of integration: $a_1 = a$, $\sigma(x) = x-a \geq 0$ for $x \in [a, b]$;
2. $m = 1$, $a_1 = b$, $\sigma(x) = x-b \leq 0$ on $[a, b]$;
3. two fixed knots are taken which coincide with the ends of the interval of integration: $m = 2$, $a_1 = a$, $a_2 = b$, $\sigma(x) = (x-a)(x-b) \leq 0$ for $x \in [a, b]$.

On the basis of the theorem proved, formula (7.1) exists in cases 1.–3. The knots $x_1, x_2, \ldots, x_n$ are roots of the polynomial of degree n, orthogonal with respect to the weight function $p(x)\sigma(x)$ on the interval $[a, b]$ to any polynomial of degree $n-1$. Moreover, $x_1, x_2, \ldots, x_n$ are real, different, and lie inside (a, b) and, consequently, conditions (7.7) are fulfilled.

On the basis of (1.5), the coefficients of the quadrature formula (7.1) may be written in the form

$$B_j = \int_a^b p(x) \frac{\sigma(x)\omega(x)}{(x-a_j)\sigma'(a_j)\omega(a_j)} dx, \qquad (7.8)$$

$$A_k = \int_a^b p(x) \frac{\sigma(x)\omega(x)}{(x-x_k)\omega'(x_k)\sigma(x_k)}\, dx. \tag{7.9}$$

We indicate the remainder term of the quadrature formula (7.1). Let $f(x)$ have a continuous derivative of order $m+2n$ and suppose the integral exists

$$\int_a^b p(x)f(x)dx.$$

We will construct the Hermitian interpolation polynomial from the conditions

$$P(x_k) = f(x_k), \qquad P'(x_k) = f'(x_k), \qquad k = 1, 2, \ldots, n.$$
$$P(a_j) = f(a_j), \qquad j = 1, 2, \ldots, m.$$

We have, by virtue of (10.24) in Chap. II,

$$f(x) = P(x) + \frac{\sigma(x)\omega^2(x)}{(m+2n)!} f^{(m+2n)}(\eta), \qquad a < \eta < b.$$

The degree of $P(x)$ is less than $m+2n$, therefore, by multiplying both sides of the last equation by $p(x)$ and integrating, we find that the remainder term of the quadrature formula (7.1) equals

$$R_n^{(m)}(f) = \frac{1}{(m+2n)!} \int_a^b p(x)\sigma(x)\omega^2(x)f^{(m+2n)}(\eta)dx.$$

Since in the cases of A. A. Markov, $\sigma(x)$ does not change sign on $[a, b]$, then according to the mean-value theorem, the last equation may be written in the form

$$R_n^{(m)}(f) = \frac{f^{(m+2n)}(\xi)}{(m+2n)!} \int_a^b p(x)\sigma(x)\omega^2(x)dx, \qquad a \leq \xi \leq b. \tag{7.10}$$

From equation (7.10), it follows that the algebraic degree of accuracy of the quadrature formula (7.1) equals $m+2n-1$. In fact, for $f(x) = x^{m+2n}$ formula (7.1) is not exact, since

$$R_n^{(m)}(x^{m+2n}) = \int_a^b p(x)\sigma(x)\omega^2(x)dx \neq 0.$$

We now examine cases 1.–3. We shall not examine case 2. separately since it may be reduced to case 1. by the linear transformation of the variable of integration $x = a+b-t$. We assume that the weight function is constant: $p(x) \equiv 1$. We consider the interval of integration to be $[-1, 1]$. We begin with case 1. We have $m = 1$, $a_1 = -1$ and formula (7.1) takes the form

$$\int_{-1}^{1} f(x)dx \cong Bf(-1)+ \sum_{k=1}^{n} A_k f(x_k).$$ (7.11)

The algebraic degree of accuracy of formula (7.11) is equal to $2n$. The knots $x_1, x_2, \ldots, x_n$ in (7.11) are the roots of the polynomial $\omega(x)$, orthogonal on $[-1, 1]$ with respect to the weight function $1+x$ to any polynomial of degree $n-1$. We note that $\omega(x)$, up to a constant multiplier, coincides with the Jacobi polynomial $P_n^{(0, 1)}(x)$.
We indicate the representation of $(1+x)\omega(x)$ by Legendre polynomials. We have

$$(1+x)\omega(x) = c_0+c_1 P_1(x)+ \ldots$$
$$\ldots +c_{n-1}P_{n-1}(x)+c_n P_n(x)+c_{n+1} P_{n+1}(x).$$

By multiplying both sides of this equation by $P_k(x)$ and integrating with respect to x from -1 to 1, we get

$$c_k = 0, \quad k = 0, 1, 2, \ldots, n-1.$$

Thus,

$$(1+x)\omega(x) = c_n P_n(x)+c_{n+1} P_{n+1}(x).$$ (7.12)

We set $x = -1$ in (7.12) and take into account that $P_k(-1) = (-1)^k$ [see (5.10)]. We get $c_n = c_{n+1}$ and, consequently,

$$(1+x)\omega(x) = c_n[P_{n+1}(x)+P_n(x)].$$ (7.13)

In order to determine c_n, we note that the coefficient of x^{n+1} on the left hand side of equation (7.13) equals 1. From the relationship [see 5.6]

$$P_{n+1}(x) = \frac{(2n+2)!}{2^{n+1}[(n+1)!]^2} x^{n+1} - \ldots$$

it follows that

$$c_n = \frac{2^{n+1}[(n+1)!]^2}{(2n+2)!}.$$ (7.14)

Thus, the knots of the quadrature formula (7.11), including the knot $a_1 = -1$, are the roots of the equation

$$P_{n+1}(x) + P_n(x) = 0.$$

The coefficients of the quadrature formula (7.11) are determined by equations (7.8) and (7.9) in which it is necessary to set $p(x) = 1$, $\sigma(x) = x+1$, $-a = b = 1$. Equations (7.8) and (7.9) are not convenient for calculation. By using the properties of Jacobi polynomials, one can show that

$$B = \frac{2}{(n+1)^2}, \qquad A_k = \frac{4}{(1+x_k)(1-x_k^2)[P_n^{(0,\,1)'}(x_k)]^2}.$$

We indicate the remainder term of formula (7.11), assuming that $f(x)$ has a continuous derivative of order $2n+1$ on $[-1, 1]$. We use formula (7.10) which, in our case, will be written thusly:

$$R_n^{(1)}(f) = \frac{f^{(2n+1)}(\xi)}{(2n+1)!} \int_{-1}^{1} (1+x)\omega^2(x)dx. \qquad (7.15)$$

We calculate the integral on the right hand side of this equation. We replace $(1+x)\omega(x)$ by the right hand side of the relationship (7.13) and perform the simple transformations:

$$\int_{-1}^{1} (1+x)\omega^2(x)dx = c_n \int_{-1}^{1} [P_{n+1}(x)+P_n(x)]\omega(x)dx =$$

$$= c_n \int_{-1}^{1} P_n(x)\omega(x)dx = c_n \int_{-1}^{1} P_n(x)x^n dx =$$

$$= c_n \int_{-1}^{1} P_n(x) \frac{2^n(n!)^2}{(2n)!} P_n(x)dx = c_n \frac{2^n(n!)^2}{(2n)!} \frac{2}{2n+1} =$$

$$= \frac{2^{2n+1}(n!)^4(n+1)}{[(2n+1)!]^2}.$$

Here, we used equalities (5.7), (5.9), and (7.14). Now, with the help of (7.15), we find that

$$R_n^{(1)}(f) = \frac{2^{2n+1}(n!)^4(n+1)}{[(2n+1)!]^3} f^{(2n+1)}(\xi), \qquad -1 \leq \xi \leq 1. \qquad (7.16)$$

The numerical values of the knots and coefficients of quadrature formula (7.11) given below are taken from V. I. Krylov's book [18]

$$n = 1$$

$$B = 0.5$$
$$x_1 = -0.33333333 \qquad A_1 = 1.5$$

$$n = 2$$

$$B = 0.22222222$$
$$x_1 = -0.28989794 \qquad A_1 = 1.02497166$$
$$x_2 = 0.68989794 \qquad A_2 = 0.75280612$$

$$n = 3$$

$$B = 0.125$$
$$x_1 = -0.5753189 \qquad A_1 = 0.6576886$$
$$x_2 = 0.1810663 \qquad A_2 = 0.7763870$$
$$x_3 = 0.8228241 \qquad A_3 = 0.4409244$$

$$n = 4$$

$$B = 0.08$$
$$x_1 = -0.7204803 \qquad A_1 = 0.4462078$$
$$x_2 = 0.1671809 \qquad A_2 = 0.6236530$$
$$x_3 = 0.4463140 \qquad A_3 = 0.5627120$$
$$x_4 = 0.8857916 \qquad A_4 = 0.2874271$$

$$n = 5$$

$$B = 0.05555556$$
$$x_1 = -0.8029298 \qquad A_1 = 0.3196408$$
$$x_2 = -0.3909286 \qquad A_2 = 0.4853872$$
$$x_3 = 0.1240504 \qquad A_3 = 0.5209268$$
$$x_4 = 0.6039732 \qquad A_4 = 0.4169013$$
$$x_5 = 0.9203803 \qquad A_5 = 0.2015884$$

$$n = 6$$

$$B = 0.04081633$$
$$x_1 = -0.8538913 \qquad A_1 = 0.2392274$$
$$x_2 = -0.5384678 \qquad A_2 = 0.3809498$$
$$x_3 = -0.1173430 \qquad A_3 = 0.4471098$$
$$x_4 = 0.3260306 \qquad A_4 = 0.4247038$$
$$x_5 = 0.7038428 \qquad A_5 = 0.3182042$$
$$x_6 = 0.9413672 \qquad A_6 = 0.1489885$$

We now examine the case 3, when the ends of the interval $[-1, 1]$ are taken as fixed knots: $m = 2$, $a_1 = -1$, $a_2 = 1$. Formula (7.1) takes the form

$$\int_{-1}^{1} f(x)dx \cong B_1 f(-1) + B_2 f(1) + \sum_{k=1}^{n} A_k f(x_k). \tag{7.17}$$

The algebraic degree of accuracy of formula (7.17) is equal to $2n+1$. Its knots $x_1, x_2, \ldots, x_n$ are the roots of the polynomial $\omega(x)$ of degree n, orthogonal with respect to the weight function $(x-1)(x+1)$ on the interval $[-1, 1]$ to all polynomials of degree $n-1$. We note that $\omega(x)$ differs from the Jacobi polynomial $P_n^{(1, 1)}(x)$ only by a constant multiplier. As in case 1, one can establish the equality

$$(x^2 - 1)\omega(x) = c_n[P_{n+2}(x) - P_n(x)], \tag{7.18}$$

where

$$c_n = \frac{2^n n!(n+2)!}{(2n+1)!(2n+3)}, \tag{7.19}$$

so that the knots of formula (7.17), including $a_1 = -1$, $a_2 = 1$, are roots of the equation

$$P_{n+2}(x) - P_n(x) = 0.$$

One can show that

$$B_1 = B_2 = \frac{2}{(n+1)(n+2)},$$

$$A_k = 8 \frac{n+1}{n+2} \frac{1}{(1-x_k^2)^2 [P_n^{(1, 1)'}(x_k)]^2}.$$

As in case 1, it is easy to indicate the representation of the remainder term of formula (7.17) with the help of (7.18) and (7.19).

$$R_n^{(2)}(f) = -\frac{2^{2n}(n!)^4(n+2)}{[(2n+1)!]^3(2n+3)} f^{(2n+2)}(\xi), \qquad -1 \leqq \xi \leqq 1. \tag{7.20}$$

The numerical values of the knots and coefficients of formula (7.17) with eight significant figures for $n = 1(1)15$ are given in V. I. Krylov's book [18]. Below, we cite the knots and coefficients of formula (7.17) for $n = 1(1)6$ from the book [18].

$$n = 1$$

$$B_1 = B_2 = 0.33333333$$
$$x_1 = 0 \qquad A_2 = 1.33333333$$

$$n = 2$$

$$B_1 = B_2 = 0.16666667$$
$$-x_1 = x_2 = 0.44721360 \qquad A_1 = A_2 = 0.83333333$$

$$n = 3$$

$$B_1 = B_2 = 0.1$$
$$-x_1 = x_3 = 0.65465367 \qquad A_1 = A_3 = 0.54444444$$
$$x_2 = 0 \qquad A_2 = 0.71111111$$

$$n = 4$$

$$B_1 = B_2 = 0.066666667$$
$$-x_1 = x_4 = 0.76505532 \qquad A_1 = A_4 = 0.37847496$$
$$-x_2 = x_3 = 0.28523152 \qquad A_2 = A_3 = 0.55485837$$

$$n = 5$$

$$B_1 = B_2 = 0.047619048$$
$$-x_1 = x_5 = 0.83022390 \qquad A_1 = A_5 = 0.27682605$$
$$-x_2 = x_4 = 0.46884879 \qquad A_2 = A_4 = 0.43174538$$
$$x_3 = 0 \qquad A_3 = 0.48761905$$

$$n = 6$$

$$B_1 = B_2 = 0.035714286$$
$$-x_1 = x_6 = 0.87174015 \qquad A_1 = A_6 = 0.21070423$$
$$-x_2 = x_5 = 0.59170018 \qquad A_2 = A_5 = 0.34112268$$
$$-x_3 = x_4 = 0.20929922 \qquad A_3 = A_4 = 0.41245881$$

These quadrature formulas find application, in particular, to the numerical solution of Fredholm integral equations of the second kind,

$$\varphi(s) = \int_a^b K(s, t)\varphi(t)dt + f(s). \tag{7.21}$$

Here, $K(s, t)$ is, on the square $a \leqq s, t \leqq b$, a given function called the *kernel* of the integral equation; $f(s)$ is a given function on $[a, b]$; and $\varphi(s)$ is the unknown function.

The method of mechanical quadratures for the solution of equation (7.21) consists of the following. By using any quadrature formula,

$$\int_a^b F(t)dt \cong \sum_{k=1} A_k F(t_k),$$
(7.22)

we replace the integral on the right hand side of (7.21) by a quadrature sum. We get

$$\varphi(s) = \sum_{k=1}^{n} A_k K(s, t_k)\varphi(t_k)+f(s)+R(s).$$
(7.23)

Here, $R(s)$ is the remainder term of the quadrature formula (7.22). We use the notation

$$K_{ik} = K(t_i, t_k), \qquad f_k = f(t_k), \qquad \varphi_k = \varphi(t_k), \qquad i, k = 1, 2, \ldots, n.$$

In (7.23), we set $s = t_1, t_2, \ldots, t_n$. We obtain a linear algebraic system for $\varphi_1, \varphi_2, \ldots, \varphi_n$

$$\varphi_i = \sum_{k=1}^{n} A_k K_{ik}\varphi_k+f_i+R(t_i), \qquad i = 1, 2, \ldots, n.$$

This system is not convenient for calculation, since the values $R(t_i)$ of the unknown remainder term $R(s)$ enter into it. By discarding these unknown values, we obtain the system

$$\tilde{\varphi}_i = \sum_{k=1}^{n} A_k K_{ik}\tilde{\varphi}_k+f_i, \qquad i = 1, 2, \ldots, n.$$
(7.24)

The numbers $\tilde{\varphi}_1, \tilde{\varphi}_2, \ldots, \tilde{\varphi}_n$, defined by the system of equations (7.24), are taken to be approximate values of the solution $\varphi(s)$ at the knots $t_1, t_2, \ldots, t_n$ of the quadrature formula.

Of real moment for the indicated method is the selection of the corresponding quadrature formula, since the number of equations of system (7.24) depends on the accuracy of the approximation of the integral sum. If the integral equation is such that the value of the unknown function $\varphi(s)$ at one or at both ends of the interval $[a, b]$ are known, then it is convenient to apply Markov's quadrature formulas.

We cite a numerical example. We examine the integral equation

$$\varphi(s) = -\int_0^1 K(s, t)\varphi(t)dt+s^2,$$
(7.25)

where

$$K(s, t) = \begin{cases} s(1-t) & \text{for} \quad 0 \leq s \leq t \leq 1, \\ t(1-s) & \text{for} \quad 0 \leq t \leq s \leq 1. \end{cases}$$

It is obvious from the integral equation (7.25), that if one sets $s = 0$ and $s = 1$ in it,

$$\varphi(0) = 0, \qquad \varphi(1) = 1.$$

It is natural to use the quadrature formula

$$\int_0^1 F(t)dt \cong \tfrac{1}{2}B_1[F(0)+F(1)]+\tfrac{1}{2}\sum_{k=1}^{n} A_k F(t_k), \tag{7.26}$$

where $t_k = \tfrac{1}{2}(x_k+1)$, x_k and A_k are the knots and coefficients of the quadrature formula (7.17).

We take $n = 2$. For $n = 2$, formula (7.26) will be written in the form

$$\int_0^1 F(t)dt \cong 0.08333[F(0)+F(1)]+0.41667[F(t_1)+F(t_2)],$$

where

$$t_1 = 0.27639, \qquad t_2 = 0.72361. \tag{7.27}$$

In our case system (7.24) will be written thusly:

$$\left. \begin{array}{l} (1+0.41667K_{11})\tilde{\varphi}_1+0.41667K_{12}\tilde{\varphi}_2 = t_1^2, \\ 0.41667K_{21}\tilde{\varphi}_1+(1+0.41667K_{22})\tilde{\varphi}_2 = t_2^2. \end{array} \right\} \tag{7.28}$$

If we substitute the numerical values $K_{ik} = K(t_i, t_k)$ and $t_k^2 = (i, k = 1, 2)$ in (7.28), where t_1 and t_2 are defined by (7.27), then we get

$$\left. \begin{array}{l} 1.08333\tilde{\varphi}_1+0.03183\tilde{\varphi}_2 = 0.07639, \\ 0.03183\tilde{\varphi}_1+1.08333\tilde{\varphi}_2 = 0.52361. \end{array} \right\}$$

By solving this system, we get

$$\tilde{\varphi}_1 = 0.05636, \qquad \tilde{\varphi}_2 = 0.48168.$$

For equation (7.25), it is not difficult to find out the exact solution

$$\varphi(s) = 2\cosh s-0.07335\sinh s-2.$$

Its values at the knots are

$$\varphi(t_1) = 0.05633, \qquad \varphi(t_2) = 0.48903,$$

so that

$$\varphi(t_1)-\tilde{\varphi}_1 = 0.00003, \qquad \varphi(t_2)-\tilde{\varphi}_2 = 0.00735.$$

8. Čebyšev's quadrature formula

P. L. Čebyšev studied the problem of constructing quadrature formulas with equal coefficients,

$$\int_a^b p(x)f(x)dx \cong C_n \sum_{k=1}^{n} f(x_k).   \tag{8.1}$$

In § 6 we examined a particular case of formula (8.1) for $[a, b] = [-1,1]$ and

$$p(x) = \frac{1}{\sqrt{1-x^2}}.$$

We assume that the moments of the function $p(x)$ exist,

$$\mu_k = \int_a^b p(x)x^k dx, \qquad k = 0, 1, 2, \ldots$$

moreover

$$\mu_0 = \int_a^b p(x)dx \neq 0.   \tag{8.2}$$

In formula (8.1), $n+1$ parameters enter into the quadrature sum: n knots and the common value C_n of the coefficients, therefore, one can hope to select these parameters so that formula (8.1) will be exact if $f(x)$ is any polynomial of degree n. As we shall see, a quadrature formula of the form (8.1) does not always exist.

We shall find the knots x_k and the constant C_n by the condition that the quadrature formula (8.1) must be exact if

$$f(x) = x^k, \qquad k = 0, 1, 2, \ldots, n.$$

Setting $f(x)$ in (8.1), we get

$$nC_n = \int_a^b p(x)dx = \mu_0,$$

from which

$$C_n = \frac{\mu_0}{n}.   \tag{8.3}$$

From the assumption (8.2), it follows that $C_n \neq 0$.

To determine the knots, we obtain the following system of non-linear equations if we take formula (8.1) to be exact for $f(x) = x^k, k = 1, 2, \ldots, n$:

$$\left.\begin{aligned} \sum_{k=1}^{n} x_k &= n \frac{\mu_1}{\mu_0}, \\[2mm] \sum_{k=1}^{n} x_k^2 &= n \frac{\mu_2}{\mu_0}, \\ \cdots\cdots\cdots\cdots \\ \sum_{k=1}^{n} x_k^n &= n \frac{\mu_n}{\mu_0}. \end{aligned}\right\} \tag{8.4}$$

Instead of finding the unknowns $x_1, x_2, \ldots, x_n$ by solving (8.4), we shall find the polynomial

$$\omega(x) = (x - x_1)(x - x_2) \ldots (x - x_n) =$$
$$= x^n + a_1 x^{n-1} + a_2 x^{n-2} + \ldots + a_n, \tag{8.5}$$

the roots of which are $x_1, x_2, \ldots, x_n$. Obviously, we have

$$\frac{\omega'(x)}{\omega(x)} = \sum_{k=1}^{n} \frac{1}{x - x_k}. \tag{8.6}$$

For $|x| > |x_k|, k = 1, 2, \ldots, n$, the following expansion is valid

$$\frac{1}{x - x_k} = \frac{1}{x} \frac{1}{1 - \dfrac{x_k}{x}} = \frac{1}{x} + \frac{x_k}{x^2} + \frac{x_k^2}{x^3} + \ldots,$$

therefore, (8.6) can be rewritten thusly:

$$\frac{\omega'(x)}{\omega(x)} = \frac{n}{x} + \frac{S_1}{x^2} + \frac{S_2}{x^3} + \ldots \tag{8.7}$$

Here, the following notation is used

$$S_j = \sum_{k=1}^{n} x_k^j, \qquad j = 1, 2, 3, \ldots \tag{8.8}$$

Multiplying both sides of equation (8.7) by $\omega(x)$, we get

$$nx^{n-1} + (n-1)a_1 x^{n-2} + (n-2)a_2 x^{n-3} + \ldots + a_{n-1} =$$
$$= (x^n + a_1 x^{n-1} + \ldots + a_n) \left(\frac{n}{x} + \frac{S_1}{x^2} + \frac{S_2}{x^3} + \ldots \right).$$

We equate the coefficients of like powers of x on the left and right hand sides of the last equation. We get Newton's formulas

$$\left.\begin{array}{l} S_1+a_1 = 0, \\ S_2+a_1 S_1+2a_2 = 0, \\ S_3+a_1 S_2+a_2 S_1+3a_3 = 0, \\ \cdots\cdots\cdots\cdots\cdots\cdots\cdots\cdots\cdots \\ S_n+a_1 S_{n-1}+a_2 S_{n-2}+ \ldots +na_n = 0. \end{array}\right\} \tag{8.9}$$

Newton's formulas and system (8.4), which gives the values of quantities (8.8) $S_j, j = 1, 2, \ldots, n$, consequently allow one to determine the coefficients $a_1, a_2, \ldots, a_n$ of the polynomial $\omega(x)$ defined by formula (8.5). Thus, the question of the possibility of constructing formula (8.1) leads to an investigation of the roots of polynomial $\omega(x)$. The roots must be real and located in the interval $[a, b]$.

P. L.Čebyšev computed the knots of formula (8.1) in the case of a finite interval $[a, b]$ which may be considered to coincide with $[-1, 1]$, and with the weight function $p(x) = 1$ for all n from 1 to 7. The knots turned out to be real, different and to belong to the interval $[-1, 1]$. Formula (8.1) for this case has the form

$$\int_{-1}^{1} f(x)dx \cong \frac{2}{n} \sum_{k=1}^{n} f(x_k). \tag{8.10}$$

Afterwards, it was discovered that for $n = 8$ there are complex roots among the roots of the polynomial $\omega(x)$. For $n = 9$ the roots again turn out to be real. S. N. Bernštein proved that for $n \geq 10$, there are always complex roots among the roots of the polynomial $\omega(x)$, so that for $n \geq 10$, Čebyšev's quadrature formula (8.10) does not exist. One can find a proof of this fact in V. I. Krylov's book [18].

The numerical values of the knots of quadrature formula (8.10) given below are taken from the article [30].

$$n = 1$$
$$x_1 = 0$$
$$n = 2$$
$$-x_1 = x_2 = 0.5773502691$$
$$n = 3$$
$$-x_1 = x_3 = 0.7071067812$$
$$x_2 = 0$$

$$n = 4$$
$$-x_1 = x_4 = 0.7946544723$$
$$-x_2 = x_3 = 0.1875924741$$

$$n = 5$$
$$-x_1 = x_5 = 0.8324974870$$
$$-x_2 = x_4 = 0.3745414096$$
$$x_3 = 0$$

$$n = 6$$
$$-x_1 = x_6 = 0.8662468181$$
$$-x_2 = x_5 = 0.4225186538$$
$$-x_3 = x_4 = 0.2666354015$$

$$n = 7$$
$$-x_1 = x_7 = 0.8838617008$$
$$-x_2 = x_6 = 0.5296567753$$
$$-x_3 = x_5 = 0.3239118105$$
$$x_4 = 0$$

$$n = 9$$
$$-x_1 = x_9 = 0.9115893077$$
$$-x_2 = x_8 = 0.6010186554$$
$$-x_3 = x_7 = 0.5287617831$$
$$-x_4 = x_6 = 0.1679061842$$
$$x_5 = 0$$

Example. We compute, by Čebyšev's formula,

$$\text{Si}(1) = \int_0^1 \frac{\sin x}{x}\, dx.$$

By the change of variable $x = \frac{1}{2}(t+1)$ we pass to the interval of integration $[-1,1]$:

$$\text{Si}(1) = \int_{-1}^1 \frac{\sin 0.5(1+t)}{1+t}\, dt.$$

Table 29

t_k	t_k+1	$0.5(t_k+1)$	$\sin 0.5(t_k+1)$	$\dfrac{\sin 0.5(t_k+1)}{t_k+1}$
—0.832497	0.167503	0.083752	0.083654	0.499418
—0.374541	0.625459	0.312730	0.307658	0.491892
0	1	0.5	0.479426	0.479426
0.374541	1.374541	0.687271	0.634430	0.461558
0.832497	1.832497	0.916249	0.793324	0.432920
				2.365214

We take $n = 5$ and compute with six decimal places.

The results of the computation are displayed in Table 29. We obtain

$$\text{Si}(1) \cong 0.946086,$$

which differs from the exact value $\text{Si}(1) = 0.946083$ by 3 units in the sixth decimal place.

9. Bernoulli numbers and polynomials

In this section, for later use we present necessary information about the numbers and polynomials of Bernoulli. We begin by defining the Bernoulli numbers.

We consider the function of the complex variable t

$$G(t) = \frac{t}{e^t - 1}. \tag{9.1}$$

The singular points of the function (9.1), with the exception of zero, are the zeros of the denominator $e^t - 1$,

$$t = 2k\pi i, \qquad k = \pm 1, \pm 2, \pm 3, \ldots$$

The point $t = 0$ is not a singular point of $G(t)$, so we may write the Maclaurin series expansion

$$G(t) = \sum_{v=0}^{\infty} \frac{B_v}{v!} t^v, \tag{9.2}$$

which converges in the disk $|t| < 2\pi$. The numbers B_v, $v = 0, 1, 2, 3, \ldots$ are also called *Bernoulli numbers*. The function $G(t)$ is called the *generating function* of the Bernoulli numbers.

In the relationship

$$G(t)(e^t - 1) = t$$

we substitute the right hand side of (9.2) for $G(t)$, and for e^t its Maclaurin series expansion.

We obtain the identity

$$\left(B_0 + \frac{B_1}{1!} t + \frac{B_2}{2!} t^2 + \ldots\right)\left(\frac{t}{1!} + \frac{t^2}{2!} + \frac{t^3}{3!} + \ldots\right) = t,$$

which is valid for $|t| < 2\pi$. By equating the coefficients of the like powers of t on the left and right hand sides, we get

$$B_0 = 1,$$

$$\frac{B_0}{k!} + \frac{B_1}{(k-1)!1!} + \frac{B_2}{(k-2)!2!} + \ldots + \frac{B_{k-1}}{1!(k-1)!} = 0, \tag{9.3}$$

for $k = 2, 3, 4, \ldots$

The recurrence relationship (9.3), consequently, allows one to determine the numbers B_k. From (9.3) one can see that the B_k are rational numbers. We write out several of the first Bernoulli numbers,

$$B_0 = 1, \quad B_1 = -\tfrac{1}{2}, \quad B_2 = \tfrac{1}{6}, \quad B_3 = 0, \quad B_4 = -\tfrac{1}{30},$$

$$B_5 = 0, \quad B_6 = \tfrac{1}{42}, \quad B_7 = 0, \quad B_8 = -\tfrac{1}{30}, \quad B_9 = 0,$$

$$B_{10} = \tfrac{5}{66}, \quad B_{11} = 0, \quad B_{12} = -\tfrac{691}{2730}, \quad B_{13} = 0, \quad B_{14} = \tfrac{7}{6},$$

$$B_{15} = 0, \quad B_{16} = -\tfrac{3617}{510}, \quad B_{17} = 0, \quad B_{18} = \tfrac{43867}{798}.$$

The relationship (9.3) may be written in a form which is easy to remember. Namely, we multiply both sides of (9.3) by $k!$, and add B_k to both sides of the equality obtained:

$$B_k = B_0 + C_k^{(1)}B_1 + C_k^{(2)}B_2 + \ldots + C_k^{(k-1)}B_{k-1} + B_k.$$

This equation may be rewritten in the symbolic form

$$B_k = (1+B)^k, \qquad k = 2, 3, 4, \ldots \tag{9.4}$$

In raising the binomial $1+B$ to the power k, one must write B_j instead of B^j.

We shall prove that the Bernoulli numbers with odd indices, with the exception of $B_1 = -\tfrac{1}{2}$, are equal to zero. By performing obvious transformations, we get

$$G(-t) = \frac{-t}{e^{-t}-1} = \frac{-te^t}{1-e^t} = \frac{te^t-t+t}{e^t-1} = t + G(t).$$

For $G(t)$ we substitute the series expansion (9.2). We get

$$\sum_{v=0}^{\infty} \frac{B_v}{v!}(-t)^v = t + \sum_{v=0}^{\infty} \frac{B_v}{v!}t^v.$$

By equating coefficients of the same powers of t, we get

$$B_0 = B_0, \quad -B_1 = 1 + B_1,$$

$$B_v(-1)^v = B_v \quad \text{where} \quad v = 2, 3, 5, \ldots$$

From these relationships, it follows that $B_v = 0$ for $v = 3, 5, 7, \ldots$ We now note the formula

$$B_{2k} = \frac{(-1)^{k-1}(2k)!}{2^{2k-1}\pi^{2k}} \sum_{m=1}^{\infty} \frac{1}{m^{2k}}, \tag{9.5}$$

from which it can be seen that for increasing k, the Bernoulli numbers increase rapidly. Formula (9.5) also gives the signs of the Bernoulli numbers,

$$\text{sign } B_{2k} = (-1)^{k-1}. \tag{9.6}$$

Formula (9.5) will be proved at the end of this paragraph.

We go on now to an examination of the Bernoulli polynomials. The function of the complex variable

$$H(x, t) = e^{xt} \frac{t}{e^t - 1} = e^{xt} G(t) \tag{9.7}$$

is regular in the disk $|t| < 2\pi$, for any complex value of the parameter x, therefore, the following expansion is valid inside of this disk,

$$H(x, t) = \sum_{v=0}^{\infty} \frac{B_v(x)}{v!} t^v. \tag{9.8}$$

The coefficients of the expansion (9.8) are functions of x. As we shall see, $B_v(x)$, $v = 0, 1, 2, 3, \ldots$, represents a polynomial of degree v. The polynomials $B_v(x)$ are called *Bernoulli polynomials*. The function (9.7) is called the *generating function of the Bernoulli polynomials*.

In equation (9.7) we substitute the expansions in a series in the neighborhood of $t = 0$ for $G(t)$, $H(x, t)$ and e^{xt}. We get the equation

$$B_0(x) + \frac{B_1(x)}{1!} t + \frac{B_2(x)}{2!} t^2 + \ldots =$$

$$= \left(B_0 + \frac{B_1}{1!} t + \frac{B_2}{2!} t^2 + \ldots \right) \left(1 + \frac{xt}{1!} + \frac{x^2 t^2}{2!} + \ldots \right),$$

which is valid in the disk $|t| < 2\pi$ for any x. By equating the coefficients of like powers on the left and right hand sides of the last equation, we get

$$\frac{B_k(x)}{k!} = \frac{B_0 x^k}{k!} = \frac{B_1 x^{k-1}}{1!(k-1)!} + \ldots + \frac{B_{k-1} x}{(k-1)!1!} + \frac{B_k}{k!}$$

or

$$B_k(x) = B_0 x^k + C_k^{(1)} B_1 x^{k-1} + C_k^{(2)} B_2 x^{k-2} + \ldots + B_k. \tag{9.9}$$

Equality (9.9) may be written in the symbolic form

$$B_k(x) = (x+B)^k, \qquad k = 0, 1, 2, \ldots, \tag{9.10}$$

where, as above, instead of B^j one must write the jth Bernoulli number B_j.

The relationship (9.10) allows one to find the polynomials $B_k(x)$. We write out several of Bernoulli polynomials:

$$B_0(x) = 1,$$
$$B_1(x) = x - \tfrac{1}{2},$$
$$B_2(x) = x^2 - x + \tfrac{1}{6},$$
$$B_3(x) = x^3 - \tfrac{3}{2}x^2 + \tfrac{1}{2}x,$$
$$B_4(x) = x^4 - 2x^3 + x^2 - \tfrac{1}{30},$$
$$B_5(x) = x^5 - \tfrac{5}{2}x^4 + \tfrac{5}{3}x^3 - \tfrac{1}{6}x,$$
$$B_6(x) = x^6 - 3x^5 + \tfrac{5}{2}x^4 - \tfrac{1}{2}x^2 + \tfrac{1}{42}.$$

From (9.9) it can be seen that

$$B_k(0) = B_k, \qquad k = 0, 1, 2, \ldots \tag{9.11}$$

We differentiate both sides of (9.8) with respect to x. We get

$$e^{xt} \frac{t^2}{e^t - 1} = \sum_{v=0}^{\infty} \frac{B_v'(x)}{v!} t^v.$$

Since the left side of this equation equals $tH(x, t)$, we get

$$t \sum_{v=0}^{\infty} \frac{B_v(x)}{v!} t^v = \sum_{v=0}^{\infty} \frac{B_v'(x)}{v!} t^v.$$

By equating coefficients of like powers of t, we find that

$$B_k'(x) = kB_{k-1}(x), \qquad k = 1, 2, 3, \ldots \tag{9.12}$$

Equation (9.12) is analogous to the rule for differentiation of the function $f(x) = x^k$.

Obviously, (9.12) can be rewritten in the integral form

$$B_k(x) = B_k + k \int_0^x B_{k-1}(t)dt. \tag{9.13}$$

We shall prove the equality

$$B_k(1-x) = (-1)^k B_k(x), \qquad k = 0, 1, 2, \ldots \tag{9.14}$$

Performing the simple transformations

$$e^{(1-x)t} \frac{t}{e^t - 1} = e^{-xt} \frac{te^t}{e^t - 1} = e^{x(-t)} \frac{-t}{e^{-t} - 1},$$

we see that

$$H(1-x, t) = H(x, -t).$$

In this relationship we substitute for the function H its expansion into the series (9.8). We get the equation

$$\sum_{v=0}^{\infty} \frac{B_v(1-x)}{v!} t^v = \sum_{v=0}^{\infty} \frac{B_v(x)}{v!} (-t)^v,$$

from which (9.14) follows.

Equation (9.14) shows that the values of $B_k(x)$ at points symmetrically situated with respect to the point $x = \frac{1}{2}$ are identical if k is even, and are identical in magnitude and have different signs, if k is odd; see figures 16 and 17, which depict the graphs of the functions $y = 4B_3(x)$ and $y = 4B_4(x)$.

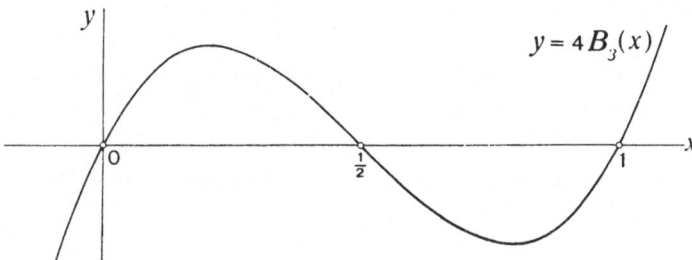

Fig. 16

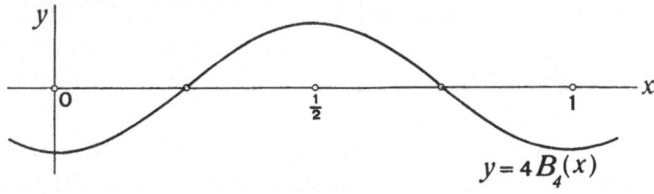

Fig. 17

We examine the polynomials which differ from the Bernoulli polynomials by the constant addends:

$$y_v(x) = B_v(x) - B_v, \qquad v = 0, 1, 2, \ldots, \tag{9.15}$$

and study the locations of the zeros of $y_v(x)$ on the interval $[0, 1]$. By virtue of (9.11), $x = 0$ is a zero of $y_v(x)$ for any $v = 0, 1, 2, \ldots$
From (9.14) we find that

$$y_v(1) = B_v(1) - B_v = (-1)^v B_v - B_v,$$

so that for any even v, $y_v(1) = 0$. For odd v, all of the Bernoulli numbers are equal to zero, except $B_1 = -\frac{1}{2}$, and, consequently, $y_v(1) = 0$ for v odd and different from one. Thus, $x = 1$ is a zero of all of the polynomials $y_v(x)$ with the exception of

$$y_1(x) = x.$$

The polynomials $y_{2m+1}(x) = B_{2m+1}(x)$ for $m = 1, 2, 3, \ldots$ have $x = \frac{1}{2}$ as a zero, which follows from the relationship (9.14):

$$B_{2m+1}(\tfrac{1}{2}) = -B_{2m+1}(\tfrac{1}{2}).$$

We shall prove that $y_{2m+1}(x)$ does not have other zeros inside the interval $(0, 1)$.
Let us assume the contrary: we suppose that $y_{2m+1}(x)$ for some fixed $m > 1$ has two zeros α and β inside $(0, 1)$, $0 < \alpha < \beta < 1$. Since 0 and 1 are also zeros of $y_{2m+1}(x)$, then, according to Rolle's theorem, $y'_{2m+1}(x)$ has at least three different roots and $y''_{2m+1}(x)$ has at least two different roots inside $(0, 1)$. But, by virtue of (9.12),

$$y''_{2m+1}(x) = B''_{2m+1}(x) = (2m+1)2mB_{2m-1}(x) =$$
$$= (2m+1)2my_{2m-1}(x),$$

and we have proved that if $y_{2m+1}(x)$ has two different zeros inside $(0, 1)$, then $y_{2m-1}(x)$ possesses the same property. Consequently, $y_3(x)$ also has two distinct zeros inside $(0, 1)$. But $y_3(x)$ is a polynomial of the third degree, the zeros of which are $0, \frac{1}{2}, 1$, and, consequently, it cannot have two distinct zeros inside $(0, 1)$. The assertion is proved.
We shall prove that $y_{2m}(x)$, $m = 1, 2, 3, \ldots$, does not have zeros inside the interval $(0, 1)$. Obviously, we can assume that $m \geq 2$. If the polynomial $y_{2m}(x)$ had a zero inside $(0, 1)$, then according to Rolle's theorem, the derivative $y'_{2m}(x)$ would have two distinct zeros inside $(0, 1)$, which is impossible in view of the equation

$$y'_{2m}(x) = B'_{2m}(x) = 2mB_{2m-1}(x) = 2my_{2m-1}(x),$$

since $y_{2m-1}(x)$ has the single zero $x = \frac{1}{2}$ inside $(0, 1)$.

We shall find the value of $B_{2m}(x)$ at $x = \frac{1}{2}$. We have:

$$e^{\frac{1}{2}t} \frac{t}{e^t-1} = \frac{e^{\frac{1}{2}t}t+t-t}{(e^{\frac{1}{2}t}+1)(e^{\frac{1}{2}t}-1)} = 2\frac{\frac{1}{2}t}{e^{\frac{1}{2}t}-1} - \frac{t}{e^t-1}.$$

From this follows the equation

$$H(\tfrac{1}{2}, t) = 2G\left(\frac{t}{2}\right) - G(t).$$

We substitute the expansions (9.8) and (9.2) for the functions H and G. We get

$$\sum_{v=0}^{\infty} \frac{B_v(\frac{1}{2})}{v!} t^v = 2\sum_{v=0}^{\infty} \frac{B_v}{v!}\left(\frac{t}{2}\right)^v - \sum_{v=0}^{\infty} \frac{B_v}{v!} t^v.$$

By equating the coefficients of t^m on the left and right hand sides, we get

$$B_m(\tfrac{1}{2}) = -B_m\left(1 - \frac{1}{2^{m-1}}\right), \qquad m = 0, 1, 2, 3, \ldots \qquad (9.16)$$

We established above that $y_{2k}(x)$ does not change sign inside $(0, 1)$. Equality (9.16) allows one to indicate this sign. In fact,

$$y_{2k}(\tfrac{1}{2}) = B_{2k}(\tfrac{1}{2}) - B_{2k} = -B_{2k}\left(2 - \frac{1}{2^{2k-1}}\right).$$

Therefore, $y_{2k}(\frac{1}{2})$ has the opposite sign to B_{2k} or, by virtue of (9.6),

$$\text{sign } y_{2k}(\tfrac{1}{2}) = (-1)^k. \qquad (9.17)$$

The sign of $y_{2k}(x)$ can also be determined directly. It coincides with the sign of the integral

$$\int_0^1 y_{2k}(x)dx = \int_0^1 [B_{2k}(x) - B_{2k}]dx =$$

$$= \left[\frac{1}{2k+1} B_{2k+1}(x) - B_{2k}x\right]_0^1 = -B_{2k},$$

and we get the previous result.

We introduce the periodic functions $B_v^*(x)$ with period 1 which coincide with $B_v(x)$ for $0 \le x < 1$. In figures 18 and 19, the graphs of the functions $y = B_0^*(x)$ and $y = B_1^*(x)$ are depicted, and in figures 20 and 21, the graphs of functions $y = 4B_2^*(x)$ and $y = 4B_3^*(x)$ are shown.

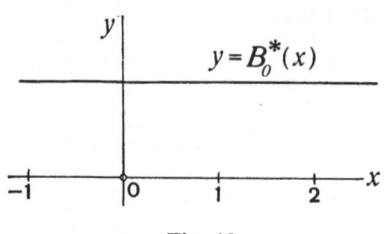

Fig. 18   Fig. 19

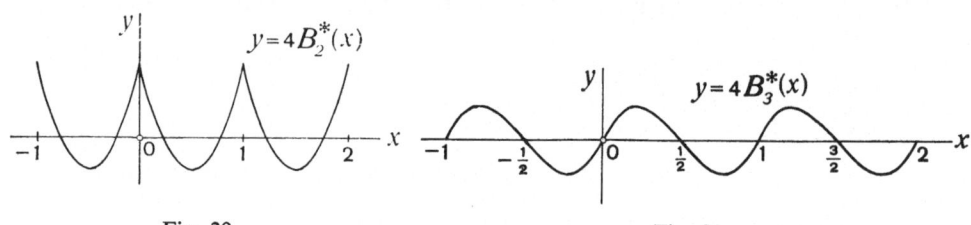

Fig. 20   Fig. 21

The function $B_1^*(x)$ is discontinuous at the integers, and has a discontinuity at them equal to minus one. The function $B_2^*(x)$ is continuous everywhere and $B_3^*(x)$ has a continuous derivative of the first order. From (9.12), it follows that

$$B_\nu^{*\prime}(x) = \nu B_{\nu-1}^*(x) \tag{9.18}$$

where

$$\nu = 2, 3, 4, \ldots,$$

so that $B_\nu^*(x)$ has continuous derivatives up to the order $\nu-2$ inclusively for $\nu \geq 2$.

We now indicate the Fourier series expansion of the functions $B_\nu^*(x)$:

$$B_\nu^*(x) = \tfrac{1}{2}a_0^{(\nu)} + \sum_{m=1}^{\infty} (a_m^{(\nu)} \cos 2\pi mx + b_m^{(\nu)} \sin 2\pi mx), \tag{9.19}$$

where

$$a_m^{(\nu)} = 2\int_0^1 B_\nu^*(x) \cos 2\pi mx\, dx = 2\int_0^1 B_\nu(x) \cos 2\pi mx\, dx,$$

$$b_m^{(\nu)} = 2\int_0^1 B_\nu^*(x) \sin 2\pi mx\, dx = 2\int_0^1 B_\nu(x) \sin 2\pi mx\, dx.$$

According to the well-known theorems on the expansion of functions in trigonometrical Fourier series, equations (9.19) are valid for all x for all of the functions $B_\nu^*(x)$, except $B_1^*(x)$. For $B_1^*(x)$, equation (9.19) does not hold for $x = 0$ (and at all integers), for at these points, the sum of the series equals

$$\frac{B_1^*(+0)+B_1^*(-0)}{2} = 0.$$

We compute the Fourier coefficients $a_m^{(\nu)}$ and $b_m^{(\nu)}$. We can suppose that $\nu \geqq 1$. We have

$$a_0^{(\nu)} = 2\int_0^1 B_\nu(x)dx = \frac{2}{\nu+1} B_{\nu+1}(x) \Big|_0^1 = 0, \qquad \nu = 1, 2, 3, \ldots$$

We compute the remaining Fourier coefficients for the case that $\nu = 2k$, $k = 1, 2, 3, \ldots$ By integrating by parts, we get

$$a_m^{(2k)} = 2\int_0^1 B_{2k}(x) \cos 2\pi mx\, dx =$$

$$= 2\frac{\sin 2\pi mx}{2\pi m} B_{2k}(x) \Big|_0^1 - 2\int_0^1 \frac{\sin 2\pi mx}{2\pi m} 2kB_{2k-1}(x)dx.$$

The boundary term becomes zero, and we continue to calculate the integral by integrating by parts:

$$a_m^{(2k)} = \frac{2k}{\pi m} \frac{\cos 2\pi mx}{2\pi m} B_{2k-1}(x) \Big|_0^1 -$$

$$- \frac{2k}{\pi m}\int_0^1 \frac{\cos 2\pi mx}{2\pi m} (2k-1)B_{2k-2}(x)dx.$$

If $k > 1$, then the boundary term becomes zero, and we get

$$a_m^{(2k)} = -\frac{2k(2k-1)}{(2\pi m)^2} a_m^{(2k-2)}. \qquad (9.20)$$

If $k = 1$, then the integral becomes zero and

$$a_m^{(2)} = \frac{1}{\pi^2 m^2}.$$

Now, from (9.20), we find

$$a_m^{(2k)} = (-1)^{k-1} \frac{2k(2k-1)(2k-2)\ldots 3}{(2\pi m)^{2(k-1)}} \, a_m^{(2)} = (-1)^{k-1} \frac{2 \cdot (2k)!}{(2\pi m)^{2k}}.$$

$$(9.21)$$

It follows from the relationship (9.14) that

$$B_{2k}(1-x) \sin 2\pi m(1-x) = -B_{2k}(x) \sin 2\pi m x$$

and, consequently,

$$b_m^{(2k)} = 0, \qquad m = 1, 2, 3, \ldots \tag{9.22}$$

Substituting the coefficients (9.21) and (9.22) in (9.19), we get

$$B_{2k}^*(x) = \frac{(-1)^{k-1}(2k)!}{2^{2k-1}\pi^{2k}} \sum_{m=1}^{\infty} \frac{\cos 2\pi m x}{m^{2k}}. \tag{9.23}$$

In the odd case, $v = 2k-1$, $k = 1, 2, \ldots$, and in the same way we find that

$$B_{2k-1}^*(x) = \frac{(-1)^k(2k-1)!}{2^{2k-2}\pi^{2k-1}} \sum_{m=1}^{\infty} \frac{\sin 2\pi m x}{m^{2k-1}}. \tag{9.24}$$

If one sets $x = 0$ in (9.23), then one obtains formula (9.5).

10. Expansion of functions in terms of Bernoulli polynomials

Let $f(x)$ have continuous derivatives up to the order $v \geq 1$, inclusive, on the interval [0,1]. Then, the following formula holds for $0 \leq x \leq 1$,

$$f(x) = \int_0^1 f(t)dt + \sum_{k=1}^{v} \frac{f^{(k-1)}(1) - f^{(k-1)}(0)}{k!} B_k(x) -$$

$$- \frac{1}{v!} \int_0^1 f^v(t) B_v^*(x-t)dt, \tag{10.1}$$

which gives an expansion of $f(x)$ in terms of Bernoulli polynomials. We shall prove formula (10.1) by performing a transformation of the following expression,

$$\rho_v(x) = \frac{1}{v!} \int_0^1 B_v^*(x-t) f^{(v)}(t)dt. \tag{10.2}$$

In the proof we assume that x is found inside the interval $(0,1)$: $0 < x < 1$. The correctness of formula (10.1) in the closed interval $0 \leq x \leq 1$ will follow from the fact that left and right hand sides of (10.1) are continuous functions of x for $0 \leq x \leq 1$.

We assume first that $v \geq 2$ and we will perform an integration by parts on the right hand side of equation (10.2):

$$\rho_v(x) = \frac{1}{v!} \int_0^1 B_v^*(x-t) f^{(v)}(t) dt =$$

$$= \frac{1}{v!} B_v^*(x-t) f^{(v-1)}(t) \Big|_0^1 - \frac{1}{v!} \int_0^1 f^{(v-1)}(t) \frac{d}{dt} [B_v^*(x-t)] dt.$$

On the basis of formula (9.18), we have that $(v \geq 2)$

$$\frac{d}{dt} B_v^*(x-t) = -v B_{v-1}^*(x-t)$$

and, by virtue of the definition of $B_v^*(x)$,

$$B_v^*(x-1) = B_v^*(x) = B_v(x),$$

therefore, the result of the integration by parts can be written in the following manner,

$$\rho_v(x) = \frac{1}{v!} [f^{(v-1)}(1) - f^{(v-1)}(0)] B_v(x) + \rho_{v-1}(x). \tag{10.3}$$

If $v-1 > 1$ then formula (10.3) can be applied to $\rho_{v-1}(x)$. Furthermore, we apply (10.3) to $\rho_{v-2}(x)$, if $v-2 > 1$, and so forth. As a result we get

$$\rho_v(x) = \sum_{k=2}^v \frac{f^{(k-1)}(1) - f^{(k-1)}(0)}{k!} B_k(x) + \rho_1(x). \tag{10.4}$$

For the transformation of

$$\rho_1(x) = \int_0^1 B_1^*(x-t) f'(t) dt,$$

one must take into account that $B_1^*(x)$ is discontinuous at the integers. We decompose the integral in the representation of $\rho_1(x)$ into the sum of the two integrals

$$I_1 = \int_0^x B_1^*(x-t) f'(t) dt, \qquad I_2 = \int_x^1 B_1^*(x-t) f'(t) dt,$$

so that

$$\rho_1(x) = I_1 + I_2. \tag{10.5}$$

We calculate I_1 by integration by parts,

$$I_1 = \int_0^x B_1^*(x-t)f'(t)dt =$$

$$= B_1^*(x-t)f(t)\Big|_0^{x-0} - \int_0^x f(t)\frac{d}{dt}[B_1^*(x-t)]dt. \tag{10.6}$$

Obviously, we have that

$$\frac{d}{dt}B_1^*(x-t) = -1 \qquad \text{for} \quad 0 \leqq t < x,$$

$$B_1^*(+0) = -\tfrac{1}{2}, \qquad B_1^*(x) = B_1(x),$$

and from (10.6) we get

$$I_1 = -\tfrac{1}{2}f(x) - f(0)B_1(x) + \int_0^x f(t)dt. \tag{10.7}$$

In an analogous manner we find that

$$I_2 = B_1^*(x-t)f(t)\Big|_{x+0}^1 - \int_x^1 f(t)\frac{d}{dt}[B_1^*(x-t)]dt =$$

$$= -\tfrac{1}{2}f(x) + f(1)B_1(x) + \int_x^1 f(t)dt. \tag{10.8}$$

We will substitute into (10.5) the right hand sides of (10.7) and (10.8) for I_1 and I_2. We obtain

$$\rho_1(x) = -f(x) + [f(1) - f(0)]B_1(x) + \int_0^1 f(t)dt. \tag{10.9}$$

Comparing (10.2), (10.4) and (10.9) we get formula (10.1). For $v = 1$, formula (10.1) becomes formula (10.9).

For $v \geqq 2$ formula (10.1) can be written in the following form,

$$f(x) = \int_0^1 f(t)dt + \sum_{k=1}^{v-1} \frac{f^{(k-1)}(1) - f^{(k-1)}(0)}{k!} B_k(x) -$$

$$- \frac{1}{v!}\int_0^1 f^{(v)}(t)[B_v^*(x-t) - B_v^*(x)]dt, \tag{10.10}$$

if one replaces the summand, corresponding to $k = v$, by the integral

$$\frac{f^{(v-1)}(1)-f^{(v-1)}(0)}{v!} B_v(x) = \frac{1}{v!} B_v^*(x) \int_0^1 f^{(v)}(t)dt.$$

We now assume that the function $f(x)$ is given on the interval $[a, a+h]$, $h > 0$, and is v times ($v \geq 2$) continuously differentiable on it. We shall write formula (10.10) for this case. We introduce the new independent variable ξ, $x = a+h\xi$, $0 \leq \xi \leq 1$. We apply formula (10.10) to the function of ξ, $\varphi(\xi) = f(a+h\xi)$, which is v times continuously differentiable for $0 \leq \xi \leq 1$.
We obtain

$$\varphi(\xi) = \int_0^1 \varphi(\tau)d\tau + \sum_{k=1}^{v-1} \frac{\varphi^{(k-1)}(1)-\varphi^{(k-1)}(0)}{k!} B_k(\xi) -$$

$$- \frac{1}{v!} \int_0^1 \varphi^{(v)}(\tau)[B_v^*(\xi-\tau)-B_v^*(\xi)]d\tau. \qquad (10.11)$$

In formula (10.11), we return to the previous notation for the independent variable and the function, and take into account the following relationships,

$$\varphi^{(k)}(\xi) = h^k f^{(k)}(a+h\xi) = h^k f^{(k)}(x),$$

$$\varphi(\tau) = f(a+h\tau) = f(t), \qquad t = a+h\tau, \qquad dt = h d\tau.$$

We get

$$f(x) = \frac{1}{h} \int_a^{a+h} f(t)dt +$$

$$+ \sum_{k=1}^{v-1} \frac{h^{k-1}[f^{(k-1)}(a+h)-f^{(k-1)}(a)]}{k!} B_k \left(\frac{x-a}{h}\right) -$$

$$- \frac{h^v}{v!} \int_0^1 f^{(v)}(a+h\tau) \left[B_v^* \left(\frac{x-a}{h}-\tau\right) - B_v^* \left(\frac{x-a}{h}\right) \right] d\tau.$$

We rewrite this formula in the following form,

$$\int_a^{a+h} f(t)dt = hf(x) - h[f(a+h)-f(a)]B_1 \left(\frac{x-a}{h}\right) -$$

$$- \sum_{k=2}^{v-1} \frac{h^k[f^{(k-1)}(a+h)-f^{(k-1)}(a)]}{k!} B_k \left(\frac{x-a}{h}\right) +$$

$$+ \frac{h^{v+1}}{v!} \int_0^1 f^{(v)}(a+h\tau) \left[B_v^* \left(\frac{x-a}{h}-\tau\right) - B_v^* \left(\frac{x-a}{h}\right) \right] d\tau.$$

$$(10.12)$$

Formula (10.12) gives the value of the integral

$$\int_a^{a+h} f(t)dt$$

in terms of the value of the function f at the point x and values of the function f and its derivatives up to the order $v-2$ at the ends of the interval $[a, a+h]$. The last summand of formula (10.12) has the significance of a remainder term, and contains the derivative of order v of the function $f(x)$.

If we keep only one summand on the right hand side of formula (10.12), then we get the approximate equation

$$\int_a^{a+h} f(t)dt \cong hf(x), \tag{10.13}$$

which represents the rectangular quadrature formula (2.11). Formula (10.13) becomes an exact equality only in the case that $f(t)$ is a constant. We write formula (10.12) for $v = 2$:

$$\int_a^{a+h} f(t)dt = hf(x) - h[f(a+h) - f(a)]B_1\left(\frac{x-a}{h}\right) +$$

$$+ \frac{h^3}{2!}\int_0^1 f''(a+h\tau)\left[B_2^*\left(\frac{x-a}{h}-\tau\right) - B_2^*\left(\frac{x-a}{h}\right)\right] d\tau. \tag{10.14}$$

If one keeps only two summands on the right hand side of this formula, then one obtains an approximate equation. It will be exact if $f(t)$ is any polynomial of the first degree, since in this case $f''(t) \equiv 0$ and the remainder term of formula (10.14) becomes zero.

If we would keep the three first terms in formula (10.12), then we would obtain an approximate equation which would be exact, if $f(t)$ is a polynomial of the second degree, and so forth.

11. The Euler-Maclaurin formula

In formula (10.12) we set $x = a$, and we take into account the equation $B_k(0) = B_k$, the periodicity of the function $B_v^*(\tau)$, and its continuity for $v \geq 2$:

$$B_v^*(-\tau) = B_v^*(1-\tau) = B_v(1-\tau), \qquad 0 \leq \tau \leq 1.$$

We get

$$\int_a^{a+h} f(t)dt = \frac{h}{2}\left[f(a)+f(a+h)\right]-$$

$$-\sum_{k=2}^{v-1}\frac{h^kB_k}{k!}\left[f^{(k-1)}(a+h)-f^{(k-1)}(a)\right]+$$

$$+\frac{h^{v+1}}{v!}\int_0^1 f^{(v)}(a+h\tau)[B_v(1-\tau)-B_v]d\tau. \tag{11.1}$$

The first summand on the right hand side of (11.1) represents the quadrature sum of the small trapezoidal rule (2.21). The following summands, as in the case of formula (10.12), are corrections. Thus, formula (11.1) makes the small trapezoidal rule precise.

The Euler-Maclaurin formula, which we shall now establish, is meant to make precise the large trapezoidal rule.

We examine the integral

$$\int_a^b f(x)dx.$$

We set $h = (b-a)/n$ and subdivide the interval $[a, b]$ into the n subintervals

$$[a+jh, a+(j+1)h], \qquad j = 0, 1, 2, \ldots, n-1.$$

We apply formula (11.1) to the integral of $f(x)$ over a subinterval:

$$\int_{a+jh}^{a+(j+1)h} f(x)dx = \frac{h}{2}\left[f(a+jh)+f(a+(j+1)h)\right]-$$

$$-\sum_{m=2}^{v-1}\frac{h^mB_m}{m!}\left[f^{(m-1)}(a+(j+1)h)-f^{(m-1)}(a+jh)\right]+$$

$$+\frac{h^{v+1}}{v!}\int_0^1 f^{(v)}(a+jh+h\tau)[B_v(1-\tau)-B_v]d\tau. \tag{11.2}$$

We introduce the notation

$$T_n = h[\tfrac{1}{2}f(a)+f(a+h)+f(a+2h)+ \cdots$$

$$\cdots +f(a+(n-1)h)+\tfrac{1}{2}f(b)] \tag{11.3}$$

and we sum both sides of formula (11.2) with respect to j from 0 to $n-1$.

We get

$$\int_a^b f(x)dx = T_n - \sum_{m=2}^{v-1} \frac{h^m B_m}{m!} [f^{(m-1)}(b) - f^{(m-1)}(a)] +$$
$$+ \frac{h^{v+1}}{v!} \int_0^1 [B_v(1-\tau) - B_v] \sum_{j=0}^{n-1} f^{(v)}(a+jh+h\tau)d\tau. \qquad (11.4)$$

This is the Euler-Maclaurin formula. Usually, in the Euler-Maclaurin formula, one takes $v = 2k$ even. Then the multiplier $B_v(1-\tau) - B_v$ under the integral sign on the right hand side, by virtue of (9.14), can be written thusly:

$$B_{2k}(1-\tau) - B_{2k} = B_{2k}(\tau) - B_{2k} = y_{2k}(\tau).$$

This multiplier does not change sign on $0 \leq \tau \leq 1$. If we now take into account that $B_m = 0$ for $m = 3, 5, 7, \ldots$, then formula (11.4) can be re-written in the following manner:

$$\int_a^b f(x)dx = T_n - \sum_{j=1}^{k-1} \frac{h^{2j} B_{2j}}{(2j)!} [f^{(2j-1)}(b) - f^{(2j-1)}(a)] +$$
$$+ \frac{h^{2k+1}}{(2k)!} \int_0^1 [B_{2k}(\tau) - B_{2k}] \sum_{j=0}^{n-1} f^{(2k)}(a+jh+h\tau)d\tau. \qquad (11.5)$$

One applies the Euler-Maclaurin formula (11.5) to the calculation of the integral $\int_a^b f(x)dx$, as well as to the calculation of the sum

$$f(a)+f(a+h)+f(a+2h)+ \ldots +f(b) = \frac{1}{h} T_n + \tfrac{1}{2}[f(a)+f(b)].$$

We are considering the Euler-Maclaurin formula as a means for the approximate calculation of integrals.

If, on the right hand side of the Euler-Maclaurin formula (11.5) one passes to the limit as $k \to \infty$ (formally), then one gets the series

$$T_n - \sum_{j=1}^{\infty} \frac{h^{2j} B_{2j}}{(2j)!} [f^{(2j-1)}(b) - f^{(2j-1)}(a)]. \qquad (11.6)$$

This series converges only for a highly restricted class of functions $f(x)$. As a rule, it diverges and its terms increase rapidly. From this it follows that the remainder term of formula (11.5),

$$R_{2k}(f) = \frac{h^{2k+1}}{(2k)!} \int_0^1 [B_{2k}(\tau) - B_{2k}] \sum_{j=0}^{n-1} f^{(2k)}(a+jh+h\tau)d\tau \qquad (11.7)$$

increases rapidly with increasing k. In calculating, it is necessary to select k so that the remainder term (11.7) has the smallest possible magnitude. We indicate two theorems on the remainder term of the Euler-Maclaurin formula.

Theorem 1. (On the representation of the remainder). If $f(x)$ has a continuous derivative of order $2k$ on $[a, b]$, then

$$R_{2k}(f) = -\frac{h^{2k}}{(2k)!}(b-a)B_{2k}f^{(2k)}(\eta), \tag{11.8}$$

where $a \leq \eta \leq b$.

Proof. We let

$$m = \min_{[a,b]} f^{(2k)}(x), \qquad M = \max_{[a,b]} f^{(2k)}(x). \tag{11.9}$$

We assume that k is even. Then,

$$y_{2k}(\tau) = B_{2k}(\tau) - B_{2k} > 0$$

for $0 < \tau < 1$ [See (9.17)], and we obtain

$$\frac{(2k)!}{h^{2k+1}} R_{2k}(f) = \int_0^1 [B_{2k}(\tau) - B_{2k}] \sum_{j=0}^{n-1} f^{(2k)}(a+jh+h\tau)d\tau \leq$$

$$\leq nM \int_0^1 [B_{2k}(\tau) - B_{2k}]d\tau = -nMB_{2k}.$$

In the same way, we obtain

$$\frac{(2k)!}{h^{2k+1}} R_{2k}(f) \geq -nmB_{2k}.$$

Thus, we have

$$-nmB_{2k} \leq \frac{(2k)!}{h^{2k+1}} R_{2k}(f) \leq -nMB_{2k}.$$

We divide through this inequality by $-nB_{2k} > 0$. We get

$$m \leq -\frac{(2k)!}{nB_{2k}h^{2k+1}} R_{2k}(f) = P \leq M.$$

Since $f^{(2k)}(x)$ is continuous on $[a, b]$, then there exists η, $a \leq \eta \leq b$, such that

$$P = f^{(2k)}(\eta).$$

This proves (11.8) in the case of even k. To prove (11.8) for odd k, one must take into account that $y_{2k}(\tau) < 0$ for $0 < \tau < 1$, and $B_{2k} > 0$.

We note that for $k = 1$ (11.8) becomes the representation of the remainder term for the trapezoidal formula (2.27).

Theorem 2. (On the evaluation of the remainder term.)

If for all x in $[a, b]$,

$$f^{(2k)}(x) \geq 0 \quad \text{and} \quad f^{(2k+2)}(x) \geq 0$$

$$[\text{or} f^{(2k)}(x) \leq 0 \quad \text{and} \quad f^{(2k+2)}(x) \leq 0], \tag{11.10}$$

then the remainder term of the Euler-Maclaurin formula $R_{2k}(f)$ has the same sign as the number

$$-\frac{h^{2k}B_{2k}}{(2k)!} \left[f^{(2k-1)}(b) - f^{(2k-1)}(a) \right] \tag{11.11}$$

and the absolute value of $R_{2k}(f)$ does not exceed the absolute value of the number (11.11).

Proof. Obviously, we have

$$R_{2k}(f) - R_{2k+2}(f) = -\frac{h^{2k}B_{2k}}{(2k)!} \left[f^{(2k-1)}(b) - f^{(2k-1)}(a) \right]. \tag{11.12}$$

By virtue of the assumptions (11.10) and the property that

$$\operatorname*{sign}_{a < \tau < b} y_{2\nu}(\tau) = (-1)^\nu,$$

the signs of $R_{2k}(f)$ and $R_{2k+2}(f)$ are opposite. This follows from the formula

$$R_{2\nu}(f) = \frac{h^{2\nu+1}}{(2\nu)!} \int_0^1 [B_{2\nu}(\tau) - B_{2\nu}] \sum_{j=0}^{n-1} f^{(2\nu)}(a+jh+h\tau)d\tau$$

for $\nu = k$ and $\nu = k+1$. But now, the conclusion of the theorem follows from the formula (11.12).

With the help of the Euler-Maclaurin formula, it is easy to indicate the representation of the remainder term of the rectangular quadrature formula (2.10) for the accurate case, that $f(x)$ is a periodic function, with period $b-a$, and has a continuous derivative of order $2k$ on the whole real axis. In fact, by virtue of the $(b-a)$-periodicity of the derivatives $f^{(j)}(x)$, the Euler-Maclaurin formula may be written in a simpler form:

$$\int_a^b f(x)dx = T_n + R_{2k}(f).$$

In view of the periodicity of $f(x)$, the quantity T_n coincides with the quadrature sum of the rectangular formula,

$$\int_a^b f(x)dx = \frac{b-a}{n}[f(a)+f(a+h)+ \ldots +f(a+(n-1)h]+R_{2k}(f),$$

and we obtain that in the case considered, the remainder term of the rectangular quadrature formula has the representation (11.8).

V. I. Krylov has constructed formulas, analogous to the Euler-Maclaurin formula, intended to make the quadrature formulas of general form precise. In particular, the formula which makes the Gaussian quadrature formula (5.1) precise is:

$$\int_{-1}^1 f(x)dx \cong \sum_{k=1}^n A_k f(x_k)+$$

$$+ \frac{1}{(2n+1)!}\left[\frac{2^n(n!)^2}{(2n)!}\right]^2 [f^{(2n-1)}(1)-f^{(2n-1)}(-1)]+$$

$$+ \frac{1}{(2n+2)!}\left[\frac{2^n(n!)^2}{(2n)!}\right]^2 \left[\frac{2n^2+2n-1}{(2n-1)(2n+1)(2n+3)}+\right.$$

$$+ \frac{n(n-1)}{(2n-1)(2n+1)} - \left.\frac{n+1}{3}\right] \times$$

$$\times [f^{(2n+1)}(1)-f^{(2n+1)}(-1)]+ \ldots$$

(see chapter 11 of the book [18]).

In calculating the integral by the Euler-Maclaurin formula, it is necessary to find the derivatives of the integrand at the ends of the interval of integration. If the integrand is given analytically, then it is possible to indicate the analytical expressions of the derivatives and to calculate their values at the ends of the interval $[a, b]$. If the integrand is given tabularly or analytically, but the expressions of its derivatives are inconvenient for calculation, then we do not apply the indicated method. We indicate how to proceed in this case.

We know the values of the integrand $f(x)$ for $x = a+kh$: $f_k = f(a+kh)$, $k = 0, 1, 2, \ldots, n$, therefore, it is possible to construct a table of finite differences of the function $f(x)$ for $x = a(h)b$. The table of finite differences of the function $f(x)$ allows one to indicate approximate values of the derivatives of $f(x)$ at the points a and b. These values are given by formulas (11.9) and (11.12) in chapter II:

$$f^{(m)}(a) \cong P_n^{(m)}(a) = \frac{m!}{h^m} \sum_{k=m}^{n} \frac{S_k^{(m)}}{k!} \Delta^k f_0,$$

$$f^{(m)}(b) \cong P_n^{(m)}(b) = \frac{m!}{h^m} \sum_{k=m}^{n} (-1)^{k-m} \frac{S_k^{(m)}}{k!} \Delta^k f_{n-k}, \qquad (11.13)$$

where the numbers $S_k^{(m)}$ are determined by equation (11.6) in chapter II. Substituting the approximate values (11.13) of the derivatives into formula (11.15), we obtain

$$\int_a^b f(x)dx \cong T_n - \frac{h}{12}(\Delta f_{n-1} - \Delta f_0) - \frac{h}{24}(\Delta^2 f_{n-2} + \Delta^2 f_0) -$$

$$- \frac{19h}{720}(\Delta^3 f_{n-3} - \Delta^3 f_0) - \frac{3h}{160}(\Delta^4 f_{n-4} + \Delta^4 f_0) -$$

$$- \frac{863h}{60\,480}(\Delta^5 f_{n-5} - \Delta^5 f_0) - \frac{275h}{24\,192}(\Delta^6 f_{n-6} + \Delta^6 f_0) - \cdots$$

$$(11.14)$$

This is Gregory's formula.

One can prove that in formula (11.14), the coefficient of

$$-h(\Delta^k f_{n-k} + (-1)^k \Delta^k f_0)$$

is equal to

$$\frac{(-1)^k}{(k+1)!} \int_0^1 t(t-1) \ldots (t-k)dt.$$

Example. The integral

$$Si(1) = \int_0^1 \frac{\sin x}{x} dx$$

was previously calculated by the trapezoidal formula in § 2 of this chapter. The value $T_{10} = 0.945831$ was obtained. We shall improve this value with the help of the Euler-Maclaurin formula.

In § 2, inequality (2.30) was established,

$$\left| \frac{d^m}{dx^m} \left(\frac{\sin x}{x} \right) \right| \leq \frac{1}{m+1}.$$

From the representation of the remainder term (11.8) of the Euler-Maclaurin formula we find

$$|R_{2k}| \leq \frac{(0.1)^{2k}}{(2k+1)!} |B_{2k}|. \tag{11.15}$$

Since for large k,

$$|B_{2k}| \cong 2 \frac{(2k)!}{(2\pi)^{2k}},$$

then it is clear that R_{2k} goes to zero rapidly, and the series (11.6) converges. For $k = 2$ we get from (11.15),

$$|R_4| < 3 \cdot 10^{-8},$$

so that one can take $k = 2$ in formula (11.5). Since

$$\frac{h^2 B_2}{2} [f'(1) - f'(0)] = -0.000251,$$

then

$$\text{Si}(1) \cong 0.945831 + 0.000251 = 0.946082,$$

which differs from the exact value by one unit in the sixth decimal place. This difference was obtained because of round off error.

12. Concluding remarks

The selection of the quadrature formula is of essential importance for the approximate calculation of integrals. This choice must be determined by many circumstances: the properties of the integrand, the manner in which it is given, the means of calculation at the disposal of the calculator, the required accuracy, and so forth.

The properties of the integrand are important, therefore, before solving the question of the selection of the quadrature formula, we must determine the representation of the graph of the function, and the behavior of its derivatives. The form of the graph of the function can, for example, tell us that it is necessary to subdivide the interval of integration into parts and to apply the quadrature formula on each subinterval.

The neglect of such preliminary research can lead to a useless waste of time and computational effort. In fact, for any quadrature formula, it is possible to construct a function, for which the remainder term of the given quadrature formula (with a given number of knots) is as large as one might desire. It is sufficient to take the square of the polynomial, the roots of which are the knots of the quadrature formula, and to multiply this polynomial by a sufficiently large constant.

On the other hand, the knowledge of the properties of the integrand and their use can, in some cases, widen the area of application of the quadrature formula. We make this idea clear in an example of the application of quadrature formulas to the computation of improper integrals and integrals of oscillating functions.

For definiteness, we consider the interval of integration to be finite. The calculation of an improper integral with the help of a quadrature formula of the form

$$\int_a^b F(x)dx \cong \sum_{k=1}^n A_k F(x_k) \tag{12.1}$$

is impossible if any knot of the quadrature formula coincides with a singular point of the integrand. If computation is possible, then, as a rule, it leads to a large error. To this we must add that also in the calculation of a proper integral by formula (12.1), we can get a large error if $F(x)$ is continuous, but has a derivative of low order which is not bounded.

Several of the quadrature formulas indicated in this chapter can be applied to the calculation of improper integrals. Such, for example, is the formula (6.2) of the Gaussian type with the weight function

$$p(x) = (1-x)^\alpha (1+x)^\beta, \qquad -1 \leq x \leq 1, \quad \alpha > -1, \quad \beta > -1.$$

In fact, if the function $F(x)$ is such that it may be represented in the form

$$F(x) = (1-x)^\alpha (1+x)^\beta f(x), \qquad \alpha > -1, \quad \beta > -1,$$

where $f(x)$ is differentiable a sufficient number of times, then to compute the integral $\int_{-1}^1 F(x)dx$ one can use the quadrature formula (6.2).

In general, if one can represent the integrand $F(x)$ in the form

$$F(x) = p(x)f(x),$$

where $p(x)$ or one of its derivatives of low order is not bounded, and $f(x)$ has a sufficient number of derivatives, then it is natural to use the interpolation quadrature formula with weight $p(x)$ for the calculation of the integral:

$$\int_a^b p(x)f(x)dx \cong \sum_{k=1}^n A_k f(x_k). \tag{12.2}$$

Such a formula, as shown in § 1, can always be constructed if the moments of the function $p(x)$ exist.

The interpolation quadrature formulas with weight functions can also be applied to the calculation of integrals of the form

$$\int_0^T \varphi(Nx)f(x)dx, \tag{12.3}$$

where $\varphi(x)$ is a T-periodic function and N is an integer; $f(x)$ is a function which is differentiable a sufficient number of times and is slowly varying. We call the function $\varphi(Nx)f(x)$ an oscillating function. The immediate application of the quadrature formula of the form (12.1) to the computation of integral (12.3) is made difficult for large N because for a sufficiently accurate approximation of $\varphi(Nx)f(x)$ by polynomials, it would be necessary to take a large number of knots.

To compute the integral (12.3), it is expedient to use the interpolation quadrature formula (12.2) with the weight function $p(x) = \varphi(Nx)$, since $f(x)$ allows a good approximation by polynomials on the whole interval of integration. Formulas of this form for $\varphi(x) = \sin x$, $\cos x$ are studied in [28].

In certain cases, one can weaken the singularity of the integrand or its derivative of a low order after having performed an integration by parts. For example, let the function $F(x)$ have the form

$$F(x) = (x-x_1)^\alpha f(x), \qquad \alpha > -1,$$

where x_1 belongs to space $[a, b]$, the number α is different from one, $f(x)$ has a sufficient number of derivatives on $[a, b]$, and $f(x_1) \neq 0$. By writing the integral in the form

$$\int_a^b F(x)dx = \int_a^{x_1} F(x)dx + \int_{x_1}^b F(x)dx$$

and by applying the formula of integration by parts to each integral on the right hand side, we get

$$\int_a^b (x-x_1)^\alpha f(x)dx =$$

$$= \frac{1}{\alpha+1}[f(b)(b-x_1)^{\alpha+1} - f(a)(a-x_1)^{\alpha+1}] -$$

$$- \frac{1}{\alpha+1}\int_a^b (x-x_1)^{\alpha+1}f'(x)dx. \tag{12.4}$$

The derivative of order $[\alpha+1]$ of the function $F(x)$ has a singularity at the point x_1, in particular, if $\alpha < 0$, then x_1 is a singular point for $F(x)$. The function under the integral sign on the right hand side of (12.4) is continuous, and x_1 is a singular point of its derivative of order $[\alpha+2]$. By applying the formula of integration by parts k times to this integral, we increase the order of differentiability of the integrand at the point x_1 by k units.

Integration by parts also turns out to be useful for the calculation of integrals of the form (12.3). We limit ourselves to the investigation of the special case

$$\int_0^1 f(x) \sin 2\pi N x\, dx.$$

By integrating by parts $2k$ times, we get

$$\int_0^1 f(x) \sin 2\pi N x\, dx = \sum_{j=0}^{k-1} \frac{(-1)^{j+1}}{(2\pi N)^{2j+1}} [f^{(2j)}(1) - f^{(2j)}(0)] +$$

$$+ (-1)^k \frac{1}{(2\pi N)^{2k}} \int_0^1 f^{(2k)}(x) \sin 2\pi N x\, dx.$$

If N is sufficiently large, then the multiplier before the integral on the right hand side will be small, and the computation of this integral can be carried out with less accuracy than the computation of the initial integral. The general case that $\varphi(x)$ in the integral (12.3) is an arbitrary T-periodic function, is examined in work [5]. The question of the calculation of the integrals of oscillating functions of more general form than in (12.3) is examined in [17].

EXERCISES FOR CHAPTER III

1.

1. Calculate the integrals

$$I_1 = \int_a^b (x-x_0)dx, \qquad I_2 = \int_a^b (x-x_0)(x-x)dx.$$

Use the results to write interpolation quadrature formulas (1.6), with $p(x) \equiv 1$, based on Newton's interpolation formula [(5.7) in Chapter II] for $n = 1, 2$. Discuss the additional calculations required to extend these formulas to $n = 3$.

2. Write interpolation quadrature formulas (1.6), with $p(x) = 1$, based on Lagrange's interpolation formula [(5.6) in Chapter II] for $n = 1, 2$. Discuss the additional calculations required to extend these formulas to $n = 3$.

3. Use the expression for $R_n(f, x)$ given by (6.2) in Chapter II to obtain the error of the interpolation quadrature formula (1.6) with $p(x) \equiv 1$ for $n = 1$. Write this expression also for the special case $x_0 = a$, $x_1 = b$.

2.

1. Prove that for the coefficients $B_k^{(n)}$ defined by (2.20),

$$B_k^{(n)} = B_{n-k}^{(n)}, \qquad k = 0, 1, 2, \ldots, n.$$

Calculate $B_3^{(6)}$.

2. The logarithmic integral is defined by

$$\text{Li}(x) = \int_1^x \frac{\ln t}{t} dt$$

for $x > 0$. Calculate $\text{Li}(2)$ by the trapezoidal rule (2.26) for $n = 2, 4, 10, 20$. Use five place tables.

3. Calculate $\text{Li}(2)$ by Simpson's rule (2.28) for $n = 2, 4, 10, 20$.

3.

1. Calculate

$$I = \int_0^{2\pi} (1 - \cos x) \cos x \, dx$$

by the rectangular formula (3.4) with $\alpha = 0$ for $n = 2, 3, 4$. Use five-place tables. Compare the results with the exact value $I = -\pi$.

2. Calculate

$$I = \int_0^{2\pi} \frac{dx}{1 + 0.5 \cos x}$$

by the rectangular formula (3.4) with $\alpha = 0$ for $n = 5, 10, 15$. Use five-place tables. Compare the results with the exact value $I = \frac{4}{3}\pi\sqrt{3}$.

3. Calculate

$$I = \int_0^{2\pi} e^{\cos x} \, dx$$

by the rectangular formula (3.4) with $\alpha = 0$ for $n = 8, 10, 12$. Use five-place tables.

4.

1. Construct the Gaussian type integration formula for $p(x) = 1, n = 1$; that is,

$$\int_a^b f(x) dx \cong A_1 f(x_1)$$

assuming that a, b are finite.

2. Construct the Gaussian type integration formula of order $n = 1$ for

$$\int_0^\infty e^{-x} f(x) dx,$$

that is, for $p(x) = e^{-x}$.
Hint: Use the fact that

$$\int_0^\infty e^{-x}(x - x_1) dx = 0$$

to determine x_1.

3. Construct the Gaussian type integration formula of order $n = 2$ for

$$\int_0^1 f(x)dx,$$

that is, for $p(x) \equiv 1$.

5.

1. Calculate

$$I = \int_1^2 \frac{dx}{x}$$

by the Gauss quadrature formula (5.16) for $n = 2, 3, 4$. Carry ten decimal places in the calculations, and round the final results to five places. Compare with the exact value $I = \ln 2 \cong 0.69315$.

2. Give an explicit representation of the remainder term (5.15) of the Gauss integration formulas for $n = 2, 3, 4$.

3. Use *Stirling's Approximation*

$$\sqrt{2n\pi} \left(\frac{n}{e}\right)^n < n! < \sqrt{2n\pi} \left(\frac{n}{e}\right)^n \left(1 + \frac{1}{12n-1}\right)$$

to obtain an upper bound for the remainder term (5.15) of the Gauss integration formula of order n.

6.

1. Use the formula (6.13) for the remainder for Mellor's formula (6.11) to obtain a correction term which will make the resulting formula exact if $f(x)$ is a polynomial of degree $2n$ with leading coefficient equal to one. Use the result for $n = 2$ to determine

$$I = \int_{-1}^1 \frac{x^4}{\sqrt{1-x^2}} dx,$$

using five significant digits in the calculations. Compare with the exact value $I = \frac{3}{8}\pi$.

2. Use the formula (6.23) for the remainder term of the quadrature formula (6.20) in the case $s = 0$ to obtain a correction term to make the

formula exact if $f(x)$ is a polynomial of degree $2n$. (Recall that $\Gamma(n+1)$ $= n!$.) For $n = 2$, compare the uncorrected and corrected results for

$$I = \int_0^\infty x^4 e^{-x} dx$$

to the exact value $I = 24$. Carry five significant figures in the calculations.

3. Calculate

$$I = \int_{-\infty}^\infty e^{-x^2} - \frac{1}{x^2} dx$$

approximately, using formula (6.15) with $n = 1$. Estimate the error by using the remainder term (6.18). Compare the estimated error with the actual error obtained by use of the exact value $I = e^{-2}\sqrt{\pi} \cong 0.239875$.

7.

1. Transform quadrature formulas (7.11) and (7.17) to obtain corresponding formulas for the approximate calculation of the integral

$$I = \int_a^b f(x) dx$$

for a, b finite.

2. Use the remainder term (7.16) to obtain a correction term which will make the quadrature formula (7.11) exact if $f(x)$ is a polynomial of degree $2n+1$. Similarly, use (7.20) to obtain a correction term which will make (7.17) exact if $f(x)$ is a polynomial of degree $2n+2$.

3. Calculate approximate values of the integral

$$I = \int_{-1}^1 \frac{dx}{1+x^2}$$

by formulas (7.11) and (7.17). Carry eight significant digits in the computations, and round the final result to six decimal places. Compare the calculated values to the exact result $I = \frac{1}{2}\pi \cong 1.570796$

8.

1. Construct the coefficients of the polynomial $\omega(x)$ defined by (8.5) for the case $p(x) \equiv 1$, $n = 3$, using (8.4) and (8.9).

2. Calculate an approximation to

$$I = \int_{-1}^{1} \frac{dx}{\sqrt{1-x^2}}$$

by formula (8.10) for $n = 4$. Use five significant digits in the computations. Compare the result with the exact value $I = \pi$.

3. Calculate the approximate value of

$$I = \int_{-1}^{1} \frac{dx}{1+x^2}$$

by using (8.10) for $n = 9$. Use five significant digits in the calculations. Compare the result with the exact result $I = \frac{1}{2}\pi$.

4. Construct quadrature formulas of Čebyšev type for the approximate calculation of integrals of the form

$$I = \int_{-1}^{1} \frac{f(x)}{1+x^2} \, dx,$$

that is, with $p(x) = 1/(1+x^2)$, of orders $n = 1, 2, 3$. Compute the knots with eight decimal place accuracy.

9.

1. Use (9.4) to compute the Bernoulli numbers B_{20}, B_{22}, B_{24}.

2. Show that

$$\frac{1}{2k-1} < \sum_{m=1}^{\infty} \frac{1}{m^{2k}} < \frac{2}{2k-1},$$

and use this result together with (9.5) to obtain upper and lower bounds for the Bernoulli numbers B_{2k}, $k = 1, 2, \ldots$.
Hint: Compare the sum $\sum_{m=1}^{\infty} 1/m^{2k}$ with the integral $\int_{1}^{\infty} dx/x^{2k}$.

3. Prove that

$$\int_{0}^{1} B_k(x)dx = 0, \qquad k = 1, 2, \ldots.$$

Hence, the Bernoulli polynomials $B_1(x)$, $B_2(x)$, ... are orthogonal to $B_0(x)$ on $[0, 1]$. Show, in general, that

$$\int_{0}^{1} B_j(x)B_k(x)dx = 0 \qquad \text{if } j \neq k.$$

Hint: Use (p. 12) and integration by parts.

10.

1. Use formula (10.1) to expand the powers x, x^2, x^3, x^4, x^5, x^6 of x in terms of Bernoulli polynomials.

2. Obtain quadrature formulas from equation (10.14) which will be exact for polynomials of degrees 2, 3, 4, respectively.

11.

1. Carry out the details of the proof of Theorem 1 for odd k.

2. Prove formula (11.12).
Hint: Apply integration by parts twice to the formula

$$R_{2k+2} = \frac{h^{2k+3}}{(2k+2)!} \int_0^1 [B_{2k+2}(\tau) - B_{2k+2}] \sum_{j=0}^{m-1} f^{(2k+2)}(a+jh+h\tau)d\tau,$$

differentiating the term in brackets and integrating the term under the summation sign.

3. Investigate the correction of the trapezoidal rule for the calculation of the logarithmic integral

$$\text{Li}(2) = \int_1^2 \frac{\ln x}{x}\, dx$$

by the Euler-Maclaurin formula (11.5) for $k = 2, 3, 4$. Use (11.11) to estimate the error.

4. Calculate the integral $\int_{250}^{260} f(x)dx$ of the function $f(x)$ with values given in the following table by Gregory's formula (11.4).

x	$f(x)$
250	0.24298
251	0.24393
252	0.24488
253	0.24584
254	0.24679
255	0.24774
256	0.24869
257	0.24964
258	0.25059
259	0.25154
260	0.25249

Chapter IV

THE NUMERICAL SOLUTION OF THE CAUCHY PROBLEM FOR ORDINARY DIFFERENTIAL EQUATIONS

1. Introduction

In the present chapter we will deal with numerical methods for solution of ordinary differential equations or of systems of such equations. To be definite, we shall speak now about one differential equation of order n

$$y^{(n)} = f(x, y, y', \ldots, y^{(n-1)}). \tag{1.1}$$

It is known that the solution of equation (1.1) is not determined uniquely by this equation. The general solution of this equation depends, generally speaking, on n arbitrary constants. Numerical methods are applicable to finding particular solutions of the equation (1.1). In order to obtain this particular solution of the differential equation (1.1), we have to impose some n additional conditions upon the sought for solution.

Two kinds or problems connected with the determination of the particular solutions of the equation (1.1) are of great significance: the *Cauchy problem*, or the problem with initial data, and the *boundary value problem*. The Cauchy problem is formulated in the following manner: to find that solution $y(x)$ of the differential equation (1.1), which at $x = x_0$ satisfies the conditions:

$$y(x_0) = y_0, y'(x_0) = y'_0, \ldots, y^{(n-1)}(x_0) = y_0^{(n-1)}. \tag{1.2}$$

Characteristic for the Cauchy problem is that the conditions (1.2) are given at the one point $x = x_0$.

In the case or boundary value problems, the conditions are given not at one but at several points. For example, in the case of the differential equation of the second order

$$y'' = f(x, y, y') \tag{1.3}$$

one may consider the two-point boundary value problem: find that solution $y(x)$ of the differential equation (1.3) on the interval $[x_0, X]$, which satisfies the conditions

$$y(x_0) = A, \qquad y(X) = B.$$

Here A and B are given constants.

In what follows, we shall concern ourselves with the numerical methods for the solution of the Cauchy problem and we shall not deal at all with the question of the solution of boundary value problems. We only note that there are methods which allow the finding of the solution of a boundary value problem which lead one to the determination of solutions of a series of Cauchy problems for the differential equation considered.

The methods examined in the present chapter allow one to construct a table of approximate values of the solution of the Cauchy problem. This table, for example, in the case of the Cauchy problem for the system of two differential equations of the first order

$$\left. \begin{array}{l} y' = f(x, y, z), \\ z' = g(x, y, z), \end{array} \right\} \qquad y(x_0) = y_0, \qquad z(x_0) = z_0$$

has the form

x	y	z
x_0	y_0	z_0
x_1	y_1	z_1
x_2	y_2	z_2
...	...	...
x_N	y_N	z_N

where $x_0 < x_1 < x_2 < \ldots < x_N$ and $y_j \cong y(x_j)$, $z_j \cong z(x_j)$ $(j = 1, 2, \ldots, N)$.

2. The Runge-Kutta method

We shall examine the Cauchy problem for the differential equation of the first order

$$y' = f(x, y), \qquad y(x_0) = y_0. \tag{2.1}$$

We assume that a solution $y(x)$ of problem (2.1) exists. The Runge-Kutta method for the differential equation and the initial condition (2.1) allows one to calculate an approximate value of the solution $y(x)$ at the point $x_1 = x_0 + h$, $h > 0$.

This is done in the following manner. Successively, four numbers k_1, k_2, k_3, k_4 are calculated according to the formula

$$k_1 = hf(x_0, y_0),$$

$$k_2 = hf\left(x_0 + \frac{h}{2}, y_0 + \frac{k_1}{2}\right),$$

$$k_3 = hf\left(x_0 + \frac{h}{2}, y_0 + \frac{k_2}{2}\right),$$

$$k_4 = hf(x_0 + h, y_0 + k_3). \tag{2.2}$$

The value

$$y_1 = y_0 + \tfrac{1}{6}(k_1 + 2k_2 + 2k_3 + k_4) \tag{2.3}$$

is taken as an approximate value of $y(x_1)$. As $y_1 \cong y(x_1)$ in the same way we calculate $y_2 \cong y(x_2)$, where $x_2 = x_1 + h$, and so forth. We note that the arguments $x_0, x_1, x_2, \ldots$ do not have to be taken to be equidistant. The approximate equality

$$y(x_0 + h) - y_0 \cong \tfrac{1}{6}(k_1 + 2k_2 + 2k_3 + k_4) \tag{2.4}$$

is the generalization of Simpson's quadrature formula in the following manner. We have the equation

$$y(x_0 + h) - y_0 = \int_{x_0}^{x_0 + h} y'(x)dx = \int_{x_0}^{x_0 + h} f(x, y(x))dx.$$

If $f(x, y) = f(x)$ does not depend on y, then we can calculate the integral on the right hand side of the last equation by Simpson's rule. We get

$$y(x_0 + h) - y_0 \cong \frac{h}{6}\left[f(x_0) + 4f\left(x_0 + \frac{h}{2}\right) + f(x_0 + h)\right], \tag{2.5}$$

which coincides with (2.4).

We assume that $f(x)$ has a Taylor series expansion in a neighborhood of the point x_0. We expand the left and right hand sides of the approximate equation (2.5) into Taylor's series in powers of h in a neighborhood of $h = 0$. It is obvious that the coefficient of h^k in the expansion on the left-hand side coincides with the coefficient of h^k on the right-hand side for $k = 0, 1, 2, 3, 4$. This assertion follows from the fact that the approximate equation (2.5) becomes an exact equation, if $f(x)$ is a polynomial of the third degree (for such $f(x)$, Simpson's rule is exact).

It turns out that an analogous assertion holds also in the general case that the right-hand side of the differential equation (2.1) depends on y. Namely, if we expand the left and right hand sides of the approximate equation (2.4) in Taylor's series in powers of h in a neighborhood of $h = 0$, then in these expansions the coefficients of h^k for $k = 0, 1, 2, 3, 4$ will be equal. We shall prove this assertion, assuming that the right hand side $f(x, y)$ of the differential equation has a sufficient number of derivatives in some region D of the (x, y) plane containing the graph of the solution $y = y(x)$.

In the proof, we shall actually find the coefficients of h^k, $k = 0, 1, 2, 3, 4$, in the indicated Taylor series expansions. These coefficients will be expressed in terms of the partial derivatives of the right hand side $f(x, y)$ of the differential equation, evaluated at the initial point (x_0, y_0). For example, the coefficient of $h^2/2!$ in the expansion of the left hand side of (2.4) is equal to

$$y''(x_0) = \frac{\partial f(x_0, y_0)}{\partial x} + f(x_0, y_0)\frac{\partial f(x_0, y_0)}{\partial y}. \qquad (2.6)$$

To simplify the notation, it is convenient to introduce the following differentiation operator

$$D = \frac{\partial}{\partial x} + f(x, y)\frac{\partial}{\partial y}.$$

The operator D can be applied to any differentiable function $u(x, y)$ of two variables, with

$$Du = \frac{\partial u}{\partial x} + f\frac{\partial u}{\partial y}.$$

We note some properties of the operator D:

    1) $D(c_1 u + c_2 v) = c_1 Du + c_2 Dv$, c_1, c_2 are constants

    2) $D(uv) = uDv + vDu$.

These properties are obvious.

We will define the nth power of the operator D:

$$D^n = \left(\frac{\partial}{\partial x} + f\frac{\partial}{\partial y}\right)^n = \sum_{v=0}^{n} C_n^{(v)} f^v \frac{\partial^n}{\partial x^{n-v} \partial y^v} \tag{2.7}$$

(here n is a positive integer). We establish the property of the operator D:

3) $\quad D(D^n f) = D^{n+1} f + nDf \cdot D^{n-1} \dfrac{\partial f}{\partial y}$.

By virtue of the definition (2.7), we have

$$D^n f = \sum_{v=0}^{n} C_n^{(v)} f^v \frac{\partial^n f}{\partial x^{n-v} \partial y^v} .$$

We apply the operator D to both sides of this equation and use properties 1) and 2):

$$D(D^n f) = \sum_{v=0}^{n} C_n^{(v)} D \left[f^v \frac{\partial^n f}{\partial x^{n-v} \partial y^v} \right] =$$

$$= \sum_{v=0}^{n} C_n^{(v)} f^v D \left(\frac{\partial^n f}{\partial x^{n-v} \partial y^v} \right) + \sum_{v=0}^{n} C_n^{(v)} \frac{\partial^n f}{\partial x^{n-v} \partial y^v} Df^v . \tag{2.8}$$

We transform the first sum on the right hand side of (2.8):

$$\sum_{v=0}^{n} C_n^{(v)} f^v D \left(\frac{\partial^n f}{\partial x^{n-v} \partial y^v} \right) =$$

$$= \sum_{v=0}^{n} C_n^{(v)} f^v \left(\frac{\partial^{n+1} f}{\partial x^{n-v+1} \partial y^v} + f \frac{\partial^{n+1} f}{\partial x^{n-v} \partial y^{v+1}} \right) =$$

$$= \sum_{v=0}^{n} C_n^{(v)} f^v \frac{\partial^{n+1} f}{\partial x^{n-v+1} \partial y^v} + \sum_{v'=1}^{n+1} C_n^{(v'-1)} f^{v'} \frac{\partial^{n+1} f}{\partial x^{n-v'+1} \partial y^{v'}} .$$

Combining in pairs the summands which are obtained for $v' = v$, and taking into account the relation

$$C_n^{(v)} + C_n^{(v-1)} = C_{n+1}^{(v)} ,$$

we get

$$\sum_{v=0}^{n} C_n^{(v)} f^v D \left(\frac{\partial^n f}{\partial x^{n-v} \partial y^v} \right) = \sum_{v=0}^{n+1} C_{n+1}^{(v)} f^v \frac{\partial^{n+1} f}{\partial x^{n+1-v} \partial y^v} = D^{n+1} f . \tag{2.9}$$

Now we transform the second sum on the right hand side of (2.8):

$$\sum_{v=0}^{n} C_n^{(v)} \frac{\partial^n f}{\partial x^{n-v} \partial y^v} Df^v = Df \sum_{v=1}^{n} C_n^{(v)} v f^{v-1} \frac{\partial^n f}{\partial x^{n-v} \partial y^v} =$$

$$= nDf \sum_{v=1}^{n} C_{n-1}^{(v-1)} f^{v-1} \frac{\partial^{n-1}}{\partial x^{n-v} \partial y^{v-1}} \left(\frac{\partial f}{\partial y}\right) =$$

$$= nDf \sum_{v'=0}^{n-1} C_{n-1}^{(v')} f^{v'} \frac{\partial^{n-1}}{\partial x^{n-1-v'} \partial y^{v'}} \left(\frac{\partial f}{\partial y}\right) =$$

$$= nDf \cdot D^{n-1} \left(\frac{\partial f}{\partial y}\right). \tag{2.10}$$

It remains to compare the relations (2.8), (2.9) and (2.10) in order to convince oneself of the validity of property 3).

Now, it is not difficult to write the derivatives of the solution $y(x)$ of the problem (2.1) at the point $x = x_0$. Using the properties of the operator D, we get successively:

$$y' = f,$$

$$y'' = \frac{d}{dx} f = Df,$$

$$y''' = \frac{d}{dx} y'' = D(Df) = D^2 f + Df \cdot \frac{\partial f}{\partial y},$$

$$y^{(IV)} = \frac{d}{dx} y''' = D\left[D^2 f + Df \cdot \frac{\partial f}{\partial y}\right] =$$

$$= D(D^2 f) + D\left[Df \cdot \frac{\partial f}{\partial y}\right] = D(D^2 f) + Df \cdot D\frac{\partial f}{\partial y} +$$

$$+ \frac{\partial f}{\partial y} D(Df) = D^3 f + 2Df \cdot D\frac{\partial f}{\partial y} + Df \cdot D\frac{\partial f}{\partial y} +$$

$$+ \frac{\partial f}{\partial y} \left[D^2 f + Df \cdot \frac{\partial f}{\partial y}\right] = D^3 f + 3Df \cdot D\frac{\partial f}{\partial y} +$$

$$+ \frac{\partial f}{\partial y} \left[D^2 f + Df \cdot \frac{\partial f}{\partial y}\right].$$

Thus, the expansion of the left hand side of the relation (2.4) in powers of h in the neighborhood of $h = 0$ has the form

$$y(x_0+h)-y_0 = hf+ \frac{h^2}{2}Df+ \frac{h^3}{6}\left[D^2f+ \frac{\partial f}{\partial y}Df\right] +$$

$$+ \frac{h^4}{24}\left\{D^3f+3Df\cdot D\frac{\partial f}{\partial y} + \frac{\partial f}{\partial y}\left[D^2f+ \frac{\partial f}{\partial y}Df\right]\right\} + \dots \qquad (2.11)$$

To simplify the notation, we do not indicate that all the values of f, Df and so forth, are taken at $x = x_0$, $y = y_0$.

We now expand in Taylor's series the quantities k_1, k_2, k_3, k_4, defined by formula (2.2) in powers of h in a neighborhood of $h = 0$. It is obvious that

$$k_1 = hf(x_0, y_0) = hf. \qquad (2.12)$$

In expanding k_2, we use Taylor's formula for a function of two variables:

$$k_2 = hf\left(x_0+ \frac{h}{2}, y_0+ \frac{k_1}{2}\right) =$$

$$= h\left\{f+ \frac{1}{1!}\left(\frac{h}{2}\frac{\partial}{\partial x} + \frac{k_1}{2}\frac{\partial}{\partial y}\right)f+ \frac{1}{2!}\left(\frac{h}{2}\frac{\partial}{\partial x} + \frac{k_1}{2}\frac{\partial}{\partial y}\right)^2f+ \right.$$

$$\left. + \frac{1}{3!}\left(\frac{h}{2}\frac{\partial}{\partial x} + \frac{k_1}{2}\frac{\partial}{\partial y}\right)^3f+ \dots\right\}.$$

We substitute $k_1 = hf$ in this and the expression obtained we write more briefly with the help of the operator D:

$$k_2 = hf+ \frac{h^2}{2}Df+ \frac{h^3}{8}D^2f+ \frac{h^4}{48}D^3f+ \dots \qquad (2.13)$$

We write the expansion of k_3:

$$k_3 = hf\left(x_0+ \frac{h}{2}, y_0+ \frac{k_2}{2}\right) = h\sum_{l=0}^{\infty}\frac{1}{l!}\left(\frac{h}{2}\frac{\partial}{\partial x} + \frac{k_2}{2}\frac{\partial}{\partial y}\right)^l f.$$

In order to get in the expansion of k_3 the terms containing h of degrees up to the fourth, inclusively, in the following expression

$$\left(\frac{h}{2}\frac{\partial}{\partial x} + \frac{k_2}{2}\frac{\partial}{\partial y}\right)^l f,$$

it is sufficient for $l = 1$, 2, 3 to substitute $4-l$ of the summands on the right hand side of (2.13) in place of k_2. We perform this substitution and make the simple transformations

$$k_3 = hf + h \left[\frac{h}{2} \frac{\partial}{\partial x} + \frac{1}{2} \left(hf + \frac{h^2}{2} Df + \frac{h^3}{8} D^2 f + \ldots \right) \frac{\partial}{\partial y} \right] f +$$

$$+ \frac{h}{2} \left[\frac{h}{2} \frac{\partial}{\partial x} + \frac{1}{2} \left(hf + \frac{h^2}{2} Df + \ldots \right) \frac{\partial}{\partial y} \right]^2 f +$$

$$+ \frac{h}{6} \left[\frac{h}{2} \frac{\partial}{\partial x} + \tfrac{1}{2}(hf + \ldots) \frac{\partial}{\partial y} \right]^3 f + \ldots =$$

$$= hf + \frac{h^2}{2} Df + \frac{h^3}{4} \frac{\partial f}{\partial y} Df + \frac{h^4}{16} \frac{\partial f}{\partial y} D^2 f + \ldots$$

$$\ldots + \frac{h}{2} \left[\frac{h}{2} D + \frac{h^2}{4} Df \frac{\partial}{\partial y} + \ldots \right]^2 f + \frac{h^4}{48} D^3 f + \ldots$$

We obtain

$$k_3 = hf + \frac{h^2}{2} Df + \frac{h^3}{4} \frac{\partial f}{\partial y} Df + \frac{h^3}{8} D^2 f +$$

$$+ \frac{h^4}{16} \frac{\partial f}{\partial y} D^2 f + \frac{h^4}{8} Df \cdot D \frac{\partial f}{\partial y} + \frac{h^4}{48} D^3 f + \ldots \qquad (2.14)$$

We pass to the expansion of k_4. We have

$$k_4 = hf(x_0 + h, y_0 + k_3) = h \sum_{l=0}^{\infty} \frac{1}{l!} \left(h \frac{\partial}{\partial x} + k_3 \frac{\partial}{\partial y} \right)^l f.$$

In place of k_3, we substitute the right-hand side of (2.14). We get

$$k_4 = hf + h \left[h \frac{\partial}{\partial x} + \left(hf + \frac{h^2}{2} Df + \frac{h^3}{4} \frac{\partial f}{\partial y} Df + \right. \right.$$

$$\left. \left. + \frac{h^3}{8} D^2 f + \ldots \right) \frac{\partial}{\partial y} \right] f +$$

$$+ \frac{h}{2} \left[h \frac{\partial}{\partial x} + \left(hf + \frac{h^2}{2} Df + \ldots \right) \frac{\partial}{\partial y} \right]^2 f +$$

$$+ \frac{h}{6} \left[h \frac{\partial}{\partial x} + (hf + \ldots) \frac{\partial}{\partial y} \right]^3 f + \ldots = hf + h^2 Df +$$

$$+ \frac{h^3}{2} \frac{\partial f}{\partial y} Df + \frac{h^4}{4} \left(\frac{\partial f}{\partial y} \right)^2 Df + \frac{h^4}{8} \frac{\partial f}{\partial y} D^2 f + \ldots$$

$$\ldots + \frac{h}{2} \left[hD + \frac{h^2}{2} Df \frac{\partial}{\partial y} + \ldots \right]^2 f + \frac{h^4}{6} D^3 f + \ldots$$

Finally, we find that

$$k_4 = hf + h^2 Df + \frac{h^3}{2} \frac{\partial f}{\partial y} Df + \frac{h^3}{2} D^2 f + \frac{h^4}{4} \left(\frac{\partial f}{\partial y}\right)^2 Df +$$

$$+ \frac{h^4}{8} \frac{\partial f}{\partial y} D^2 f + \frac{h^4}{2} Df \cdot D \left(\frac{\partial f}{\partial y}\right) + \frac{h^4}{6} D^3 f + \dots \tag{2.15}$$

Now, we can write the first four terms of the expansion in Taylor series of the quantity

$$\tfrac{1}{6}(k_1 + 2k_2 + 2k_3 + k_4) =$$
$$= A_1 h + A_2 h^2 + A_3 h^3 + A_4 h^4 + \dots, \tag{2.16}$$

forming the linear combination indicated on the left-hand side of (2.16), with the help of formulas (2.12), (2.13), (2.14), (2.15). It is easy to convince oneself that the A_l, $l = 1, 2, 3, 4$, coincide with the corresponding coefficients of the expansion (2.11). For example,

$$A_3 = \frac{1}{6} \left[2 \cdot \tfrac{1}{8} D^2 f + 2 \left(\frac{1}{4} \frac{\partial f}{\partial y} Df + \tfrac{1}{8} D^2 f \right) + \right.$$

$$\left. + \frac{1}{2} \frac{\partial f}{\partial y} Df + \tfrac{1}{2} D^2 f \right] = \frac{1}{6} \left(D^2 f + \frac{\partial f}{\partial y} Df \right),$$

which coincides with the coefficient of h^3 in the expansion (2.11).

We have already pointed out above that if $f(x, y)$ does not depend on y and is a polynomial in x of degree not higher than three (in the case examined, this is equivalent to the solution $y(x)$ being a polynomial of a degree not higher than four), then the Runge-Kutta method gives an exact value of the solution $y(x)$ for any h. In general, when $f(x, y)$ depends on y and the solution $y(x)$ is a polynomial of degree not higher than four, not coinciding with a polynomial of the first degree, this situation does not occur.

The calculations for the Runge-Kutta method are shown in the diagram:

x	y	$hf(x, y)$	k
x_0	y_0	$k_1 = hf(x_0, y_0)$	
$x_0 + \dfrac{h}{2}$	$y_0 + \dfrac{k_1}{2}$	$k_2 = hf\left(x_0 + \dfrac{h}{2}, y_0 + \dfrac{k_1}{2}\right)$	$k = \tfrac{1}{6}(k_1 + 2k_2 + 2k_3 + k_4)$
$x_0 + \dfrac{h}{2}$	$y_0 + \dfrac{k_2}{2}$	$k_3 = hf\left(x_0 + \dfrac{h}{2}, y_0 + \dfrac{k_2}{2}\right)$	
$x_0 + h$	$y_0 + k_3$	$k_4 = hf(x_0 + h, y_0 + k_3)$	
$x_1 = x_0 + h$	$y_1 = y_0 + k$	$\dots$	$\dots$

The main advantage of the Runge-Kutta method in comparison with the difference methods presented below consists of the fact that its application does not require the construction of the so-called starting table. Another advantage of the Runge-Kutta method consists of the fact that one can change the side of the step h in the course of the calculation without this resulting in additional calculations.

At the same time, the Runge-Kutta method possesses essential deficiencies. It is more laborious than the method of differences, since at each step one has to calculate four values of the right hand side $f(x, y)$ of the differential equation. As we will see, in the case of difference methods, it is necessary at each step to calculate one value of $f(x, y)$ (the extrapolation method of Adams) or, what is practically as much, two or three values (interpolation methods). Furthermore, the computational scheme of the Runge-Kutta method does not contain in itself the elements of the control of the validity of the calculations.

The Runge-Kutta method can be applied to the Cauchy problem for systems of ordinary differential equations. We introduce the corresponding formulas for the case of the system of two equations:

$$\left.\begin{array}{l} y' = f(x, y, z), \\ z' = g(x, y, z), \end{array}\right\} \qquad y(x_0) = y_0, \qquad z(x_0) = z_0.$$

The number pairs $(k_j, l_j), j = 1, 2, 3, 4$ are calculated successively:

$$k_1 = hf(x_0, y_0, z_0),$$

$$k_2 = hf\left(x_0 + \frac{h}{2}, \; y_0 + \frac{k_1}{2}, \; z_0 + \frac{l_1}{2}\right),$$

$$k_3 = hf\left(x_0 + \frac{h}{2}, \; y_0 + \frac{k_2}{2}, \; z_0 + \frac{l_2}{2}\right),$$

$$k_4 = hf(x_0 + h, \; y_0 + k_3, \; z_0 + l_3).$$

$$l_1 = hg(x_0, y_0, z_0),$$

$$l_2 = hg\left(x_0 + \frac{h}{2}, \; y_0 + \frac{k_1}{2}, \; z_0 + \frac{l_1}{2}\right),$$

$$l_3 = hg\left(x_0 + \frac{h}{2}, \; y_0 + \frac{k_2}{2}, \; z_0 + \frac{l_2}{2}\right),$$

$$l_4 = hg(x_0 + h, \; y_0 + k_3, \; z_0 + l_3).$$

As the approximate values for $y(x_0+h)$ and $z(x_0+h)$ the following numbers are taken:

$$y(x_0+h) \cong y_0+k, \qquad z(x_0+h) \cong z_0+l,$$

where

$$k = \tfrac{1}{6}(k_1+2k_2+2k_3+k_4), \qquad l = \tfrac{1}{6}(l_1+2l_2+2l_3+l_4).$$

In the book [3] the generalized Runge-Kutta method is presented for the case of the Cauchy problem for the differential equation of the order n:

$$y^{(n)} = f(x, y, y', \ldots, y^{(n-1)}).$$

Example. We shall find the numerical solution of the Cauchy problem

$$y' = 0.25y^2+x^2, \qquad y(0) = -1 \tag{2.17}$$

on the interval [0, 1] by the Runge-Kutta method.

First of all, it is necessary to convince oneself that the solution of problem (2.17) exists on the interval [0,1].

We shall use the following theorem from the theory of the ordinary differential equations.

Picard's theorem. We assume that in the rectangle R

$$|x-x_0| \leqq a, \qquad |y-y_0| \leqq b$$

the function $f(x, y)$ is continuous as a function of the variables (x, y) and it satisfies the Lipschitz condition in y

$$|f(x, y')-f(x, y'')| \leqq L|y'-y''|.$$

Let

$$M = \max_{R} |f(x, y)|.$$

Then, the Cauchy problem

$$y' = f(x, y), \qquad y(x_0) = y_0$$

has a unique solution $y = y(x)$ on the interval

$$|x-x_0| \leqq c,$$

where $c = \min [a, b/M]$, and the graph of the solution is contained in the rectangle R.

Table 30

x	y	$0.25\, y$	$hf(x, y)$	$\dfrac{h}{2} f(x, y)$	$k_1 + 2k_2 + {} \atop {} + 2k_3 + k_4$	$\frac{1}{6}(k_1 + 2k_2 + {} \atop {} + 2k_3 + k_4)$
0	—1	—0.25	0.025	0.01250		
0.05	—0.98750	—0.21688	0.024629	0.01231	0.148317	0.02472
0.05	—0.98769	—0.24692	0.024638			
0.1	—0.97536	—0.24384	0.024783			
0.1	—0.97528	—0.24382	0.024779	0.01239		
0.15	—0.96289	—0.24072	0.025429	0.01271	0.153020	0.02550
0.15	—0.96257	—0.24064	0.025413			
0.2	—0.94987	—0.23747	0.026557			
0.2	—0.94978	—0.23745	0.026553	0.01328		
0.25	—0.93650	—0.23413	0.028176	0.01409	0.169417	0.02824
0.25	—0.93569	—0.23392	0.028138			
0.3	—0.92164	—0.23041	0.030236			
0.3	—0.92154	—0.23039	0.030231	0.01512		
0.35	—0.90642	—0.22661	0.032790	0.01640	0.197024	0.03284
0.35	—0.90514	—0.22629	0.032732			
0.4	—0.88881	—0.22220	0.035749			
0.4	—0.88870	—0.22218	0.035745	0.01787		
0.45	—0.87083	—0.21771	0.039209	0.01960	0.235475	0.03925
0.45	—0.86910	—0.21728	0.039134			
0.5	—0.84957	—0.21239	0.043044			
0.5	—0.84945	—0.21236	0.043039	0.02152		
0.55	—0.82793	—0.20698	0.047386	0.02369	0.284491	0.04742
0.55	—0.82576	—0.20644	0.047297			
0.6	—0.80215	—0.20054	0.052086			
0.6	—0.80203	—0.20051	0.052082	0.02604		
0.65	—0.77599	—0.19400	0.057304	0.02865	0.343965	0.05733
0.65	—0.74483	—0.18621	0.062869			
0.7	—0.74483	—0.18621	0.062869			
0.7	—0.74470	—0.18618	0.062865	0.03143		
0.75	—0.71327	—0.17832	0.068969	0.03448	0.413944	0.06899
0.75	—0.71022	—0.17756	0.068861			
0.8	—0.67584	—0.16896	0.075419			
0.8	—0.67571	—0.16893	0.075415	0.03771		
0.85	—0.63800	—0.15950	0.082426	0.04121	0.494700	0.08245
0.85	—0.63450	—0.15863	0.082315			
0.9	—0.59339	—0.14835	0.089803			
0.9	—0.59326	—0.14832	0.089800	0.04490		
0.95	—0.54836	—0.13709	0.097767	0.04888	0.586792	0.09780
0.95	—0.54438	—0.13610	0.097659			
1	—0.49560	—0.12390	0.106140			
1	—0.49546					

Picard's theorem allows to assert that problem (2.17) has a solution on the interval $|x| \leq 0.7$. In fact, let us take $a = 0.7, b = 1.2$. In the rectangle

$$|x| \leq 0.7 \qquad |y+1| \leq 1.2$$

the greatest value of the function $f(x, y) = 0.25y^2 + x^2$ is equal to

$$M = 0.25(1+b)^2 + a^2 = 1.7.$$

Since $b/M = 1.2/1.7 > 0.7$, then $c = 0.7$. We note that for a different choice of the parameters a and b, we do not obtain an essential increase in c.

However, one can show that the successive approximations, determined in the proof of Picard's theorem

$$y_{n+1}(x) = -1 + \int_0^x [0.25y_n^2(t) + t^2]dt, \qquad y_0(x) = -1,$$

for $x \geq 0$ belong to the rectangle

$$0 \leq x \leq 1.2 \qquad -1 \leq y \leq 0.$$

From this, it is not difficult to conclude the existence of a solution of the problem (2.17) on the interval $[0; 1.2]$.

We construct the numerical solution of problem (2.17) for the equidistant values of the argument with the step $h = 0.1$ to five decimal places. We take $h = 0.1$ for the following reason. We write the desired solution $y(x)$ in the form of a series in powers of x:

$$y(x) = y(0) + \frac{y'(0)}{1!}x + \frac{y''(0)}{2!}x^2 + \ldots$$

In order to obtain an approximate value of $y(0.1)$ with five significant decimal places by the help of this series, it is sufficient to retain the terms of the series containing the powers of x up to the fourth inclusively (see § 5, example 1). This gives the basis for assuming that the Runge-Kutta method, at least at the beginning of the calculation, will give values accurate to the fifth decimal place. The results of the calculations are shown in table 30.

3. On difference methods for the solution of the Cauchy problem

We present the main idea behind difference methods for the solution of the Cauchy problem in the case of one first order differential equation

$$y' = f(x, y), \qquad y(x_0) = y_0. \tag{3.1}$$

Let it be necessary to find a numerical solution of the problem (3.1) on the finite interval $[x_0, X]$, where $X > x_0$. We assume, of course, that the solution $y(x)$ of the problem (3.1) exists on the interval $[x_0, X]$. We also assume that in some closed region D of the (x, y) plane, containing the graph of the solution $y = y(x)$ on $[x_0, X]$, the function $f(x, y)$ is continuous as a function of the variables (x, y) and has continuous partial derivatives (including also mixed derivatives) up to the required order l. From this assumption it follows that the solution $y(x)$ has a continuous derivative of the order $l+1$ on the interval $[x_0, X]$.

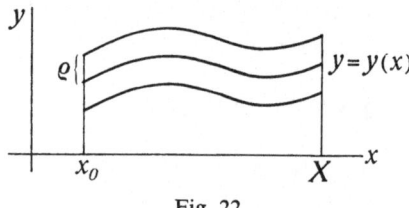

Fig. 22

We assume the region D to be convex in y. This means that if any two points (x, y') and (x, y'') belong to the region D, then the line segment connecting them belongs to D. For D we can take, for example, the curvilinear band (see fig. 22),

$$x_0 \leqq x \leqq X, \qquad -\rho \leqq y(x)-y \leqq \rho, \qquad \rho > 0$$

We consider that the assumptions about problem (3.1) are always satisfied.

Let $h > 0$ and

$$x_j = x_0+jh, \qquad j = 0, 1, 2, \ldots, N, \tag{3.2}$$

be equidistant points with spacing h, located in the interval $[x_0, X]$:

$$x_0 + Nh \leqq X < x_0+(N+1)h.$$

We call the number h the *step* of integration.*)

We assume that approximate values of the solution $y(x)$ are known at the points

$$x_0, x_1, \ldots, x_k,$$

---

* In speaking about the equidistant values of the argument, we do not take into account the possibility of a change in the value of the step h in the course of the numerical integration.

where $k < N$. We denote these values by

$$y_0, y_1, \ldots, y_k,$$ (3.3)

so that

$$y_j \cong y(x_j), \qquad j = 1, 2, \ldots, k.$$

The value y_0 is given and it coincides with $y(x_0)$. If $k \geq 1$, then $y_1, y_2, \ldots, y_k$ have to be calculated by some other method, for example, by the Runge-Kutta method.

The numbers (3.3), which can be written in the form of the table

x	x_0	x_1	$\cdots$	x_k
y	y_0	y_1	$\cdots$	y_k

are called the initial values or the start of the table. We shall examine methods for construction of the start of the table in § 5.

Difference methods allow the extension of the start of the table: to calculate approximate values of $y(x)$ at the points (3.2) for $j \geq k+1$. These approximate values, as well as the initial values (3.3), we denote by y_j:

$$y_j \cong y(x_j), \qquad j = k+1, k+2, \ldots, N.$$

Let the already calculated values by

$$y_0, y_1, \ldots, y_n$$ (3.4)

with $n \geq k$. The basis of the calculation of y_{n+1} by difference methods is algebraic interpolation. We construct the interpolation polynomial for the function $y'(x)$ — the derivative of the solution $y(x)$ of the problem (3.1) — from its values at the following $k+p+1$ knots:

$$x_{n-k}, x_{n-k+1}, \ldots, x_n, x_{n+1}, \ldots, x_{n+p}.$$ (3.5)

Here

$$n+p < N.$$

We take the interpolation polynomial in the form of Lagrange

$$P(x) = \sum_{j=-p}^{k} \frac{\omega(x)}{(x-x_{n-j})\omega'(x_{n-j})} y'(x_{n-j}),$$ (3.6)

where

$$\omega(x) = (x-x_{n-k})(x-x_{n-k+1})\ldots(x-x_{n+p}).$$

We have

$$y'(x) = P(x)+r(x), \tag{3.7}$$

where $r(x)$ is the remainder term of interpolation.

If $f(x, y)$ has continuous partial derivatives in the region D up to the order $l = k+p+1$ inclusive, then the remainder term can be represented in the form

$$r(x) = \frac{\omega(x)}{(k+p+1)!} y^{(k+p+2)}(\eta). \tag{3.8}$$

Since we will use the representation (3.8) on the interval $[x_n, x_{n+1}]$, then we consider that $x_{n-k} \leqq \eta \leqq x_{n+p}$ for $p \geqq 1$ and $x_{n-k} \leqq \eta \leqq x_{n+1}$ for $p = 0$.

In the formula

$$y(x_{n+1}) = y(x_n)+\int_{x_n}^{x_{n+1}} y'(x)dx \tag{3.9}$$

we replace $y'(x)$ by the right-hand side of equation (3.7). We get

$$y(x_{n+1}) = y(x_n)+\int_{x_n}^{x_{n+1}} P(x)dx+\int_{x_n}^{x_{n+1}} r(x)dx. \tag{3.10}$$

We transform the first of the integrals on the right-hand side of the equation (3.10). By virtue of (3.6), we have

$$\int_{x_n}^{x_{n+1}} P(x)dx = \sum_{j=-p}^{k} I_j y'(x_{n-j}), \tag{3.11}$$

where

$$I_j = \int_{x_n}^{x_n+h} \frac{\omega(x)}{(x-x_{n-j})\omega'(x_{n-j})} dx. \tag{3.12}$$

We make the change of the variable of integration $x = x_n+ht$ in the integral (3.12). If we set

$$\sigma(t) = (t+k)(t+k-1)\ldots t(t-1)\ldots(t-p), \tag{3.13}$$

then we get

$$\omega(x) = h^{k+p+1}\sigma(t),$$
$$\omega'(x_{n-j}) = h^{k+p}\sigma'(-j),$$
$$x - x_{n-j} = h(t+j).$$

Thus, we find that

$$I_j = h\int_0^1 \frac{\sigma(t)}{(t+j)\sigma'(-j)}\,dt. \tag{3.14}$$

We set

$$b_{kj}^{(p)} = \int_0^1 \frac{\sigma(t)}{(t+j)\sigma'(-j)}\,dt. \tag{3.15}$$

We note that the numbers $b_{kj}^{(p)}$ do not depend on n and h. On the basis of (3.14) and (3.15), equation (3.11) can be written in the form

$$\int_{x_n}^{x_{n+1}} P(x)dx = h\sum_{j=-p}^{k} b_{kj}^{(p)}y'(x_{n-j}). \tag{3.16}$$

We transform the second integral of the right-hand side of (3.10):

$$R_{n,k}^{(p)} = \int_{x_n}^{x_{n+1}} r(x)dx. \tag{3.17}$$

For $r(x)$ we substitute the right-hand side of equation (3.8), and we make the change of the variable of integration $x = x_n + ht$. We obtain

$$R_{n,k}^{(p)} = \frac{h^{k+p+2}}{(k+p+1)!}\int_0^1 \sigma(t)y^{(k+p+2)}(\eta)dt.$$

Since $\sigma(t)$ does not change sign on the interval of integration and $y^{(k+p+2)}$ is continuous on $[x_0, X]$, then the last equation can be rewritten thus:

$$R_{n,k}^{(p)} = \frac{h^{k+p+2}}{(k+p+1)!}y^{(k+p+2)}(\xi)\int_0^1 \sigma(t)dt, \tag{3.18}$$

where $x_{n-k} \leq \xi \leq x_{n+p}$ with $p \geq 1$ and $x_{n-k} \leq \xi \leq x_{n+1}$ with $p = 0$. Now, formula (3.10), by virtue of (3.16) and (3.17), is written in the following manner:

$$y(x_{n+1}) = y(x_n) + h\sum_{j=-p}^{k} b_{kj}^{(p)}y'(x_{n-j}) + R_{n,k}^{(p)}, \tag{3.19}$$

where the numbers $b_{kj}^{(p)}$ are defined by equation (3.15) and the remainder term $R_{n,k}^{(p)}$ has the representation (3.18), if $f(x, y)$ has continuous partial derivatives up to the order $l = k+p+1$ on D.

From formula (3.19), it is possible to obtain different difference methods for the solution of the Cauchy problem. The methods obtained here for the case $p \geq 1$ are called interpolated. The method corresponding to $p = 0$ is called extrapolated. The reason for these names is the following. The smallest interval containing the knots (3.5), on the basis of which the interpolation polynomial $P(x)$ is constructed, is $[x_{n-k}, x_{n+p}]$. In the case $p = 0$, this interval is outside the interval $[x_n, x_{n+1}]$, on which the integral (3.11) of $P(x)$ is calculated. The values of the polynomial are extrapolated onto the interval $[x_n, x_{n+1}]$. In the case $p \geq 1$, the interval $[x_n, x_{n+1}]$ is contained inside $[x_{n-k}, x_{n+p}]$, consequently, here we deal with interpolation in the proper sense of this word.*

We examine the extrapolation method first. We write formula (3.19) for $p = 0$:

$$y(x_{n+1}) = y(x_n) + h \sum_{j=0}^{k} b_{kj} y'(x_{n-j}) + R_{n,k}. \tag{3.20}$$

The coefficients $b_{kj} = b_{kj}^{(0)}$ are determined by formula (3.15) with $p = 0$:

$$b_{kj} = (-1)^j \int_0^1 \frac{t(t+1)\ldots(t+k)}{(t+j)j!(k-j)!} dt, \qquad j = 0, 1, 2, \ldots, k. \tag{3.21}$$

The remainder term $R_{n,k} = R_{n,k}^{(0)}$ can be written by formula (3.18) with $p = 0$:

$$R_{n,k} = \frac{h^{k+2}}{(k+1)!} y^{(k+2)}(\xi) \int_0^1 t(t+1)\ldots(t+k)dt,$$

$$x_{n-k} \leq \xi \leq x_{n+1}. \tag{3.22}$$

We cite the numerical values of the coefficients b_{kj} for $k = 0, 1, 2, 3$:

$b_{00} = 1,$

$b_{10} = \frac{3}{2}, \qquad b_{11} = -\frac{1}{2},$

$b_{20} = \frac{23}{12}, \qquad b_{21} = -\frac{4}{3}, \qquad b_{22} = \frac{5}{12},$

$b_{30} = \frac{55}{24}, \qquad b_{31} = -\frac{59}{24}, \qquad b_{32} = \frac{37}{24}, \qquad b_{33} = -\frac{3}{8}.$

---

* Extrapolation methods are frequently called *explicit* methods, and interpolation methods are also referred to as *implicit* methods. TR.

It is obvious from formula (3.21) that the numbers b_{kj} are positive for even j, and are negative for odd j. In particular,

$$b_{k0} > 0, \qquad k = 0, 1, 2, \ldots \tag{3.23}$$

Formula (3.20) is useless for calculation, since the unknown remainder term $R_{n,k}$ enters into it, the values of the first derivative of the desired solution

$$y'(x_{n-k}), \qquad y'(x_{n-k+1}), \ldots, y'(x_n) \tag{3.24}$$

and $y(x_n)$. If the exact values

$$y(x_{n-k}), \qquad y(x_{n-k+1}), \ldots, y(x_n),$$

were known, then with the help of the differential equation (3.1) we could find the exact values (3.24):

$$y'(x_{n-j}) = f(x_{n-j}, y(x_{n-j})), \qquad j = 0, 1, 2, \ldots, k.$$

However, only the approximate values (3.24) of the desired solution are known to us

$$y_{n-k}, y_{n-k+1}, \ldots, y_n \qquad (n \geq k),$$

and with their help we can find the approximate values of the derivative. We denote these approximate values by y'_{n-j}:

$$y'_{n-j} = f(x_{n-j}, y_{n-j}), \qquad j = 0, 1, 2, \ldots, k. \tag{3.25}$$

We substitute into (3.20) for the derivatives (3.24), their approximate values (3.25), for the value $y(x_n)$, its approximate value y_n, and we discard the remainder term $R_{n,k}$. We get the approximate equation

$$y(x_{n+1}) \cong y_n + h \sum_{j=0}^{k} b_{kj} f(x_{n-j}, y_{n-j}).$$

The right-hand side of the equation we accept as y_{n+1}:

$$y_{n+1} = y_n + h \sum_{j=0}^{k} b_{kj} f(x_{n-j}, y_{n-j}). \tag{3.26}$$

In (3.26) we set $n = k$ and substituting into the right hand side the known values (3.3) $y_0, y_1, \ldots, y_k$, we find y_{k+1}. In (3.26) we set $n = k+1$ and substituting $y_1, y_2, \ldots, y_k$ and the already found value y_{k+1}, we obtain y_{k+2}, and so forth. Thus, formula (3.26) allows one to find y_n successively for $n = k+1, k+2, \ldots, N$. The method of the numerical solution of the Cauchy problem defined by formula (3.26) is called the *Adams extrapolation method*.

Formula (3.26) is a so-called difference equation or an equation in finite differences (on the subject of difference equations see [7], chapter V). The Adams extrapolation method can be interpreted as a method for construction of solutions y_n of this difference equation, having for $n = 0, 1, 2, \ldots, k$ the given values (3.3).

If the solution $y(x)$ of problem (3.1) is a polynomial of degree not higher than $k+1$, then the values $y(x_n)(n = 0, 1, 2, \ldots, N)$ satisfy the difference equation (3.26). In fact, by virtue of (3.22), $R_{n,k} = 0$, and this assertion follows from equation (3.20).

We examine the interpolation method, corresponding to $p = 1$. We write formula (3.19) with $p = 1$:

$$y(x_{n+1}) = y(x_n) + h \sum_{j=-1}^{k} b_{kj}^* y'(x_{n-j}) + R_{n,k}^*. \tag{3.27}$$

The coefficients $b_{kj}^* = b_{kj}^{(1)}$ are given by formula (3.15) with $p = 1$:

$$b_{kj}^* = (-1)^{j+1} \int_0^1 \frac{(t-1)t(t+1) \ldots (t+k)}{(t+j)(j+1)!(k-j)!} \, dt,$$

$$j = -1, 0, 1, 2, \ldots, k. \tag{3.28}$$

The representation of the remainder term $R_{n,k}^* = R_{n,k}^{(1)}$ is obtained from (3.18) with $p = 1$:

$$R_{n,k}^* = \frac{h^{k+3}}{(k+2)!} y^{(k+3)}(\xi) \int_0^1 (t-1)t(t+1) \ldots (t+k)dt,$$

$$x_{n-k} \leqq \xi \leqq x_{n+1}. \tag{3.29}$$

We remark that formulas (3.27), (3.28) and (3.29) also make sense for $k = -1$, if one agrees to consider that $x_{n-k} = x_n$ for $k = -1$ and

$$t(t+1) \ldots (t+k) = 1 \quad \text{for} \quad k = -1.$$

In the case $k = -1$ the interpolation polynomial of the function $y'(x)$ is constructed using the one knot x_{n+1}.

We cite the numerical values of coefficients b_{kj}^* for $k = -1, 0, 1, 2$:

$$b^*_{-1,-1} = 1,$$

$$b^*_{0,-1} = \tfrac{1}{2}, \qquad b^*_{00} = \tfrac{1}{2},$$

$$b^*_{1,-1} = \tfrac{5}{12}, \qquad b^*_{10} = \tfrac{2}{3}, \qquad b^*_{11} = -\tfrac{1}{12},$$

$$b^*_{2,-1} = \tfrac{3}{8}, \qquad b^*_{20} = \tfrac{19}{24}, \qquad b^*_{21} = -\tfrac{5}{24}, \qquad b^*_{22} = \tfrac{1}{24}.$$

From formula (3.28) we get

$$b^*_{k,-1} > 0, \qquad k = -1, 0, 1, 2, \ldots \tag{3.30}$$

Furthermore, for $j \geqq 0$ the coefficients b^*_{kj} are positive, if j is even, and they are negative, if j is odd. In particular,

$$b^*_{k0} > 0, \qquad k = 0, 1, 2, \ldots \tag{3.31}$$

In order to make formula (3.27) useful for calculation, we discard the remainder term $R^*_{n,k}$, we replace $y'(x_{n-j})$ by

$$y'_{n-j} = f(x_{n-j}, y_{n-j})$$

for $j = -1, 0, 1, 2, \ldots k$ and $y(x_n)$ by y_n. We get

$$y_{n+1} = y_n + h \sum_{j=-1}^{k} b^*_{kj} f(x_{n-j}, y_{n-j}). \tag{3.32}$$

Formula (3.32) does not give an explicit expression for y_{n+1}, since y_{n+1} enters into the term on the right-hand side, corresponding to $j = -1$. Thus, (3.32) represents an equation for y_{n+1}. Usually, this equation is solved by the method of iteration. The method of numerical solution of the Cauchy problem, defined by formula (3.32), is called the *Adams interpolation method*.

If $y(x)$ is an polynomial of degree not higher than $k+2$, then, as follows from (3.29) and (3.27), the values $y(x_n)$ satisfy the difference equation (3.32).

We compare the Adams extrapolation and interpolation methods. Here, one should compare formula (3.26) with the formula

$$y_{n+1} = y_n + h \sum_{j=-1}^{k-1} b^*_{k-1,j} f(x_{n-j}, y_{n-j}), \tag{3.33}$$

which is obtained from (3.32) by the substitution of k for $k-1$, since in constructing these formulas, the same number $k+1$ of knots is used for the interpolation of the function $y'(x)$. Namely, in the case of formula

(3.26) the following knots are used

$$x_{n-k}, \qquad x_{n-k+1}, \ldots, x_n, \tag{3.34}$$

and in the case of formula (3.33),

$$x_{n-k+1}, \qquad x_{n-k+2}, \ldots, x_n, \qquad x_{n+1}. \tag{3.35}$$

The approximation of the function $y'(x)$ at the points of the interval $[x_n, x_{n+1}]$ by the interpolation polynomial, constructed according to the knots (3.35), will be better than the approximation by the polynomial, constructed according to the knots (3.34) (see § 6 chapter II). In this case the Adams interpolation method is more accurate than the extrapolation one.

The most accurate, in the sense indicated, is the interpolation method which we obtain for $p = k+1$. In formula (3.19) set $p = k+1$.

$$y(x_{n+1}) = y(x_n) + h \sum_{j=-k-1}^{k} \tilde{b}_{kj} y'(x_{n-j}) + \tilde{R}_{n,k}. \tag{3.36}$$

The coefficients $\tilde{b}_{kj} = b_{kj}^{(k+1)}$ are determined by equation (3.15) with $p = k+1$. The remainder term $\tilde{R}_{n,k} = R_{n,k}^{(k+1)}$ is written according to formula (3.18):

$$\tilde{R}_{n,k} = \frac{h^{2k+3}}{(2k+2)!} y^{(2k+3)}(\xi) \int_0^1 (t+k)(t+k-1) \ldots$$

$$\ldots t(t-1) \ldots (t-k-1)dt, \tag{3.37}$$

$$x_{n-k} \leqq \xi \leqq x_{n+k+1}.$$

From (3.26) we get the formula

$$y_{n+1} = y_n + h \sum_{j=-k-1}^{k} \tilde{b}_{kj} f(x_{n-j}, y_{n-j}). \tag{3.38}$$

Formula (3.38) represents an equation with $k+1$ unknowns

$$y_{n+1}, \qquad y_{n+2}, \ldots, y_{n+k+1}.$$

The method of the numerical solution of the Cauchy problem, defined by formula (3.38), is called the Cowell type method. We shall examine this method in § 7.

4. The Adams extrapolation method

The formula for the difference methods, which we obtained in § 3, are often applied in modified form in calculations. The modifications reduce to the fact that the values of the function $f(x_j, y_j)$ are replaced by finite differences. In the present paragraph, the modified formula for the Adams extrapolation method is introduced and the scheme of calculations according to this formula is indicated.

We write formula (3.26) for the Adams extrapolation method

$$\Delta y_n = h \sum_{j=0}^{k} b_{kj} f(x_{n-j}, y_{n-j}). \tag{4.1}$$

This formula was established in the following manner. The interpolation polynomial $P(x)$ was constructed from the values of function $y'(x)$ at the points

$$x_{n-k}, \; x_{n-k+1}, \ldots, x_n.$$

Furthermore, the integral of $P(x)$ was calculated over the interval $[x_n, x_{n+1}]$ and in the expression obtained, values of $y'(x_{n-j})$ were replaced by the following values

$$y'_{n-j} = f(x_{n-j}, y_{n-j}), \qquad j = 0, 1, 2, \ldots k.$$

Obviously, we obtain the same result if we calculate the integral over the interval $[x_n, x_{n+1}]$ of the interpolation polynomial $Q(x)$, constructed by the following conditions

$$Q(x_{n-j}) = y'_{n-j}, \qquad j = 0, 1, 2, \ldots, k. \tag{4.2}$$

Thus, formula (4.1) can be written in the form

$$\Delta y_n = \int_{x_n}^{x_{n+1}} Q(x) dx$$

or, if one makes the change of the variable of integration $x = x_n + ht$,

$$\Delta y_n = h \int_0^1 Q(x_n + ht) dt, \tag{4.3}$$

where $Q(x)$ is the interpolation polynomial, constructed according to the conditions (4.2).

Since the interval $[x_n, x_{n+1}]$, on which we use the values of the interpolation polynomial $Q(x)$, occurs at the end of the table of values

$$y'_{n-k}, \qquad y'_{n-k+1}, \ldots, y'_n,$$

then it is natural to write $Q(x)$ according to Newton's formula for interpolation at the end of the table [see (7.12), chapter II]

$$Q(x_n+ht) = y'_n + \frac{t}{1!} \Delta y'_{n-1} + \frac{t(t+1)}{2!} \Delta^2 y'_{n-2} + \ldots$$

$$\ldots + \frac{t(t+1)\ldots(t+k-1)}{k!} \Delta^k y'_{n-k}. \tag{4.4}$$

We introduce the notation

$$a_j = \frac{1}{j!} \int_0^1 t(t+1)\ldots(t+j-1)dt, \qquad j = 0, 1, 2, \ldots \tag{4.5}$$

The numbers a_j do not depend on k. We write out some of the first of the numbers a_j:

$$a_0 = 1, \qquad a_1 = \tfrac{1}{2}, \qquad a_2 = \tfrac{5}{12}, \qquad a_3 = \tfrac{3}{8},$$

$$a_4 = \tfrac{251}{720}, \qquad a_5 = \tfrac{95}{288}, \qquad a_6 = \tfrac{19\,087}{60\,480}.$$

We integrate both sides of equation (4.4) over the interval $[0, 1]$, and use the notation (4.5). We get

$$\int_0^1 Q(x_n+ht)dt = y'_n + \tfrac{1}{2}\Delta y'_{n-1} + \tfrac{5}{12}\Delta^2 y'_{n-2} + \ldots + a_k \Delta^k y'_{n-k}.$$

Comparing this equality with equality (4.3), we find that

$$\Delta y_n = h(y'_n + \tfrac{1}{2}\Delta y'_{n-1} + \tfrac{5}{12}\Delta^2 y'_{n-2} + \ldots + a_k \Delta^k y'_{n-k}). \tag{4.6}$$

This is the formula for the Adams extrapolation method, written in terms of finite differences.
We set

$$\eta_j = hf(x_j, y_j), \qquad j = 0, 1, 2, \ldots \tag{4.7}$$

Formula (4.6) is written in the following manner in this notation:

$$\Delta y_n = \eta_n + \tfrac{1}{2}\Delta \eta_{n-1} + \tfrac{5}{12}\Delta^2 \eta_{n-2} + \tfrac{3}{8}\Delta^3 \eta_{n-3} + \ldots + a_k \Delta^k \eta_{n-k}. \tag{4.8}$$

Formula (4.8) is usually applied in calculations.

For $k = 0$, formula (4.8) takes the form

$$\Delta y_n = hf(x_n, y_n). \tag{4.9}$$

The application of this formula does not require the construction of the beginning of the table. The method of the numerical solution of the Cauchy problem, defined by formula (4.9), is known as the *Euler method*.

The calculations by formula (4.8) are assumed to be laid out according to the following scheme (for the sake of definiteness, we assume $k = 4$):

x	y	Δy	$\eta = hf(x, y)$	$\Delta \eta$	$\Delta^2 \eta$	$\Delta^3 \eta$	$\Delta^3 \eta$
x_0	y_0		η_0				
		Δy_0		$\Delta \eta_0$			
x_1	y_1		η_1		$\Delta^2 \eta_0$		
		Δy_1		$\Delta \eta_1$		$\Delta^3 \eta_0$	
x_2	y_2		η_2		$\Delta^2 \eta_1$		$\Delta^4 \eta_0$
		Δy_2		$\Delta \eta_2$		$\Delta^3 \eta_1$	
x_3	y_3		η_3		$\Delta^2 \eta_2$		
		Δy_3		$\Delta \eta_3$			
x_4	y_4		η_4				
x_5							

All the quantities entered in this table are known, since it is supposed that the beginning of the table has already been constructed.

Formula (4.8) for $k = 4$ is written thusly:

$$\Delta y_n = \eta_n + \tfrac{1}{2}\Delta\eta_{n-1} + \tfrac{5}{12}\Delta^2\eta_{n-2} + \tfrac{3}{8}\Delta^3\eta_{n-3} + \tfrac{251}{720}\Delta^4\eta_{n-4}. \tag{4.10}$$

In (4.10), we set $n = 4$:

$$\Delta y_4 = \eta_4 + \tfrac{1}{2}\Delta\eta_3 + \tfrac{5}{12}\Delta^2\eta_2 + \tfrac{3}{8}\Delta^3\eta_1 + \tfrac{251}{720}\Delta^4\eta_0.$$

All the quantities entering on the right-hand side of this formula are located on the lower diagonal of the table of finite differences, so we can find Δy_4 and, consequently, y_5. From y_5, we calculate $\eta_5 = hf(x_5, y_5)$ and we add a new diagonal in the table of finite differences. Assuming that $n = 5$ in (4.10), we find Δy_5, and so forth.

5. The construction of the beginning of the table

We shall study the question of the calculation of the starting values of the solution of the Cauchy problem (3.1),

$$y' = f(x, y), \qquad y(x_0) = y_0. \tag{5.1}$$

It is necessary to find the approximate values of the solution $y(x)$ at the points $x_j = x_0 + jh, j = 1, 2, \ldots, m$:

$$y_1, y_2, \ldots, y_m, \tag{5.2}$$

where $m \geq k \geq 1$.

Let $f(x, y)$ be expanded in the series

$$f(x, y) = \sum_{i, j} A_{ij}(x-x_0)^i(y-y_0)^j, \tag{5.3}$$

convergent for

$$|x-x_0| < R_1, \qquad |y-y_0| < R_2 \qquad (R_1 > 0, R_2 > 0). \tag{5.4}$$

Then, according to Cauchy's theorem in the theory of differential equations, the problem (5.1) has the solution $y(x)$, represented by the series

$$y(x) = \sum_{i=0}^{\infty} \frac{y^{(i)}(x_0)}{i!} (x-x_0)^i. \tag{5.5}$$

convergent in some neighborhood of the point x_0. The derivatives $y^{(i)}(x_0)$ can be found successively with the help of the differential equation and the initial condition:

$$y'(x_0) = f(x_0, y_0),$$

$$y''(x_0) = \frac{\partial f(x_0, y_0)}{\partial x} + \frac{\partial f(x_0, y_0)}{\partial y} f(x_0, y_0)$$

$$\cdots \cdots \cdots \cdots \cdots \cdots \cdots \cdots \cdots \cdots \cdots$$

Cauchy's theorem allows one also to indicate the neighborhood of the point x_0, in which the series (5.5) converges. Namely, by M we denote a constant such that

$$|f(x, y)| \leq M$$

with

$$|x-x_0| \leq r_1 < R_1, \qquad |y-y_0| \leq r_2 < R_2,$$

where R_1 and R_2 are the numbers, defining the region (5.4) of the convergence of the series (5.3), and r_1 and r_2 are some positive numbers. Then one can assert that series (5.5) converges for

$$|x-x_0| < r,$$

where

$$r = r_1(1-e^{-(1/2M)(r_2/r_1)}). \tag{5.6}$$

The indicated value r, as a rule, turns out to be considerably smaller than the radius of convergence of the series (5.5).

The construction of the beginning of the table in the case considered is thus realized. A segment of the series (5.5) is constructed,

$$y(x) \cong \sum_{i=0}^{l} \frac{y^{(i)}(x_0)}{i!} (x-x_0)^i$$

and with its help the numbers (5.2) are found:

$$y_j = \sum_{i=0}^{l} \frac{y^{(i)}(x_0)}{i!} (jh)^i, \qquad j = 1, 2, \ldots, m.$$

Of course, the number l should be taken so that the error

$$y(x_j)-y_j = \sum_{i=l+1}^{\infty} \frac{y^{(i)}(x_0)}{i!} (x_j-x_0)^i, \qquad j = 1, 2, \ldots, m,$$

does not exceed in absolute value half a unit in the last decimal place retained in the value of y_j.

In constructing the beginning of the table, the problem of the selection of the value of the step h should be solved. The step should be selected so that the highest difference in formula (4.8), $\Delta^k \eta_i (i = 0, 1, 2, \ldots, m-k)$ is constant to within a few units in the last place. In order to convince oneself of the constancy of $\Delta^k \eta_i$, one should take $m > k$, i.e. construct the beginning of the table with more values (by one or two) than is required for the application of the formula (4.8). One should also take into account that application of formula (4.8) is suitable with small $k(k = 2, 3, 4, 5)$.

Example 1. To select the step of integration and to construct the beginning of the table for the Cauchy problem (2.17)

$$y' = 0.25y^2+x^2, \qquad y(0) = -1. \tag{5.7}$$

We calculate the values y_j with five correct decimal places.
We find the derivatives from the differential equation:

$$y'' = \tfrac{1}{2}yy' + 2x,$$

$$y''' = \tfrac{1}{2}(y'^2 + yy'') + 2,$$

$$y^{(IV)} = \tfrac{3}{2}y'y'' + \tfrac{1}{2}yy''',$$

$$y^{(V)} = \tfrac{3}{2}y''^2 + 2y'y''' + \tfrac{1}{2}yy^{(IV)}.$$

Successively, we determine

$$y'(0) = \tfrac{1}{4}, \qquad y''(0) = -\tfrac{1}{8}, \qquad y'''(0) = \tfrac{67}{32}, \qquad y^{(IV)}(0) = -\tfrac{35}{32},$$
$$y^{(V)}(0) = \tfrac{207}{128}$$

and we write the first six terms of the series

$$y(x) = -1 + \tfrac{1}{4}x - \tfrac{1}{16}x^2 + \tfrac{67}{192}x^3 - \tfrac{35}{768}x^4 + \tfrac{69}{5120}x^5 - \dots \tag{5.8}$$

The Cauchy theorem guarantees the convergence of the series (5.8) on the interval $|x| \le 0.27$. In fact, for x and y satisfying the inequalities

$$|x| \le 0.5|y+1| \le 1,$$

$$|f(x, y)| = |0.25[(y+1)-1]^2 + x^2| \le 0.25(|y+1|+1)^2 + |x|^2 \le 1.25$$

so that from formula (5.6) with $r_1 = 0.5, r_2 = 1$ and $M = 1.25$, we obtain

$$r = 0.5(1 - e^{-0.8}) \cong 0.275.$$

It is natural to select a number h with one or two significant figures, for example 0.02, 0.05, 0.1, 0.25, 0.3, 0.5. We calculate $y(0.1)$ with the help of (5.8). We get

$$y(0.1) = -1 + 0.025 - 0.000625 + 0.000349 - 0.000005 = -0.97528.$$

It is obvious that it is sufficient to take four terms of the series in order to obtain five decimal placec in the value of $y(0.1)$. From here we conclude that $y(x)$ is approximated by a polynomial of the third degree on the interval $[-0.1; 0.1]$ to an accuracy of $5 \cdot 10^{-6}$.

Differentiating the series (5.8), we find that the derivative $y'(x)$ can be approximated by a polynomial of third degree on the interval $[-0.1; 0.1]$ with the same accuracy, here the leading term of the polynomial, $\tfrac{35}{192}x^3$, influences only the fourth decimal place.

Furthermore, one can conduct the following heuristic reasoning. It is highly probable that $y'(x)$ can be approximated with an accuracy up to $5 \cdot 10^{-6}$ by a polynomial of a third degree on the interval $[0; 0.2]$; here, the leading term ax^3 of the polynomial will be of the order of magnitude of a few units in the fourth decimal place. But then, $hy'(x)$ on the interval $[0; 0.2]$, if one takes $h = 0.05$, will be given accurately up to some units in the fifth decimal place by a polynomial of second degree. Therefore, we are right to have confidence that by selecting $h = 0.05$ in the table of finite differences of the quantity (4.7)

$$\eta_j = hf(x_j, y_j), \quad j = 0, 1, 2, 3, 4,$$

the second differences will be constant to five decimal places.

The values $y(j \cdot 0.05)$, $j = 1, 1, 3, 4$, calculated to five decimal places by (5.8), are shown in table 31. In the table the values $\eta_j = hf(x_j, y_j)$ and their differences are also indicated. The values η_j are given with six decimal places, since in what follows, in the numerical solution of problem (5.7) by Adams method, we shall need an extra place. It is obvious from the table that the second differences $\Delta^2 \eta_j (j = 0, 1, 2)$ are constant to five decimal places.

Table 31

x	y	η	$\Delta\eta$	$\Delta^2\eta$	$\Delta^3\eta$
0	—1	0.012500			
			—183		
0.05	—0.98761	0.012317		256	
			73		—8
0.1	—0.97528	0.012390		248	
			321		—4
0.15	—0.96275	0.012711		244	
			565		
0.2	—0.94798	0.013276			

We conclude with the method of construction of the beginning of the table, proposed by A. N. Krylov [16]. A. N. Krylov's method is also useful in the case that the right-hand side $f(x, y)$ of the differential equation is given tabularly.

We assume that we have already selected the step of integration h and that the highest difference $\Delta^k \eta_j$ in formula (4.8) is constant. For the sake of simplicity we assume that $k = 3$. We rewrite formula (4.8) for this case

$$\Delta y_n = \eta_n + \tfrac{1}{2}\Delta\eta_{n-1} + \tfrac{5}{12}\Delta^2\eta_{n-2} + \tfrac{3}{8}\Delta^3\eta_{n-3}.$$

We set here $n = 0, 1, 2$:

$$\Delta y_0 = \eta_0 + \tfrac{1}{2}\Delta\eta_{-1} + \tfrac{5}{12}\Delta^2\eta_{-2} + \tfrac{3}{8}\Delta^3\eta_{-3},$$
$$\Delta y_1 = \eta_1 + \tfrac{1}{2}\Delta\eta_0 + \tfrac{5}{12}\Delta^2\eta_{-1} + \tfrac{3}{8}\Delta^3\eta_{-2}, \tag{5.9}$$
$$\Delta y_2 = \eta_2 + \tfrac{1}{2}\Delta\eta_1 + \tfrac{5}{12}\Delta^2\eta_0 + \tfrac{3}{8}\Delta^3\eta_{-1}.$$

From the condition of constancy of the third differences,

$$\Delta^3\eta_{-3} = \Delta^3\eta_{-2} = \Delta^3\eta_{-1} = \Delta^3\eta_0$$

We find

$$\Delta^2\eta_{-1} = \Delta^2\eta_0 - \Delta^3\eta_{-1} = \Delta^2\eta_0 - \Delta^3\eta_0,$$
$$\Delta^2\eta_{-2} = \Delta^2\eta_{-1} - \Delta^3\eta_{-2} = \Delta^2\eta_0 - 2\Delta^3\eta_0,$$
$$\Delta\eta_{-1} = \Delta\eta_0 - \Delta^2\eta_{-1} = \Delta\eta_0 - \Delta^2\eta_0 + \Delta^3\eta_0.$$

Substituting the values found for the differences with negative indices into equations (5.9), we get

$$\Delta y_0 = \eta_0 + \tfrac{1}{2}\Delta\eta_0 - \tfrac{1}{12}\Delta^2\eta_0 + \tfrac{1}{24}\Delta^3\eta_0,$$
$$\Delta y_1 = \eta_1 + \tfrac{1}{2}\Delta\eta_0 + \tfrac{5}{12}\Delta^2\eta_0 - \tfrac{1}{24}\Delta^3\eta_0, \tag{5.10}$$
$$\Delta y_2 = \eta_2 + \tfrac{1}{2}\Delta\eta_1 + \tfrac{5}{12}\Delta^2\eta_0 + \tfrac{3}{8}\Delta^3\eta_0.$$

One may consider equations (5.10) to be a system of, generally speaking, non-linear equations for the unknown y_1, y_2, y_3. Krylov's method is an original method of iteration for the solution of this system. Let y_0 and $\eta_0 = hf(x_0, y_0)$ be known to us. The first of equations (5.10) gives an approximate equation

$$\Delta y_0 \cong \eta_0,$$

from which we find

$$y_1^{(1)} = y_0 + \eta_0.$$

From $y_1^{(1)}$, one can find an approximate value of η_1 for this approximate value of y_1,

$$\eta_1^{(1)} = hf(x_1, y_1^{(1)})$$

and, consequently, an approximate value of the difference $\Delta\eta_0$

$$\Delta\eta_0^{(1)} = \eta_1^{(1)} - \eta_0.$$

Thus, we have at our disposal the quantities

$$
\begin{array}{l}
\eta_0 \\
\qquad \Delta\eta_0^{(1)} \\
\eta_1^{(1)}
\end{array}
\tag{5.11}
$$

From the first two equations of (5.10), keeping two terms in each of them and substituting into these terms the approximate values (5.11), we find

$$\Delta y_0^{(2)} = \eta_0 + \tfrac{1}{2}\Delta\eta_0^{(1)},$$
$$\Delta y_1^{(2)} = \eta_1^{(1)} + \tfrac{1}{2}\Delta\eta_0^{(1)}.$$

From this,

$$y_1^{(2)} = y_0 + \Delta y_0^{(2)}, \qquad y_2^{(2)} = y_1^{(2)} + \Delta y_1^{(2)}.$$

From $y_1^{(2)}$ and $y_2^{(2)}$ we obtain

$$\eta_1^{(2)} = hf(x_1, y_1^{(2)}), \qquad \eta_2^{(2)} = hf(x_2, y_2^{(2)}),$$

and now we display the quantities

$$
\begin{array}{l}
\eta_0 \\
\qquad \Delta\eta_0^{(2)} \\
\eta_1^{(2)} \qquad\qquad \Delta^2\eta_0^{(2)} \\
\qquad \Delta\eta_1^{(2)} \\
\eta_2^{(2)}
\end{array}
\tag{5.12}
$$

With the help of quantities (5.12), if one keeps three first terms in each of the equation (5.10), then one gets

$$\Delta y_0^{(3)} = \eta_0 + \tfrac{1}{2}\Delta\eta_0^{(2)} - \tfrac{1}{12}\Delta^2\eta_0^{(2)},$$
$$\Delta y_1^{(3)} = \eta_1^{(2)} + \tfrac{1}{2}\Delta\eta_0^{(2)} + \tfrac{5}{12}\Delta^2\eta_0^{(2)},$$
$$\Delta y_2^{(3)} = \eta_2^{(2)} + \tfrac{1}{2}\Delta\eta_1^{(2)} + \tfrac{5}{12}\Delta^2\eta_0^{(2)},$$
$$y_1^{(3)} = y_0 + \Delta y_0^{(3)}, \qquad y_2^{(3)} = y_1^{(3)} + \Delta y_1^{(3)}, \qquad y_3^{(3)} = y_2^{(3)} + \Delta y_2^{(3)}.$$
$$\tag{5.13}$$

Now we have the approximate values (5.13) for all of the unknowns y_1, y_2, y_3. Now, the usual method of iteration is applied for the solution of system (5.10) with initial approximations (5.13), $y_1^{(3)}$, $y_2^{(3)}$, $y_3^{(3)}$. Thus, the process of obtaining the quantities (5.13) can be considered to be a method of construction of the initial approximations to the solutions of the system (5.10).

The usual method of iteration is carried out in the following manner. We calculate the numbers

$$\eta_j^{(3)} = hf(x_J, y_j^{(3)}), \qquad j = 1, 2, 3,$$

and we form the finite difference table

η_0			
	$\Delta\eta_0^{(3)}$		
$\eta_1^{(3)}$		$\Delta^2\eta_0^{(3)}$	
	$\Delta\eta_1^{(3)}$		$\Delta^3\eta_0^{(3)}$
$\eta_2^{(3)}$		$\Delta^2\eta_1^{(3)}$	
	$\Delta\eta_2^{(3)}$		
$\eta_3^{(3)}$			

with the help of the table of the system (5.10), we find $y_1^{(4)}$, $y_2^{(4)}$, $y_3^{(4)}$. We continue the calculations until two successive approximations coincide to the required accuracy

$$y_1^{(j)} = y_1^{(j+1)}, \qquad y_2^{(j)} = y_2^{(j+1)}, \qquad y_3^{(j)} = y_3^{(j+1)}.$$

With the aim of simplifying the computational scheme, the initial approximations to the solutions of system (5.10) can be found by the Euler method, using formula (4.9):

$$y_{n+1} = y_n + hf(x_n, y_n).$$

We cite without proof the system, used in Krylov's method for the case $k = 5$:

$$\Delta y_0 = \eta_0 + \tfrac{1}{2}\Delta\eta_0 - \tfrac{1}{12}\Delta^2\eta_0 + \tfrac{1}{24}\Delta^3\eta_0 - \tfrac{19}{720}\Delta^4\eta_0 + \tfrac{3}{160}\Delta^5\eta_0,$$

$$\Delta y_1 = \eta_1 + \tfrac{1}{2}\Delta\eta_0 + \tfrac{5}{12}\Delta^2\eta_0 - \tfrac{1}{24}\Delta^3\eta_0 + \tfrac{11}{720}\Delta^4\eta_0 - \tfrac{11}{1440}\Delta^5\eta_0,$$

$$\Delta y_2 = \eta_2 + \tfrac{1}{2}\Delta\eta_1 + \tfrac{5}{12}\Delta^2\eta_0 + \tfrac{3}{8}\Delta^3\eta_0 - \tfrac{19}{720}\Delta^4\eta_0 + \tfrac{11}{1440}\Delta^5\eta_0, \qquad (5.14)$$

$$\Delta y_3 = \eta_3 + \tfrac{1}{2}\Delta\eta_2 + \tfrac{5}{12}\Delta^2\eta_1 + \tfrac{3}{8}\Delta^3\eta_0 + \tfrac{251}{720}\Delta^4\eta_0 - \tfrac{3}{160}\Delta^5\eta_0,$$

$$\Delta y_4 = \eta_4 + \tfrac{1}{2}\Delta\eta_3 + \tfrac{5}{12}\Delta^2\eta_2 + \tfrac{3}{8}\Delta^3\eta_1 + \tfrac{251}{720}\Delta^4\eta_0 + \tfrac{95}{288}\Delta^5\eta_0.$$

This system contains as particular cases the systems employed in Krylov's method for any k, $1 \leq k \leq 5$. In order to obtain the necessary system for such k, we take from system (5.14) the first k equations and set the differences $\Delta^i\eta_j$ equal to zero for $i \geq k+1$. In particular, for $k = 3$, we obtain system (5.10).

Table 32

x	y	Δy	0.25 y	f(x, y)	η	Δη	Δ²η	Δ³η
0	—1		—0.25	0.25	0.012500			
0	—1				0.012500			
		0.01250				—186		
0.05	—0.98750		—0.24688	0.24629	0.012314			
0	—1				0.012500			
		0.01241				—183		
0.05	—0.98759		—0.24690	0.24634	0.012317		258	
		0.01222				75		
0.10	—0.97537		—0.24384	0.24783	0.012392			
0	—1				0.012500			
		0.01239				—183		
0.05	—0.98761		—0.24690	0.24634	0.012317		256	
		0.01233				73		—9
0.10	—0.97528		—0.24382	0.24779	0.012390		247	
		0.01254				320		
0.15	—0.96274		—0.24068	0.25421	0.012710			
0	—1				0.012500			
		0.01239				—183		
0.05	—0.98761				0.012317		256	
		0.01233				73		—8
0.10	—0.97528				0.012390		248	
		0.01253				321		
0.15	—0.96275		—0.24069	0.25422	0.012711			

Table 33

x	y	Δy	0.25 y	f(x, y)	η	Δη	Δ²η	Δ³η
0	—1		—0.25	0.25	0.012500			
		0.01250				—186		
0.05	—0.98750		—0.24688	0.24629	0.012314		260	
		0.01231				74		—10
0.10	—0.97519		—0.24380	0.24775	0.012388		250	
		0.01239				324		
0.15	—0.96280		—0.24070	0.25425	0.012712			
0	—1				0.012500			
		0.01239				—183		
0.05	—0.98761		—0.24690	0.24634	0.012317		256	
		0.01233				73		—8
0.10	—0.97528		—0.24382	0.24779	0.012390		248	
		0.01253				321		
0.15	—0.96275		—0.24069	0.25422	0.012711			

Table 34

x	y	Δy	$0.25\,y$	$f(x, y)$	η	$\Delta\eta$	$\Delta^2\eta$	$\Delta^3\eta$
0	—1		—0.25	0.25	0.012500			
						—183		
0.05	—0.98761		—0.24690	0.24634	0.012317		256	
						73		—8
0.10	—0.97528		—0.24382	0.24779	0.012390		248	
						321		—4
0.15	—0.96275		—0.24069	0.25422	0.012711		244	
						565		—6
0.20	—0.94978		—0.23744	0.26552	0.013276		238	
		0.01366				803		—5
0.25	—0.93612		—0.23403	0.28158	0.014079		233	
		0.01458				1036		—3
0.30	—0.92154		—0.23038	0.30230	0.015115		230	
		0.01573				1266		—5
0.35	—0.90581		—0.22645	0.32762	0.016381		225	
		0.01711				1491		—2
0.40	—0.88870		—0.22218	0.35745	0.017872		223	
		0.01871				1714		—3
0.45	—0.86999		—0.21750	0.39172	0.019586		220	
		0.02054				1934		—2
0.50	—0.84945		—0.21236	0.43039	0.021520		218	
		0.02258				2152		—1
0.55	—0.82687		—0.20672	0.47343	0.023672		217	
		0.02484				2369		1
0.60	—0.80203		—0.20051	0.52082	0.026041		218	
		0.02732				2587		—1
0.65	—0.77471		—0.19368	0.57255	0.028628		217	
		0.03001				2804		5
0.70	—0.74470		—0.18618	0.62865	0.031432		222	
		0.03292				3026		2
0.75	—0.71178		—0.17794	0.68915	0.034458		224	
		0.03607				3250		4
0.80	—0.67571		—0.16893	0.75415	0.037708		228	
		0.03943				3478		8
0.85	—0.63628		—0.15907	0.82371	0.041186		236	
		0.04302				3714		7
0.90	—0.59326		—0.14832	0.89799	0.044900		243	
		0.04686				3957		
0.95	—0.54640		—0.13660	0.97714	0.048857			
		0.05091						
1	—0.49546							

Example 2. We construct the beginning of the table for the Cauchy problem (5.7) by Krylov's method. We take $h = 0.05$. We use formulas (5.10) in the calculation. The results of the calculations are cited in table 32.

The beginning of the table of the same problem (5.7) is given in table 33, using the Euler method for the construction of the initial approximation to the solution of the system (5.10). We see that the computational scheme in this case is simpler than for the method of A. N. Krylov. In order to obtain five accurate decimal places it was necessary to make only one iteration.

The Runge-Kutta method for the numerical solution of the Cauchy problem can also be applied to the construction of the beginning of the table.

Example 3. We shall find the numerical solution of problem (5.7) by the Adams extrapolation method. The beginning of the table has already been constructed. In the calculations we use formula (4.8) for $k = 3$:

$$\varDelta y_n = \eta_n + \tfrac{1}{2}\varDelta\eta_{n-1} + \tfrac{5}{12}\varDelta^2\eta_{n-2} + \tfrac{3}{8}\varDelta^3\eta_{n-3}. \tag{5.15}$$

We calculate the quantities η_j with one extra place (in the same way as above when constructing the beginning of the table) in order to decrease the influence of the error of rounding off, arising in the calculation of $\varDelta y_n$ by formula (5.15). The results of the calulations are cited in table 34.

6. The Adams interpolation method

The Adams interpolation method is defined by formula (3.32),

$$\varDelta y_n = h \sum_{j=-1}^{k} b_{kj}^* f(x_{n-j}, y_{n-j}). \tag{6.1}$$

It is clear from the method of construction of this formula that its right-hand side represents the integral over the interval $[x_n, x_{n+1}]$ of the interpolation polynomial $Q(x)$, constructed from the conditions

$$Q(x_{n-j}) = y'_{n-j} = f(x_{n-j}, y_{n-j}), \qquad j = -1, 0, 1, 2, \ldots, k.$$

Thus, formula (6.1) can be represented in the form

$$\varDelta y_n = \int_{x_n}^{x_{n+1}} Q(x)dx. \tag{6.3}$$

The interval $[x_n, x_{n+1}]$, over which the integration of $Q(x)$ extends, is found at the end of the table of values

$$y'_{n-k}, y'_{n-k+1}, \ldots, y'_n, y'_{n+1},$$

therefore, we write $Q(x)$ by Newton's formula for interpolation at the end of the table:

$$Q(x_{n+1}+ht) = y'_{n+1} + \frac{t}{1!} \Delta y'_n + \frac{t(t+1)}{2!} \Delta^2 y'_{n-1} + \ldots$$

$$\ldots + \frac{t(t+1)\ldots(t+k)}{(k+1)!} \Delta^{k+1} y'_{n-2}. \tag{6.4}$$

We make the change $x = x_{n+1}+ht$ of the variable of integration in the integral (6.3):

$$\Delta y_n = h \int_{-1}^{0} Q(x_{n+1}+ht)dt. \tag{6.5}$$

In order to write the integral over the interval $[-1, 0]$ of the interpolation polynomial (6.4), we use the notation

$$a_j^* = \frac{1}{j!} \int_{-1}^{0} t(t+1)\ldots(t+j-1)dt, \qquad j = 0, 1, 2, \ldots \tag{6.6}$$

We write out the first few of the a_j^*:

$$a_0^* = 1, \qquad a_1^* = -\tfrac{1}{2}, \qquad a_2^* = -\tfrac{1}{12}, \qquad a_3^* = -\tfrac{1}{24},$$

$$a_4^* = -\tfrac{19}{720}, \qquad a_5^* = -\tfrac{3}{160}, \qquad a_6^* = -\tfrac{863}{60\,480}.$$

Integrating (6.4) over $[-1, 0]$, we get

$$\int_{-1}^{0} Q(x_{n+1}+ht)dt =$$

$$= y'_{n+1} - \tfrac{1}{2}\Delta y'_n - \tfrac{1}{12}\Delta^2 y'_{n-1} - \ldots + a_{k+1}\Delta^{k+1} y'_{n-k}.$$

We replace the integral in (6.5) by the right-hand side of the last equation. We obtain a modified formula for the Adams interpolation method:

$$\Delta y_n = h(y'_{n+1} - \tfrac{1}{2}\Delta y'_n - \tfrac{1}{12}\Delta^2 y'_{n-1} - \ldots + a_{k+1}^* \Delta^{k+1} y'_{n-k}). \tag{6.7}$$

In § 3 we have already noted that the Adams interpolation method is more accurate in some sense than the extrapolation method. This situation becomes apparent from the fact that the coefficients of formula (6.7) decrease with the growth of n faster than the coefficients of formula (4.6).

For calculation, formula (6.7) is applied in the form

$$\Delta y_n = \eta_{n+1} - \tfrac{1}{2}\Delta\eta_n - \tfrac{1}{12}\Delta^2\eta_{n-1} - \tfrac{1}{24}\Delta^3\eta_{n-2} - \ldots + a^*_{k+1}\Delta^{k+1}\eta_{n-k},$$
(6.8)

where the η_j are determined by equation (4.7):

$$\eta_j = hf(x_j, y_j), \qquad j = 0, 1, 2, \ldots$$

Formula (6.8) represents an equation for y_{n+1} [see (6.1)]:

$$y_{n+1} = hb^*_{k,-1} f(x_{n+1}, y_{n+1}) + F(y_{n-k}, \ldots, y_n),$$

which is usually solved by the method of iteration. It is obvious from the form of this equation that the method of iteration converges for it, if the step h is chosen sufficiently small and the initial approximation $y^{(0)}_{n+1}$ is taken sufficiently close to the solution y_{n+1}. Of course, we assume here that $f(x, y)$ has a partial derivative $\partial f(x, y)/\partial y$, continuous in some band

$$x_0 \leqq x \leqq X, \qquad -\rho \leqq y(x) - y \leqq \rho, \qquad \rho > 0,$$

containing $y(x)$ and y_j for $j = 1, 2, \ldots, N$. (See theorem 1 on the convergence of the method of iteration, chapter I, § 3).

The numerical solution of the Cauchy problem by the Adams interpolation method according to formula (6.8) is performed in the following manner.

From the values $y_0, y_1, \ldots, y_k, y_{k+1}$, a table similar to the one for the Adams extrapolation method is constructed. We assume, that in addition to the initial values $y_0, y_1, \ldots, y_k$, necessary for the deduction of formula (6.8), still another value y_{k+1} is known. We write the table for $k = 3$:

x	y	Δy	$n = hf(x, y)$	$\Delta\eta$	$\Delta^2\eta$	$\Delta^3\eta$	$\Delta^3\eta$
x_0	y_0		η_0				
		Δy_0		$\Delta\eta_0$			
x_1	y_1		η_1		$\Delta^2\eta_0$		
		Δy_1		$\Delta\eta_1$		$\Delta^3\eta_0$	
x_2	y_2		η_2		$\Delta^2\eta_1$		$\Delta^4\eta_0$
		Δy_2		$\Delta\eta_2$		$\Delta^3\eta_1$	
x_3	y_3		η_3		$\Delta^2\eta_2$		$\Delta^4\eta_1$
		Δy_3		$\Delta\eta_3$		$\Delta^3\eta_2$	
x_4	y_4		η_4		$\Delta^2\eta_3$		
				$\Delta\eta_4$			
x_5			η_5				

In the case $k = 3$, formula (6.8) has the form

$$\Delta y_n = \eta_{n+1} - \tfrac{1}{2}\Delta\eta_n - \tfrac{1}{12}\Delta^2\eta_{n-1} - \tfrac{1}{24}\Delta^3\eta_{n-2} - \tfrac{19}{720}\Delta^4\eta_{n-3}.$$

We set $n = 4$:

$$\Delta y_4 = \eta_5 - \tfrac{1}{2}\Delta\eta_4 - \tfrac{1}{12}\Delta^2\eta_3 - \tfrac{1}{24}\Delta^3\eta_2 - \tfrac{19}{720}\Delta^4\eta_1. \tag{6.9}$$

The differences

$$\eta_5, \Delta\eta_4, \Delta^2\eta_3, \Delta^3\eta_2, \Delta^4\eta_1, \tag{6.10}$$

situated in the table below the heavy stepped line enter into the right-hand side of this formula. These differences are unknown and we have to indicate initial approximations for them.

In order to indicate the initial approximations to the differences (6.10), it is sufficient to indicate an approximation to one of them. Simplest of all is to indicate the initial approximation to the fourth difference $\Delta^4\eta_1$. In fact, if the step h is correctly selected, then up to a few units in the last significant figure

$$\Delta^4\eta_1 \cong \Delta^4\eta_0,$$

and we can take the quantity $\Delta^4\eta_1^{(0)} = \Delta^4\eta_0$ as the initial approximation for $\Delta^4\eta_1$. This allows one to calculate approximations for all the differences (6.10)

$$\eta_5^{(0)}, \Delta\eta_4^{(0)}, \Delta^2\eta_3^{(0)}, \Delta^3\eta_2^{(0)}, \Delta^4\eta_1^{(0)} \tag{6.11}$$

successively according to the formulas

$$\Delta^3\eta_2^{(0)} = \Delta^3\eta_1 + \Delta^4\eta_1^{(0)},$$
$$\Delta^2\eta_3^{(0)} = \Delta^2\eta_2 + \Delta^3\eta_2^{(0)},$$
$$\Delta\eta_4^{(0)} = \Delta\eta_3 + \Delta^2\eta_3^{(0)},$$
$$\eta_5^{(0)} = \eta_4 + \Delta\eta_4^{(0)}.$$

Substituting the initial approximations (6.11) into formula (6.9), we find $\Delta y_4^{(1)}$ and consequently,

$$y_5^{(1)} = y_4 + \Delta y_4^{(1)}.$$

Now, we can calculate

$$\eta_5^{(1)} = hf(x_5, y_5^{(1)}).$$

If $\eta_5^{(1)} = \eta_5^{(0)}$, then we assume that $y_5 = y_5^{(1)}$ and we consider the calculation of y_5 to be finished.

If $\eta_5^{(1)} \neq \eta_5^{(0)}$, then from $\eta_5^{(1)}$ we calculate new values of the differences (6.10)

$$\eta_5^{(1)}, \Delta\eta_4^{(1)}, \Delta^2\eta_3^{(1)}, \Delta^3\eta_2^{(1)}, \Delta^4\eta_1^{(1)} \tag{6.12}$$

successively according to the formulas

$$\Delta\eta_4^{(1)} = \eta_5^{(1)} - \eta_4,$$
$$\Delta^2\eta_3^{(1)} = \Delta\eta_4^{(1)} - \Delta\eta_3,$$
$$\Delta^3\eta_2^{(1)} = \Delta^2\eta_3^{(1)} - \Delta^2\eta_2,$$
$$\Delta^4\eta_1^{(1)} = \Delta^3\eta_2^{(1)} - \Delta^3\eta_1.$$

Substituting the approximations (6.12) into the formula (6.9), one finds $\Delta y_4^{(2)}$ and, consequently, $y_5^{(2)}$. If it turns out that $y_5^{(2)} = y_5^{(1)}$, then we consider that $y_5 = y_5^{(2)}$. In the contrary case, one should continue the iteration. Of course, it is not convenient to make many iterations. The step should be such that one or two iterations are sufficient. After y_5 is found, we calculate y_6 in the same way, and so forth.

Sometimes the initial approximations for the unknown differences

$$\eta_{n+1}, \Delta\eta_n, \Delta^2\eta_{n-1}, \ldots, \Delta^{k+1}\eta_{n-k}$$

are calculated with the help of the formula of the Adams extrapolation method. Namely, we calculate $y_{n+1}^{(0)}$ by formula (4.8). This permits us to find $\eta_{n+1}^{(0)}$ and, consequently, to indicate the initial approximations for all differences.

Example. We shall find the numerical solution of the Cauchy problem

$$y' = \frac{x}{y} + 0.5y, \qquad y(0) = 1 \tag{6.13}$$

by the interpolation method of Adams. The solution of this problem exists on the whole x-axis:

$$y(x) = \sqrt{3e^x - 2 - 2x}.$$

The beginning of the table we will construct by the method of expansion of $y(x)$ in series of powers of x. For the calculation of the derivatives, we write the differential equation in the form

$$y'y - 0.5y^2 - x = 0.$$

Calculating the successive derivatives with respect to x from the left hand side of this equation, we get

$$y''y + y'^2 - yy' - 1 = 0,$$
$$y'''y + 3y'y'' - y'^2 - yy'' = 0, \qquad (6.14)$$
$$\dots\dots\dots\dots\dots\dots\dots$$

From the differential equation, assuming that $x = 0$, we find that $y'(0) = 0.5$. From the first of the relations (6.14), setting $x = 0$ and knowing $y(0)$ and $y'(0)$, we find $y''(0)$; from the second of the relations (6.14), we find $y'''(0)$, etc. As a result of the calculations we obtain

$$y(0) = 1, \qquad y'(0) = \tfrac{1}{2}, \qquad y''(0) = \tfrac{5}{4}, \qquad y'''(0) = -\tfrac{3}{8},$$
$$y^{(IV)}(0) = -\tfrac{39}{16}, \qquad y^{(V)}(0) = \tfrac{393}{32}, \qquad y^{(VI)}(0) = \tfrac{573}{64}.$$

We have

$$y(x) = 1 + \tfrac{1}{2}x + \tfrac{5}{8}x^2 - \tfrac{1}{16}x^3 - \tfrac{13}{128}x^4 + \tfrac{131}{1280}x^5 + \tfrac{191}{15360}x^6 + \dots$$
$$(6.15)$$

We select $h = 0.1$ as the step of integration. We intend to obtain values of $y(x)$ to five decimal places. As the basis for the selection of the step $h = 0.1$ is the fact that in order to obtain $y(0.3)$ with five accurate decimal places, it is sufficient to keep terms written out in the expansion (6.15) (up to x^6 inclusively) and, consequently, on the interval $[-0.3, 0.3]$, $y(x)$ can be considered to be a polynomial of the sixth degree with the accepted accuracy. The same can be said also about the quantity $hy'(x)$.

In fact, $y(x)$ and also $hy'(x)$ can obviously be represented on the interval considered by a polynomial of a lower degree — fourth or fifth — to five decimal places (since the segment of the Taylor series is not the best polynomial in the sense of a uniform approximation). It is clear from the reasoning above, that the fourth or the fifth difference in the table of values of the quantity

$$\eta_i = hf(x_i, y_i) \cong hy'(x_i)$$

at least for x_i close to $x = 0$, will be constant to five decimal places. We need to calculate five or six values of $y(x)$ for the equidistant values of x with the step 0.1. It is convenient to calculate $y(\pm 0.1)$, $y(\pm 0.2)$, $y(\pm 0.3)$ in order not to require additional terms of the series, beyond the ones written in (6.15).

Table 35

x	y	Δy	x/y	$0.5\,y$	$f(x, y)$	η	$\Delta\eta$	$\Delta^2\eta$	$\Delta^3\eta$	$\Delta^4\eta$
−0.3	0.90689		−0.33080	0.45345	0.12265	0.012265				
							12386			
−0.2	0.92531		−0.21614	0.46265	0.24651	0.024651		321		
							12707		−386	
−0.1	0.95630		−0.10457	0.47815	0.37358	0.037358		−65		86
							12642		−300	
0	1		0	0.5	0.5	0.050000		−365		117
							12277		−183	
0.1	1.05618		0.09468	0.52809	0.62277	0.062277		−548		112
							11729		−71	
										86
0.2	1.12437		0.17788	0.56218	0.74006	0.074006		−619		107
									15	
							11110		36	
								−604		54
0.3	1.20398		0.24917	0.60199	0.85116	0.085116		−583		60
							10506		69	
		0.09042					10527		75	
						0.095622		−535		29
0.4	1.29440		0.30902	0.64720	0.95622	0.095643		−529		22
							9971		98	
		0.10065					9977		91	
						0.105593		−437		10
0.5	1.39505		0.35841	0.69752	1.05593	0.105599		−444		4
							9534		108	
		0.11039					9527		102	
						0.115127		−329		−5
0.6	1.50544		0.39855	0.75272	1.15127	0.115120		−335		−9
							9205		103	
		0.11975					9199		99	
						0.124332		−226		−5
0.7	1.62519		0.43072	0.81260	1.24332	0.124326		−230		−20
		0.12884					8979		98	
		0.12883					8975		83	
	1.75403		0.45609	0.87702	1.33311	0.133311		−128		−6
0.8	1.75402		0.45610	0.87701	1.33311	0.133307		−143		−5
							8851		92	
		0.13774					8836		93	
						0.142162		−36		−18
0.9	1.89177		0.47574	0.94588	1.42162	0.142147		−35		−6
							8815		74	
		0.14657					8816		86	
						0.150977		38		−3
1.0	2.03834		0.49060	1.01917	1.50977	0.150978		50		−30
							8853		71	
		0.15540					8865		44	
						0.159830		109		
1.1	2.19374		0.50143	1.09687	1.59830	0.159842		82		
							8962			
		0.16430					8935			
		0.16429								
	2.35804					0.168792				
1.2	2.35803		0.50890	1.17902	1.68792	0.168765				

The calculated values $y(i \cdot 0, 1)$, the quantities η_i, $i = 0, \pm 1, \pm 2, \pm 3$ and their differences are written in table 35 in bold face. The quantity η_i we write with one spare digit in order to decrease the influence of the round off errors arising in the calculation of Δy_n. We see that the fourth differences (which in units of the sixth decimal place are equal to 86, 117, 112) differ from each other by about three units in the fifth place. We carry out the calculations for the continuation of the beginning of the table by formula (6.8) for $k = 3$; here we will write the coefficients with six decimal places:

$$\Delta y_n = \eta_{n+1} - 0.5\Delta\eta_n - 0.083333\Delta^2\eta_{n-1} -$$
$$- 0.041667\Delta^3\eta_{n-2} - 0.026389\Delta^4\eta_{n-3} . \tag{6.16}$$

We set $x_l = 0.1l$, $l = 0, \pm 1, \pm 2, \ldots$.

We shall explain table 35. To calculate Δy_3, we take $\Delta^4\eta_0^{(0)} = 107$, assuming that the fifth difference is constant and equal to $\Delta^4\eta_{-1} - \Delta^4\eta_{-2} = 112 - 117 = -5$. From the difference $\Delta^4\eta_0^{(0)} = 107$ we calculate the differences standing on the diagonal. These differences are written in the table in regular type. Now, by formula (6.16) we find $\Delta y_3^{(1)} = 0.09042$ and, consequently,

$$y_4^{(1)} = y_3 + \Delta y_3^{(1)} = 1.29440.$$

Furthermore, we calculate $f(x_4, y_4^{(1)}) = 0.95622$. The numbers $\Delta y_3^{(1)}$, $y_4^{(1)}$ and $f(x_4, y_4^{(1)})$ are written in the table in ordinary type.

We calculate $\eta_4^{(1)} = 0.1f(x_4, y_4^{(1)})$ and we recalculate all differences on the diagonal. The numbers $\Delta^k\eta_{4-k}^{(1)}$ are written in the table in bold face above the corresponding numbers $\Delta^k\eta_{4-k}^{(0)}(k = 0, 1, 2, 3, 4)$. By formula (6.16), we find $\Delta y_3^{(2)}$, which turns out to be equal to 0.09042, and therefore we consider $y_4^{(1)} = 1.29440$ to be the final value and we pass to the calculation of Δy_4.

The numbers on the diagonal

$$\eta_4^{(0)}, \Delta\eta_3^{(0)}, \Delta^2\eta_2^{(0)}, \Delta^3\eta_1^{(0)}, \Delta^4\eta_0^{(0)},$$

are written in ordinary type, and in further calculations they are not taken into account. In the calculation of Δy_4 we take $\Delta^4\eta_1^{(0)} = 60$, etc. At some steps of the calculation it turns out that $\Delta y_m^{(1)} \neq \Delta y_m^{(2)}$ (the difference is one unit in the fifth decimal place), for example, for $m = 7$, but each time the value $\Delta y_m^{(2)}$ turns out to be final.

We remark that the initial differences, which we use for obtaining the first approximations and which are written in ordinary type, are written in pencil in calculations, and the differences, used for obtaining the second approximation, are written in ink, correcting the modified last figures of the initial differences.

7. Methods of Cowell type

We shall express the method (3.38) of Cowell type

$$\Delta y_n = h \sum_{j=-k-1}^{k} \bar{b}_{kj} f(x_{n-j}, y_{n-j}) \tag{7.1}$$

in the form in which it is usually applied in calculations.
Let $Q(x)$ be an interpolating polynomial constructed according to the conditions

$$Q(x_{n-j}) = y'_{n-j} = f(x_{n-j}, y_{n-j}),$$
$$j = -k-1, \quad -k, \ldots, 0, \quad 1, \ldots, k. \tag{7.2}$$

From the method of construction of formula (7.1), it is clear that

$$\Delta y_n = \int_{x_n}^{x_{n+1}} Q(x)dx = h \int_0^1 Q(x_n + ht)dt. \tag{7.3}$$

The knots used in the construction of the interpolating polynomial $Q(x)$,

$$x_{n-k}, x_{n-k+1}, \ldots, x_n, x_{n+1}, \ldots, x_{n+k+1},$$

are situated symmetrically with respect to the midpoint of the interval $[x_n, x_{n+1}]$, over which the integration of $Q(x)$ is performed. The number of knots is equal to $2k+2$. It is natural to express the polynomial $Q(x)$ by means of Bessel's formula [see (8.11) chapter II]:

$$Q(x_n + ht) = \frac{y'_{n+1} + y'_n}{2} = \frac{t - \frac{1}{2}}{1!} \Delta y'_n +$$

$$+ \frac{t(t-1)}{2!} \frac{\Delta^2 y'_n + \Delta^2 y'_{n-1}}{2} + \ldots$$

$$\ldots + \frac{(t+k-1)(t+k-2) \ldots t(t-1) \ldots (t-k)}{(2k)!} \times$$

$$\times \frac{\Delta^{2k} y'_{n-k+1} + \Delta^{2k} y'_{n-k}}{2} +$$

$$+ \frac{(t-\frac{1}{2})(t+k-1)(t+k-2) \ldots t(t-1) \ldots (t-k)}{(2k+1)!} \Delta^{2k+1} y'_{n-k}. \tag{7.4}$$

We introduce the notation

$$\tilde{a}_{2j} = \frac{1}{(2j)!} \int_0^1 (t+j-1)(t+j-2) \ldots t(t-1) \ldots (t-j)dt,$$

$$\tilde{a}_{2j+1} = \frac{1}{(2j+1)!} \int_0^1 (t-\tfrac{1}{2})(t+j-1)(t+j-2) \ldots$$

$$\ldots t(t-1) \ldots (t-j)dt, \qquad j = 0, 1, 2, \ldots \qquad (7.5)$$

We assume that

$$(t+j-1)(t+j-2) \ldots t(t-1) \ldots (t-j) = 1 \qquad \text{for} \quad j = 0.$$

The numbers $\tilde{a}_{2j+1}$ are equal to zero. In fact, if one denotes the integrand, a polynomial of degree $2j+1$, by

$$F(t) = (t-\tfrac{1}{2})(t+j-1)(t+j-2) \ldots t(t-1) \ldots (t-j)$$

and performs the change of variables $t=1-z$, then one gets

$$F(t) = (-z+\tfrac{1}{2})(-z+j) \ldots (-z+1)(-z) \ldots (-z-j+1).$$

By reversing the sign inside each parentheses (there are $2j+1$) parentheses), we find

$$F(t) = -F(z) = -F(1-t).$$

From this it can be seen that

$$(2j+1)!\tilde{a}_{2j+1} = \int_0^1 F(t)dt = 0.$$

The numbers $\tilde{a}_{2j}$ are all distinct from zero since the integrand is a polynomial, the only zeros of which are integers. We write out several of the first values of $\tilde{a}_{2j}$:

$$\tilde{a}_0 = 1, \qquad \tilde{a}_2 = -\tfrac{1}{12}, \qquad \tilde{a}_4 = \tfrac{11}{720}, \qquad \tilde{a}_6 = -\tfrac{191}{60480}.$$

We integrate both sides of (7.4) over the interval [0, 1], and in so doing we use the notation of (7.5):

$$\int_0^1 Q(x_n + ht)dt = \frac{y'_{n+1} + y'_n}{2} - \frac{1}{12}\frac{\Delta^2 y'_n + \Delta^2 y'_{n-1}}{2} + \ldots$$

$$\ldots + \tilde{a}_{2k}\frac{\Delta^{2k} y'_{n-k+1} + \Delta^{2k} y'_{n-k}}{2}.$$

Comparing this equation with (7.3), we get

$$\Delta y_n = h\left(\frac{y'_{n+1} + y'_n}{2} - \frac{1}{12}\frac{\Delta^2 y_n + \Delta^2 y'_{n-1}}{2} + \ldots\right.$$

$$\left.\ldots + \tilde{a}_{2k}\frac{\Delta^{2k} y'_{n-k+1} + \Delta^{2k} y'_{n-k}}{2}\right). \tag{7.6}$$

This is the required formula for the method of Cowell type.

The placement of knots occurring in the resulting formula (7.6) is the best with respect to the minimization of the remainder of integration on the points of the interval $[x_n, x_{n+1}]$ (see § 6, chap. II), therefore, the coefficients of formula (7.6) decrease the most rapidly (by comparison with the corresponding formula for any difference method derived by the method of § 3 for $p \neq k+1$). For this reason, formula (7.6) is also called the *formula with most rapidly decreasing coefficients*.

By using the notation (4.7),

$$\eta_j = hf(x_j, y_j), \qquad j = 0, 1, 2, \ldots,$$

We rewrite (7.6) in the form

$$\Delta y_n = \frac{\eta_{n+1} + \eta_n}{2} - \frac{1}{12}\frac{\Delta^2\eta_n + \Delta^2\eta_{n-1}}{2} +$$

$$+ \frac{11}{720}\frac{\Delta^4\eta_{n-1} + \Delta^4\eta_{n-2}}{2} - \ldots + \tilde{a}_{2k}\frac{\Delta^{2k}\eta_{n-k+1} + \Delta^{2k}\eta_{n-k}}{2}.$$

$$\tag{7.7}$$

Formula (7.7) is also applied for computation. Its right-hand side depends on $y_{n+1}, y_{n+2}, \ldots, y_{n+k+1}$, so that this formula represents an equation with $k+1$ unknowns.

We now explain how to conduct calculations according to formula (7.7). We suppose that the following values are known:

$$y_0, y_1, y_2, \ldots, y_{2k}.$$

For definiteness, we assume that $k = 2$. We construct a table as in the case of the Adams methods:

x	y	Δy	$\eta = hf(x, y)$	$\Delta\eta$	$\Delta^2\eta$	$\Delta^3\eta$	$\Delta^4\eta$
x_0	y_0		η_0				
		Δy_0		$\Delta\eta_0$			
x_1	y_1		η_1		$\Delta^2\eta_0$		
		Δy_1		$\Delta\eta_1$		$\Delta^3\eta_0$	
x_2	y_2		η_2		$\Delta^2\eta_1$		$\Delta^4\eta_0$
		Δy_2		$\Delta\eta_2$		$\Delta^3\eta_1$	
x_3	y_3		η_3		$\Delta^2\eta_2$		$\Delta^4\eta_1$
		Δy_3		$\Delta\eta_3$		$\Delta^3\eta_2$	
x_4	y_4		η_4		$\Delta^2\eta_3$		$\Delta^4\eta_2$
				$\Delta\eta_4$		$\Delta^3\eta_3$	
x_5			η_5		$\Delta^2\eta_4$		$\Delta^4\eta_3$
				$\Delta\eta_5$		$\Delta^3\eta_4$	
x_6			η_6		$\Delta^2\eta_5$		
				$\Delta\eta_6$			
x_7			η_7				

In the table, all of the quantities above the heavy stepped line are known. Formula (7.7) is written thusly for $k = 2$:

$$\Delta y_n = \frac{\eta_{n+1}+\eta_n}{2} - \frac{1}{12}\frac{\Delta^2\eta_n+\Delta^2\eta_{n-1}}{2} + \frac{11}{720}\frac{\Delta^4\eta_{n-1}+\Delta^4\eta_{n-2}}{2}. \quad (7.8)$$

In this formula, we set $n = 4$:

$$\Delta y_4 = \frac{\eta_5+\eta_4}{2} - \frac{1}{12}\frac{\Delta^2\eta_4+\Delta^2\eta_3}{2} + \frac{11}{720}\frac{\Delta^4\eta_3+\Delta^1\eta_2}{2}. \quad (7.9)$$

On the right-hand side of formula (7.9), differences occur which depend on the three unknown quantities η_5, η_6, η_7. In the table, these differences are placed below the heavy stepped line in three diagonals containing η_5, η_6, η_7. We take initial approximations to the fourth differences.

$$\Delta^4\eta_1^{(0)}, \Delta^4\eta_2^{(0)}, \Delta^4\eta_3^{(0)},$$

Supposing that they are, for example, all equal to $\Delta^4\eta_0$. This allows one to indicate initial approximations for all the unknown differences in the diagonals. By formula (7.9), we find

$$\Delta y_4^{(1)} = \frac{\eta_5^{(0)}+\eta_4^{(0)}}{2} - \frac{1}{12}\frac{\Delta^2\eta_4^{(0)}+\Delta^2\eta_3^{(0)}}{2} + \frac{11}{720}\frac{\Delta^4\eta_3^{(0)}+\Delta^4\eta_2^{(0)}}{2}$$

and determine $y_5^{(1)}$.

We calculate

$$\eta_5^{(1)} = hf(x_5, y_5^{(1)}).$$

If $\eta_5^{(1)} = \eta_5^{(0)}$, then we assume that $y_5 = y_5^{(1)}$ and go on to the calculation of y_6. If $\eta_5^{(1)} \neq \eta_5^{(0)}$, then we recompute all the differences in the diagonal after replacing $\eta_5^{(0)}$ by $\eta_5^{(1)}$. We obtain a new value for the fourth order difference $\Delta^4 \eta_1^{(1)}$ and can indicate new approximations for the two following differences of the fourth order: $\Delta^4 \eta_2^{(1)}$, $\Delta^4 \eta_3^{(1)}$. Further, we find the required elements of the diagonals and calculate $y_5^{(2)}$ by formula (7.9). If $y_5^{(2)} = y_5^{(1)}$, then $y_5^{(2)}$ is taken to be the desired value of y_5. If $y_5^{(2)} \neq y_5^{(1)}$, then we find $\eta_5^{(2)}$. In case $\eta_5^{(2)} = \eta_5^{(1)}$ we consider that $y_5 = y_5^{(2)}$, in the contrary case, we correct the fourth difference $\Delta^4 \eta_1^{(2)}$, we indicate new approximations for the two following fourth differences, and so forth. In the calculations, the step h must be chosen so that $y_5^{(1)}$ or $y_5^{(2)}$ require no further correction.

We remark that in calculating by formula (7.7) in contrast to the Adams interpolation method, we do not have full control by use of the differential equation: only the first of the unknown diagonals which contains η_{n+1} is controlled. But this is not very dangerous since the influence on the right-hand side of formula (7.7) of errors in the differences of the following diagonals is insignificant. For example, from the second diagonal (containing η_{n+2}), the second difference $\Delta^2 \eta_n$ enters into the third term of formula (7.7). The second differences are significantly less in absolute value than the values of η_i, and the coefficient in the third term is $\frac{1}{12}$, while the coefficient in the second term, containing η_{n+1}, is 1. The influence of the following diagonals is still less.

Example. We investigate the Cauchy problem (6.13),

$$y' = \frac{x}{y} + 0.5y, \qquad y(0) = 1.$$

In § 6, the beginning of the table was constructed for this problem (with the step $h = 0.1$). In tables 36, 37, and 38, the numerical integration of this problem by the method of Cowell type is shown. In table 36 the values of y_4, y_5 and y_6 are calculated, in table 37 the values of y_7, y_8, and y_9 are calculated, and in table 38, the values of y_{10} through y_{15}. The results of the calculations are distributed over three tables in order to simplify the notation in the tables.

We shall explain table 36. The values of y_i, η_i, $i = 0, \pm 1, \pm 2, \pm 3$ are known. These values and their differences are printed in bold face. In order to calculate Δy_3, we must indicate initial approximations for the three differences of fourth order, following $\Delta^4 \eta_{-1} = 112$ (in units in the

Table 36

x	y	Δy	x/y	0.5 y	f(x, y)	η	Δη	Δ²η	Δ³η	Δ⁴η
—0.3	**0.90689**		—0.33080	0.45345	0.12265	0.012265				
							12386			
—0.2	**0.92531**		—0.21614	0.46265	0.24651	0.024651		321		
							12707		—386	
—0.1	**0.95630**		—0.10457	0.47815	0.37358	0.037358		—65		86
							12642		—300	
0	**1**		0	0.5	0.5	0.050000		—365		117
							12277		—183	
0.1	**1.05618**		0.09468	0.52809	0.62277	0.062277		—548		112
							11729		—71	
										86
0.2	**1.12437**		0.17788	0.56218	0.74006	0.074006		—619		107
									15	
							11110		36	
										54
								—604		*56*
0.3	**1.20398**		0.24917	0.60199	0.85116	0.085116		—583		102
									69	
						0.090369	*10506*	*—568*	*71*	*42*
		0.09042				0.090380	10527	—514	138	100
								—535		29
						0.095622		*—533*		*29*
0.4	1.29440		0.30902	0.64720	0.95622	0.095643		—445		97
						0.100608	**9971**	**—486**	**98**	**16**
		0.10065				*0.100608*	*9973*	*—483*	*100*	*16*
		0.10071				0.100684	10082	—328	235	94
						0.105593		**—437**		**4**
	1.39505		*0.35841*	*0.69752*	*1.05593*	*0.105595*		*—433*		*2*
0.5	1.39511		0.35839	0.69756	1.05595	0.105725		—210		92
		0.11039				**0.110360**	**9534**	**—386**	**102**	**—8**
		0.11040				*0.110365*	*9540*	*—382*	*102*	*—12*
		0.11067				0.110661	9872	—46	327	90
	1.50544		0.39855	0.75272	1.15127	0.115127		—335		—21
	1.50545		*0.39855*	*0.75272*	*1.15127*	*0.115135*		*—331*		*—25*
0.6	1.50578		0.39846	0.75289	1.15135	0.115597		117		87

Table 37

x	y	Δy	x/y	0.5 y	f(x, y)	η	$\Delta\eta$	$\Delta^2\eta$	$\Delta^3\eta$	$\Delta^4\eta$
0.4	1.29440					0.095622		—535		29
		0.10065					9971		98	
										10
0.5	1.39505					0.105593		—437		4
									108	
		0.11039					9534		102	
										—5
								—329		—6
0.6	1.50544					0.115127		—335		—21
									103	
						0.119730	9205	—278	102	—4
		0.11975				0.119726	9199	—294	81	—34
								—226		—5
								—227		—3
						0.124332				
0.7	1.62519		0.43072	0.81260	1.24332	0.124326		—254		—46
						0.128822	8979	—177	98	—5
		0.12884				0.128821	8978	—178	99	—2
		0.12882				0.128798	8945	—236	35	—58
						0.133311		—128		—5
	1.75403		0.45609	0.87702	1.33311	0.133310		—128		0
0.8	1.75401		0.45610	0.87700	1.33310	0.133271		—219		—71
						0.137736	8851	—82	93	—5
		0.13774				0.137735	8850	—78	99	2
		0.13765				0.137634	8726	—237	—36	—84
						0.142162		—35		—5
	1.89177		0.47574	0.94588	1.42162	0.142160		—29		3
0.9	1.89166		0.47577	0.94583	1.42160	0.141997		—255		—96

sixth decimal place). In order to calculate all three values y_4, y_5, y_6, we indicate the initial approximations for the five differences of the fourth order: $\Delta^4\eta_0^{(0)} = 107$, $\Delta^4\eta_1^{(0)} = 102$, $\Delta^4\eta_2^{(0)} = 97$, $\Delta^4\eta_3^{(0)} = 92$, $\Delta^4\eta_4^{(0)} = 87$. We complete the table of the finite differences for the quantity $\eta_i^{(0)}$ up to the horizontal row containing $x_6 = 0.6$. Namely, by consecutive summations we find the differences of third order

$$\Delta^3\eta_1^{(0)} = \Delta^3\eta_0 + \Delta^4\eta_0^{(0)} = -71 + 107 = 36,$$
$$\Delta^3\eta_2^{(0)} = \Delta^3\eta_1^{(0)} + \Delta^4\eta_1^{(0)} = 36 + 102 = 138,$$
$$\Delta^3\eta_3^{(0)} = \Delta^3\eta_2^{(0)} + \Delta^4\eta_2^{(0)} = 138 + 97 = 235,$$
$$\Delta^3\eta_4^{(0)} = \Delta^3\eta_3^{(0)} + \Delta^4\eta_3^{(0)} = 235 + 92 = 327.$$

Table 38

x	y	Δy	x/y	0.5 y	f(x, y)	η	Δη	Δ²η	Δ³η	Δ⁴η
0.6	1.50544					0.115127		—329		—5
							9205		103	
0.7	1.62519					0.124332		—226		—5
							8979		98	
										—6
0.8	1.75403					0.133311		—128		—5
									92	
							8851		93	
								—36		—18
0.9	1.89177					0.142162		—35		—5
						0.146570	*8815*	*1*	*74*	*—10*
		0.14657				0.146570	8816	9	88	—5
										—3
						0.150977		*38*		*—2*
1.0	1.03834		0.49060	1.01917	1.50977	0.150978		53		—5
									71	
						0.155404	*8853*	*74*	*72*	*—6*
		0.15540				0.155412	8869	94	83	—5
								109		—10
						0.159830		*110*		*—9*
1.1	2.19374		0.50143	1.09687	1.59830	0.159847		136		—5
						0.164311	**8962**	**140**	**61**	**—7**
		0.16430				*0.164312*	*8963*	*142*	*63*	*—8*
		0.16434				0.164350	9005	175	78	—5
						0.168792		**170**		**—4**
	2.35804		*0.50890*	*1.17902*	*1.68792*	*0.168793*		*173*		*—7*
1.2	2.35808		0.50889	1.17904	1.68793	0.168852		214		—5
		0.17334				*0.173358*	*9132*	*198*	*57*	*—6*
		0.17344				*0.173361*	*9136*	*201*	*56*	*—4*
						0.173462	9219	250	73	—5
						0.177924		**227**		**—7**
	2.53138		*0.51355*	*1.26569*	*1.77924*	*0.177929*		*229*		*—1*
1.3	2.53152		0.51353	1.26576	1.77929	0.178071		287		—5
		0.18258				**0.182604**	**9359**	**252**	**50**	**—8**
		0.18259				*0.182612*	*9365*	*256*	*55*	*2*
		0.18280				0.182824	9506	321	68	—5
	2.71396		**0.51585**	**1.35698**	**1.87283**	**0.187283**		**277**		**—10**
	2.71397		*0.51585*	*1.35698*	*1.87283*	*0.187294*		*284*		*5*
1.4	2.71432		0.51578	1.35716	1.87294	0.187577		355		—5
		0.19208				**0.192101**	**9636**	**297**	**40**	**—12**
		0.19209				*0.192118*	*9649*	*314*	*60*	*8*
		0.19148				0.192508	9861	386	63	—5
	2.90604		**0.51617**	**1.45302**	**1.96919**	**0.196919**		**317**		**—13**
	2.90606		*0.51616*	*1.44303*	*1.96919*	*0.196943*		*344*		*11*
1.5	2.90680		0.51603	1.45340	1.96043	0.197438		418		—5

In the same way we find the differences of second order:

$$\Delta^2\eta_2^{(0)} = \Delta^2\eta_1 + \Delta^3\eta_1^{(0)} = -619 + 36 = -583,$$

$$\Delta^2\eta_3^{(0)} = \Delta^2\eta_2^{(0)} + \Delta^3\eta_2^{(0)} = -583 + 138 = -445,$$

$$\Delta^2\eta_4^{(0)} = \Delta^2\eta_3^{(0)} + \Delta^3\eta_3^{(0)} = -445 + 235 = -210,$$

$$\Delta^2\eta_5^{(0)} = \Delta^2\eta_4^{(0)} + \Delta^3\eta_4^{(0)} = -210 + 327 = 117;$$

the differences of first order:

$$\Delta\eta_3^{(0)} = \Delta\eta_2 + \Delta^2\eta_2^{(0)} = 11110 - 583 = 10527,$$

$$\Delta\eta_4^{(0)} = \Delta\eta_3^{(0)} + \Delta^2\eta_3^{(0)} = 10527 - 445 = 10082,$$

$$\Delta\eta_5^{(0)} = \Delta\eta_4^{(0)} + \Delta^2\eta_4^{(0)} = 10082 - 210 = 9872$$

and the values of η:

$$\eta_4^{(0)} = \eta_3 + \Delta\eta_3^{(0)} = 85116 + 10527 = 95643,$$

$$\eta_5^{(0)} = \eta_4^{(0)} + \Delta\eta_4^{(0)} = 95643 + 10082 = 105725,$$

$$\eta_6^{(0)} = \eta_5^{(0)} + \Delta\eta_5^{(0)} = 105725 + 9872 = 115597.$$

We note the averages of the differences of even order

$$\frac{\eta_i^{(0)} + \eta_{i+1}^{(0)}}{2}, \quad \frac{\Delta^2\eta_{i-1}^{(0)} + \Delta^2\eta_i^{(0)}}{2}, \quad \frac{\Delta^4\eta_{i-2}^{(0)} + \Delta^4\eta_{i-1}^{(0)}}{2},$$

$$i = 3, 4, 5 (\eta_3^{(0)} = \eta_3).$$

Each average of the differences is written in the row placed between the differences of which it is composed and in the same column from which the differences are taken.

Further, according to formula (7.8),

$$\Delta y_i^{(1)} = \frac{\eta_i^{(0)} + \eta_{i+1}^{(0)}}{2} - \frac{1}{12}\frac{\Delta^2\eta_{i-1}^{(0)} + \Delta^2\eta_i^{(0)}}{2} + \frac{11}{720}\frac{\Delta^4\eta_{i-2}^{(0)} + \Delta^4\eta_{i-1}^{(0)}}{2}$$

We find $\Delta y_i^{(1)}$, $i = 3, 4, 5$. The averages of the differences needed for the calculation of $\Delta y_i^{(1)}$ are found in the same horizontal row in which one would write $\Delta y_i^{(1)}$. Knowing $\Delta y_i^{(1)}$, we can calculate

$$y_{i+1}^{(1)} = y_i^{(1)} + \Delta y_i^{(1)}, \quad i = 3, 4, 5 \quad (y_3^{(1)} = y_3),$$

and find

$$f(x_{i+1}, y_{i+1}^{(1)}), \quad i = 3, 4, 5.$$

All the quantities that we are talking about now: the initial values of the differences, the averages of the differences of even order,

$$\Delta y_i^{(1)}, \; y_{i+1}^{(1)}, \; f(x_{i+1}, y_{i+1}^{(1)}), \qquad i = 3, 4, 5,$$

are written in the table in ordinary numbers.
Further, we find $\eta_i^{(1)}$ by the formulas

$$\eta_i^{(1)} = 0.1 f(x_i, y_i^{(1)}), \qquad i = 4, 5, 6.$$

We calculate the differences of the first order

$$\Delta \eta_i^{(1)} = \eta_{i+1}^{(1)} - \eta_i^{(1)}, \qquad i = 3, 4, 5 \quad (\eta_3^{(1)} = \eta_3),$$

the differences of the second order

$$\Delta^2 \eta_i^{(1)} = \Delta \eta_{i+1}^{(1)} - \Delta \eta_i^{(1)}, \qquad i = 2, 3, 4 \quad (\Delta \eta_2^{(1)} = \Delta \eta_2),$$

the differences of the third order

$$\Delta^3 \eta_i^{(1)} = \Delta^2 \eta_{i+1}^{(1)} - \Delta^2 \eta_i^{(1)}, \qquad i = 1, 2, 3 \quad (\Delta^2 \eta_1^{(1)} = \Delta^2 \eta_1),$$

and the differences of the fourth order

$$\Delta^4 \eta_i^{(1)} = \Delta^3 \eta_{i+1}^{(1)} - \Delta^3 \eta_i^{(1)}, \qquad i = 0, 1, 2 \quad (\Delta^3 \eta_0^{(1)} = \Delta^3 \eta_0).$$

In order to use formula (7.8) and find $\Delta y_i^{(2)}$, the values found for the $\eta_i^{(1)}$ and their differences are insufficient. We obtain the missing differences in the following manner. We have already found

$$\Delta^4 \eta_0^{(1)} = 86, \qquad \Delta^4 \eta_1^{(1)} = 56, \qquad \Delta^4 \eta_2^{(1)} = 29.$$

assuming that the fifth difference is constant and equal to -27, we propose

$$\Delta^4 \eta_3^{(1)} = 2, \qquad \Delta^4 \eta_4^{(1)} = -25.$$

This allows one to find $\Delta^3 \eta_4^{(1)}$ and $\Delta^2 \eta_5^{(1)}$. Each of the numbers $\Delta^k \eta_i^{(1)}$ is written in italics above the number $\Delta^k \eta_i^{(0)}$ in the same column in the table that $\Delta^k \eta_i^{(0)}$ is found.
From the differences $\Delta^k \eta_i^{(1)}$, we find the second approximation $\Delta y_i^{(2)}$ ($i = 3, 4, 5$) just as we found the first approximation to $\Delta y_i^{(1)}$ ($i = 3, 4, 5$) above from the initial differences $\Delta^k \eta_i^{(0)}$. Namely, we calculate the averages of the even differences

$$\frac{\eta_i^{(1)} + \eta_{i+1}^{(1)}}{2}, \quad \frac{\Delta^3 \eta_{i-1}^{(1)} + \Delta^2 \eta_i^{(1)}}{2}, \quad \frac{\Delta^4 \eta_{i-2}^{(1)} + \Delta^4 \eta_{i-1}^{(1)}}{2},$$

$$i = 3, 4, 5 \quad (\eta_3^{(1)} = \eta_3).$$

and write it in italics above the corresponding averages of the initial differences. By the formula:

$$\Delta y_i^{(2)} = \frac{\eta_i^{(1)}+\eta_{i+1}^{(1)}}{2} - \frac{1}{12}\frac{\Delta^2\eta_{i-1}^{(1)}+\Delta^2\eta_i^{(1)}}{2} + \frac{11}{720}\frac{\Delta^4\eta_{i-2}^{(1)}+\Delta^4\eta_{i-1}^{(1)}}{2}$$

we find $\Delta y_i^{(2)}$, $i = 3, 4, 5$. Moreover, it turns out that

$$\Delta y_3^{(2)} = \Delta y_3^{(1)} = 0.09042,$$

therefore the value of $\Delta y_3^{(2)}$ is not written in the table, and we consider the value of $\Delta y_3^{(1)}$ and, that means, of $y_4^{(1)}$ to be final; $\Delta y_i^{(2)} \neq \Delta y_i^{(1)}$ for $i = 4, 5$, therefore $\Delta y_i^{(2)}$, $y_{i+1}^{(2)}$, and also $f(x_{i+1}, y_{i+1}^{(2)})$, $i = 4, 5$, are written (in italics) in the table above the corresponding values of the first approximation.
We calculate

$$\eta_i^{(2)} = 0.1f(x_i, y_i^{(2)}), \qquad i = 5, 6,$$

and their differences. In the table these numbers, and also the numbers obtained from further calculations, are written in bold face (above the corresponding number written in italics). In particular, we get

$$\Delta^4\eta_1^{(2)} = 54, \qquad \Delta^4\eta_2^{(2)} = 29,$$

considering the fifth difference to be constant and equal to -25, we propose

$$\Delta^4\eta_3^{(2)} = 4, \qquad \Delta^4\eta_4^{(2)} = -21.$$

We find $\Delta^3\eta_4^{(2)}$ and $\Delta^2\eta_5^{(2)}$.
From the differences $\Delta^k\eta_i^{(2)}$, one can find the third approximations: $\Delta y_4^{(3)}$ and $\Delta y_5^{(3)}$. Here, we get

$$\Delta y_4^{(3)} = \Delta y_4^{(2)} = 0.10065,$$

therefore, $\Delta y_4^{(3)}$ is not written in the table and $\Delta y_4^{(2)}$ and, consequently, $y_5^{(2)}$ are considered to be final values. Further, we will write $\Delta y_5^{(3)} = 0.11039$ and $y_6^{(3)} = 1.50544$, which differ from $y_6^{(2)}$ by one unit in the fifth decimal place. In calculating $f(x_6, y_6^{(3)})$, we get:

$$f(x_6, y_6^{(3)}) = f(x_6, y_6^{(2)}) = 1.15127$$

and, consequently, $\eta_6^{(3)} = \eta_6^{(2)}$. For this reason, $y_6^{(3)}$ is also considered to be final.

8. Numerical integration of systems of equations of the first order

We shall examine the Cauchy problem for a system of ordinary differential equations of the first order. In order not to complicate the notation, we limit ourselves to the case of a system of two equations:

$$\left.\begin{array}{l} y' = f(x, y, z), \\ z' = g(x, y, z), \end{array}\right\} \tag{8.1}$$

$$y(x_0) = y_0, \qquad z(x_0) = z_0.$$

It is necessary to construct tables of values of the functions $y(x)$ and $z(x)$ with the step h and initial values y_0 and z_0, respectively. We assume, as in the case of the Cauchy problem for one first order equation, that we know the beginning of the table, i.e. approximate values of the functions $y(x)$ and $z(x)$ are given for $x = x_0 + jh$, $j = 0, 1, 2, \ldots, k$:

$$y_0, y_1, y_2, \ldots, y_k, \tag{8.2}$$

$$z_0, z_1, z_2, \ldots, z_k.$$

The beginning of the table can be constructed by expanding the solutions $y(x)$ and $z(x)$ in a series in powers of $x - x_0$, and also by the Runge-Kutta method.

For the continuation of the tables (8.2) any of the methods examined above will serve for equations of the first order. We will show, for example, how one can apply the computational formula for the Adams extrapolation method (4.8) for the continuation of the tables (8.2).

We will introduce the quantities

$$\eta_l = hy'_l = hf(x_l, y_l, z_l),$$

$$\zeta_l = hz'_l = hg(x_l, y_l, z_l), \qquad l = 0, 1, 2, \ldots, \tag{8.3}$$

and construct two finite differences tables:

x	y	η	$\Delta\eta$	$\Delta^2\eta$	$\Delta^3\eta$	$\Delta^4\eta$	x	z	ζ	$\Delta\zeta$	$\Delta^2\zeta$	$\Delta^3\zeta$	$\Delta^4\zeta$
x_0	y_0	η_0					x_0	z_0	ζ_0				
			$\Delta\eta_0$							$\Delta\zeta_0$			
x_1	y_1	η_1		$\Delta^2\eta_0$			x_1	z_1	ζ_1		$\Delta^2\zeta_0$		
			$\Delta\eta_1$		$\Delta^3\eta_0$					$\Delta\zeta_1$		$\Delta^3\zeta_0$	
x_2	y_2	η_2		$\Delta^2\eta_1$		$\Delta^4\eta_0$	x_2	z_2	ζ_2		$\Delta^2\zeta_1$		$\Delta^4\zeta_0$
			$\Delta\eta_2$		$\Delta^3\eta_1$					$\Delta\zeta_2$		$\Delta^3\zeta_1$	
x_3	y_3	η_3		$\Delta^2\eta_2$			x_3	z_3	ζ_3		$\Delta^2\zeta_2$		
			$\Delta\eta_3$							$\Delta\zeta_3$			
x_4	y_4	η_4					x_4	z_4	ζ_4				

We took $k = 4$ here for definiteness. The values of y_5 and z_5 can be found by the formulas

$$y_5 = y_4 + \eta_4 + \tfrac{1}{2}\Delta\eta_3 + \tfrac{5}{12}\Delta^2\eta_2 + \tfrac{3}{8}\Delta^3\eta_1 + \tfrac{251}{720}\Delta^4\eta_0,$$
$$z_5 = z_4 + \zeta_4 + \tfrac{1}{2}\Delta\zeta_3 + \tfrac{5}{12}\Delta^2\zeta_2 + \tfrac{3}{8}\Delta^3\zeta_1 + \tfrac{251}{720}\Delta^4\zeta_0.$$

These formulas are obtained on the basis of (4.8). By knowing y_5 and z_5, we can find η_5 and ζ_5 by formula (8.3) and, consequently, can add a diagonal to each of the tables. In the same way, we find y_6 and z_6, etc.

In the case of a system of n equations it is necessary to construct simultaneously n finite difference tables.

It is known that the Cauchy problem for an ordinary differential equation of order n

$$\begin{aligned}
&y^{(n)} = f(x, y, y', y'', \ldots, y^{(n-1)}),\\
&y(x_0) = y_0,\, y'(x_0) = y_0',\, \ldots,\, y^{(n-1)}(x_0) = y_0^{(n-1)}
\end{aligned} \tag{8.4}$$

leads to a Cauchy problem for a system of first order differential equations. Namely, we set

$$y' = y_1, \qquad y'' = y_2, \ldots, y^{(n-1)} = y_{n-1}.$$

Obviously, the Cauchy problem (8.4) is equivalent to the Cauchy problem for a system of differential equations of the first order

$$\begin{aligned}
&y' = y_1,\\
&y_1' = y_2,\\
&\cdots\cdots\\
&y_{n-2}' = y_{n-1},\\
&y_{n-1}' = f(x, y, y_1, y_2, \ldots, y_{n-1})
\end{aligned}$$

with the initial conditions

$$y(x_0) = y_0, \qquad y_1(x_0) = y_0', \qquad y_2(x_0) = y_0'', \ldots, y_{n-1}(x_0) = y_0^{(n-1)}.$$

Thus, we obtain a method for solving the Cauchy problem (8.4) for a differential equation of order n.

9. Störmer's extrapolation method

In this and the next two sections we shall examine difference methods for solving the Cauchy problem for an ordinary differential equation of the second order of the particular form

$$y'' = f(x, y), \qquad y(x_0) = y_0, \qquad y'(x_0) = y'_0. \tag{9.1}$$

We suppose that the solution of problem (9.1) exists on the finite interval $[x_0, X], X > x_0$. We consider that in a certain closed area D of the (x, y) plane containing the graph of the solution $y = y(x)$ on $[x_0, X]$, the function $f(x, y)$ is continuous and has continuous partial derivatives up to the required order.

Of course, problem (9.1) could have been solved by the method of introducing a system of first order equations as shown in § 8, then however, it would be necessary to fill out two tables of finite differences. The fact that $f(x, y)$ does not depend on y' allows one to indicate methods whose realization requires filling out only one table of finite differences.

Let $h > 0$ and

$$x_j = x_0 + jh, \qquad j = 0, 1, 2, \ldots, N,$$

where N satisfies the inequality

$$x_0 + Nh \leq X < x_0 + (N+1)h.$$

We suppose that the beginning of the table exists, i.e., approximate values of the solution $y(x)$ are indicated for $x = x_j, j = 0, 1, 2, \ldots, k$:

$$y_0, y_1, \ldots, y_k. \tag{9.2}$$

The beginning of the table can be constructed by the method of expanding the solution in a series in powers of $x - x_0$, by the Runge-Kutta method, or the method analogous to A. N. Krylov's method (§ 5).

The difference method for solving the problem (9.1) allows one to calculate approximate values of the solution $y(x)$ at the points x_j for $j \geq k+1$. These approximate values, like the initial values (9.2), will be denoted by y_j.

Let the already calculated values be

$$y_0, y_1, \ldots, y_n, \tag{9.3}$$

where $n \geq k$. It is necessary to calculate y_{n+1}. According   to   Taylor's formula with integral form of the remainder term, we have

$$y(x) = y(x_n) + y'(x_n)(x - x_n) + \int_{x_n}^{x} y''(z)(x - z)dz. \tag{9.4}$$

In (9.4), we first set $x = x_n + h$, and then $x = x_n - h$. We get the two equations

$$y(x_{n+1}) = y(x_n) + hy'(x_n) + \int_{x_n}^{x_n+h} y''(z)(x_n+h-z)dz,$$

$$y(x_{n-1}) = y(x_n) - hy'(x_n) + \int_{x_n}^{x_n-h} y''(z)(x_n-h-z)dz.$$

We eliminate the quantity $hy'(x_n)$ from these equations, by performing a term by term summation:

$$y(x_{n+1}) + y(x_{n-1}) = 2y(x_n) + \int_{x_n}^{x_n+h} y''(x)(h+x_n-x)dx +$$

$$+ \int_{x_n}^{x_n-h} y''(x)(x_n-h-x)dx. \tag{9.5}$$

In the second integral we interchange the limits of integration and change the sign of the integrand. Then, this integral is written thusly:

$$\int_{x_n-h}^{x_n} y''(x)(h+x-x_n)dx.$$

Obviously, both integrals in the right-hand side of (9.5) may now be combined and formula (9.5) will be rewritten in the following form:

$$\Delta^2 y(x_{n-1}) = \int_{x_n-h}^{x_n+h} y''(x)(h-|x-x_n|)dx. \tag{9.6}$$

As we started from formula (3.9)

$$\Delta y(x_n) = \int_{x_n}^{x_n+h} y'(x)dx,$$

in the presentation of the difference methods for the solution of the Cauchy problem for first order equations, in the presentation of difference methods for the solution of (9.1) we shall start from formula (9.6). We construct the interpolation polynomial $P(x)$ for the function $y''(x)$ from its values at the following $k+p+1$ knots:

$$x_{n-k}, x_{n-k+1}, \ldots, x_n, x_{n+1}, \ldots, x_{n+p}. \tag{9.7}$$

We have

$$y''(x) = P(x) + r(x),$$

where $r(x)$ is the remainder term of interpolation. By substituting the right hand side of the last equation for $y''(x)$ in (9.6), we get, as in § 3, various difference methods for the solution of the problem (9.1).

For $p = 0$ we get the so-called *Störmer extrapolation method*. The difference methods obtained for $p \geq 1$ are called *interpolation methods*. In the following we shall examine two such methods: the *Störmer interpolation method*, corresponding to $p = 1$, and *Cowell's method*, corresponding to $p = k$.

We establish the formula for Störmer's extrapolation method. The interpolating polynomial $P(x)$, constructed from the conditions

$$P(x_{n-j}) = y''(x_{n-j}), \qquad j = 0, 1, 2, \ldots, k,$$

is naturally written according to Newton's formula for interpolation at the end of the table since the values of this polynomial will be used on the interval $[x_{n-1}, x_{n+1}]$. We have:

$$y''(x_n + ht) = y''(x_n) + \frac{t}{1!} \Delta y''(x_{n-1}) + \frac{t(t+1)}{2!} \Delta^2 y''(x_{n-2}) + \ldots$$

$$\ldots + \frac{t(t+1) \ldots (t+k-1)}{k!} \Delta^k y''(x_{n-k}) + r(x_n + ht), \qquad (9.8)$$

where $r(x_n + ht)$ is the remainder term of interpolation.

In the integral in (9.6) will make the change $x = x_n + ht$ of the variable of integration:

$$\Delta^2 y(x_{n-1}) = h^2 \int_{-1}^{1} y''(x_n + ht)(1 - |t|) dt.$$

Here, we substitute the right-hand side of formula (9.8) for $y''(x_n + ht)$. We get

$$\Delta^2 y(x_{n-1}) = h^2 \int_{-1}^{1} \left[y''(x_n) + \frac{t}{1!} \Delta y''(x_{n-1}) + \ldots \right.$$

$$\left. \ldots + \frac{t(t+1) \ldots (t+k-1)}{k!} \Delta^k y''(x_{n-k}) \right] (1 - |t|) dt +$$

$$+ h^2 \int_{-1}^{1} r(x_n + ht)(1 - |t|) dt. \qquad (9.9)$$

We introduce the notation

$$\alpha_j = \frac{1}{j!}\int_{-1}^{1} t(t+1)\ldots(t+j-1)(1-|t|)dt, \qquad j = 0, 1, 2, \ldots \quad (9.10)$$

We write out several of the first of the numbers α_j:

$$\alpha_0 = 1, \qquad \alpha_1 = 0, \qquad \alpha_2 = \tfrac{1}{12}, \qquad \alpha_3 = \tfrac{1}{12}, \qquad \alpha_4 = \tfrac{19}{240},$$

$$\alpha_5 = \tfrac{3}{40}, \alpha_6 = \tfrac{863}{12\,096}.$$

With the help of this notation, formula (9.9) may be written thusly:

$$\Delta^2 y(x_{n-1}) = h^2[y''(x_n)+\tfrac{1}{12}\Delta^2 y''(x_{n-2})+\tfrac{1}{12}\Delta^3 y''(x_{n-3})+ \cdots$$
$$\cdots +\alpha_k \Delta^k y''(x_{n-k})]+R_{n,k}. \qquad (9.11)$$

Here

$$R_{n,k} = h^2 \int_{-1}^{1} r(x_n+ht)(1-|t|)dt. \qquad (9.12)$$

Supposing that $y^{(k+3)}(x)$ exists on the interval $[x_{n-k}, x_{n+1}]$, one may use the representation of the remainder term of interpolation of (7.13) from chapter II

$$r(x_n+ht) = h^{k+1}\frac{t(t+1)\ldots(t+k)}{(k+1)!} y^{(k+3)}(\xi), \qquad x_{n-k} < \xi < x_{n+1}.$$

We substitute this expression for $r(x_n+ht)$ into (9.12) We get

$$R_{n,k} = \frac{h^{k+3}}{(k+1)!}\int_{-1}^{1} t(t+1)\ldots(t+k)(1-|t|)y^{(k+3)}(\xi)dt. \qquad (9.13)$$

The multiplier of $y^{(k+3)}(\xi)$ under the integral sign changes sign on the interval of integration, therefore, one may not apply the theorem of the mean.

Formula (9.11) is exact, but it cannot be used in calculating since the remainder term and the unknown values of $y''(x_{n-j})$ enter into it. In (9.11), we neglect the remainder term $R_{n,k}$, and substitute the approximate values y_i'', determined from the differential equation

$$y_i'' = f(x_i, y_i), \qquad i = 0, 1, 2, \ldots \qquad (9.14)$$

for the quantity $y''(x_i)$.

The values (9.3), $y_0, y_1, \ldots, y_n$, are known. We get

$$\Delta^2 y_{n-1} = h^2(y_n''+\tfrac{1}{12}\Delta^2 y_{n-2}''+\tfrac{1}{12}\Delta^3 y_{n-3}''+$$
$$+\tfrac{19}{240}\Delta^4 y_{n-4}''+\tfrac{3}{40}\Delta^5 y_{n-5}''+ \ldots +\alpha_k \Delta^k y_{n-k}''). \qquad (9.15)$$

We introduce the notation

$$\eta_i = h^2 y_i'' = h^2 f(x_i, y_i), \qquad i = 0, 1, 2, \ldots, \tag{9.16}$$

and rewrite formula (9.15) in the form

$$\Delta^2 y_{n-1} = \eta_n + \tfrac{1}{12}\Delta^2 \eta_{n-2} + \tfrac{1}{12}\Delta^3 \eta_{n-3} +$$
$$+ \tfrac{19}{240}\Delta^4 \eta_{n-4} + \tfrac{3}{40}\Delta^5 \eta_{n-5} + \ldots + \alpha_k \Delta^k \eta_{n-k}. \tag{9.17}$$

This is the formula for the Störmer extrapolation method.
The calculations by formula (9.17) are displayed in the following table
(we assume $k = 3$):

x	y	Δy	$\Delta^2 y$	$\eta = h^2 f(x, y)$	$\Delta \eta$	$\Delta^2 \eta$	$\Delta^3 \eta$
x_0	y_0			η_0			
		Δy_0			$\Delta \eta_0$		
x_1	y_1		$\Delta^2 y_0$	η_1		$\Delta^2 \eta_0$	
		Δy_1			$\Delta \eta_1$		$\Delta^3 \eta_0$
x_2	y_2		$\Delta^2 y_1$	η_2		$\Delta^2 \eta_1$	
		Δy_2			$\Delta \eta_2$		
x_3	y_3			η_3			
x_4							

The quantities indicated in the table are known. We need to calculate y_4
or, what is the same, $\Delta^2 y_2$. We set $k = 3$ and $n = 3$ in (9.17). We get

$$\Delta^2 y_2 = \eta_3 + \tfrac{1}{12}\Delta^2 \eta_1 + \tfrac{1}{12}\Delta^3 \eta_0.$$

All the quantities on the right-hand side of this formula are located on the
lower diagonal of the table of difference of η_i. Having determined y_4,
we calculate $\eta_4 = h^2 f(x_4, y_4)$ and add the new diagonal that allows one
to find $\Delta^2 y_3$, etc.
Example. We examine the Cauchy problem

$$y'' = y \cos x, \qquad y(0) = 1, \qquad y'(0) = 0. \tag{9.18}$$

Its solution exists on the whole real axis $-\infty < x < +\infty$.
To construct the beginning of the table, we use the expansion of the
solution $y(x)$ in a series in powers of x:

$$y(x) = 1 + \tfrac{1}{2}x^2 - \tfrac{1}{144}x^6 + \tfrac{1}{4480}x^8 + \ldots \tag{9.19}$$

For the step of integration, we choose $h = 0.2$ and calculate the values
of $y(x)$ for $x = x_j = j \cdot 0.2$; $j = \pm 1, \pm 2, \pm 3$ to five decimal places.

Table 39

x	y	Δy	$\Delta^2 y$	$\cos x$	$y\cos x$	η	$\Delta\eta$	$\Delta^2\eta$	$\Delta^3\eta$	$\Delta^4\eta$	$\Delta^5\eta$	$\Delta^6\eta$
-0.6	1.17968			0.82534	0.97364	0.038946						
-0.4	1.07997	-0.09971	0.03974	0.92106	0.99472	0.039789	843	-645				
-0.2	1.02000	-0.05997	0.03997	0.98007	0.99967	0.039987	198	-185	460	-301		
0	1	-0.02000	0.04000	1	1	0.040000	13	-26	159	-318	-17	34
0.2	1.02000	0.02000	0.03997	0.98007	0.99967	0.039987	-13	-185	-159	-301	17	23
0.4	1.07997	0.05997	0.03974	0.92106	0.99472	0.039789	-198	-645	-460	-261	40	54
0.6	1.17968	0.09971	0.03883	0.82534	0.97364	0.038946	-843	-1366	-721	-167	94	50
0.8	1.31822	0.13854	0.03655	0.69671	0.91842	0.036737	-2209	-2254	-888	-23	144	71
1	1.49331	0.17509	0.03201	0.54030	0.80684	0.032274	-4463	-3165	-911	192	215	32
1.2	1.70041	0.20710	0.02432	0.36236	0.61616	0.024646	-7628	-3884	-719	439	247	7
1.4	1.93183	0.23142	0.01279	0.16997	0.32835	0.013134	-11512	-4164	-280	693	254	
1.6	2.17604	0.24421	-0.00286	-0.02920	-0.06354	-0.002542	-15676	-3751	413			
1.8	2.41739	0.24135	-0.02217	-0.22720	-0.54923	-0.021969	-19427					
2.0	2.63657	0.21918										

To get five correct decimal places, the terms of the series which are written out in (9.19) are sufficient.

We calculate the values of $y_j, j = 4, 5, \ldots, 10$ by Störmer's extrapolation method. We use formula (9.17) for $k = 6$:

$$\Delta^2 y_{n-1} = \eta_n + \tfrac{1}{12}\Delta^2\eta_{n-2} + \tfrac{1}{12}\Delta^3\eta_{n-3} +$$
$$+ \tfrac{19}{240}\Delta^4\eta_{n-4} + \tfrac{3}{40}\Delta^5\eta_{n-5} + \tfrac{863}{120\,96}\Delta^6\eta_{n-6}.$$

The results of the calculations are shown in table 39.

10. Störmer's interpolation method

We shall represent the function $y''(x)$ with the help of the interpolation polynomial constructed from its values at the knots (9.7) for $p = 1$:

$$x_{n-k}, x_{n-k+1}, \ldots, x_{n-1}, x_n, x_{n-1}. \tag{10.1}$$

We write the interpolation polynomial by Newton's formula for interpolation at the end of the table:

$$y''(x_{n+1} + ht) = y''(x_{n+1}) +$$
$$+ \frac{t}{1!}\Delta y''(x_n) + \frac{t(t+1)}{2!}\Delta^2 y''(x_{n-1}) + \ldots$$
$$\ldots + \frac{t(t+1)\ldots(t+k)}{(k+1)!}\Delta^{k+1}y''(x_{n-k}) + r(x_{n+1} + ht). \tag{10.2}$$

Here $r(x_{n+1} + ht)$ is the remainder term of interpolation.
In formula (9.6),

$$\Delta^2 y(x_{n-1}) = \int_{x_n-h}^{x_n+h} y''(x)(h - |x - x_n|)dx,$$

we will make the change $x = x_{n+1} + ht$ of the variable of integration:

$$\Delta^2 y(x_{n-1}) = h^2 \int_{-2}^{0} y''(x_{n+1} + ht)(1 - |1 + t|)dt. \tag{10.3}$$

We introduce the notation

$$\alpha_j^* = \frac{1}{j!}\int_{-2}^{0} t(t+1)\ldots(t+j-1)(1 - |1 + t|)dt, \qquad j = 0, 1, 2, \ldots \tag{10.4}$$

We write out several of the first of the numbers α_j^*:

$$\alpha_0^* = 1, \quad \alpha_1^* = -1, \quad \alpha_2^* = \tfrac{1}{12}, \quad \alpha_3^* = 0, \quad \alpha_4^* = -\tfrac{1}{240},$$

$$\alpha_5^* = -\tfrac{1}{240}, \quad \alpha_6^* = -\tfrac{221}{60\,480}.$$

In (10.3), we will substitute the right-hand side of formula (10.2) for $y''(x_{n+1}+ht)$, and take into consideration the notation (10.4). The first two terms obtained in this way which contain $y''(x_{n+1})$ and $\Delta y''(x_n)$ may be combined since

$$y''(x_{n+1}) - \Delta y''(x_n) = y''(x_n).$$

We get

$$\Delta^2 y(x_{n-1}) = h^2[y''(x_n) + \tfrac{1}{12}\Delta^2 y''(x_{n-1}) - \tfrac{1}{240}\Delta^4 y''(x_{n-3}) - \cdots$$
$$\cdots + \alpha_{k+1}^* \Delta^{k+1} y''(x_{n-k})] + R_{n,k}^*. \tag{10.5}$$

Here

$$R_{n,k}^* = h^2 \int_{-2}^{0} r(x_{n+1}+ht)(1-|1+t|)dt.$$

We suppose that $y(x)$ has a derivative of order $k+4$. Then, the following representation is correct for the remainder term of interpolation,

$$r(x_{n+1}+ht) = \frac{h^{k+2}t(t+1)\dots(t+k+1)}{(k+2)!} y^{(k+4)}(\xi),$$

$$x_{n-k} < \xi < x_{n+1},$$

and $R_{n,k}^*$ may be written in the form

$$R_{n,k}^* = \frac{h^{k+4}}{(k+2)!}\int_{-2}^{0} t(t+1)\dots$$
$$\dots(t+k+1)(1-|1+t|)y^{(k+4)}(\xi)dt. \tag{10.6}$$

In formula (10.5) we throw away the remainder term and replace the quantities $y''(x_i)$ by their approximate values

$$y_i'' = f(x_i, y_i), \quad i = 0, 1, 2, \dots$$

We get

$$\Delta^2 y_{n-1} = h^2(y_n'' + \tfrac{1}{12}\Delta^2 y_{n-1}'' - \tfrac{1}{240}\Delta^4 y_{n-3}'' -$$
$$- \tfrac{1}{240}\Delta^5 y_{n-4}'' - \cdots + \alpha_{k+1}^* \Delta^{k+1} y_{n-k}''). \tag{10.7}$$

We rewrite (10.7), using the notation (9.16),

$$\eta_i = h^2 f(x_i, y_i), \qquad i = 0, 1, 2, \ldots$$

We get the formula for Störmer's interpolation method

$$\Delta^2 y_{n-1} = \eta_n + \tfrac{1}{12}\Delta^2 \eta_{n-1} - \tfrac{1}{240}\Delta^4 \eta_{n-3} -$$
$$- \tfrac{1}{240}\Delta^5 \eta_{n-4} - \ldots + \alpha^*_{k+1}\Delta^{k+1}\eta_{n-k}. \tag{10.8}$$

This formula is an equation for y_{n+1}.

The calculations by formula (10.8) are carried out as in the case of the Adams interpolation method. We assume that the following values are known

$$y_0, y_1, \ldots, y_k, y_{k+1}.$$

Let $k = 3$ for definiteness. We construct the table:

x	y	Δy	$\Delta^2 y$	$\eta = h^2 f(x, y)$	$\Delta \eta$	$\Delta^2 \eta$	$\Delta^3 \eta$	$\Delta^4 \eta$
x_0	y_0			η_0				
		Δy_0			$\Delta \eta_0$			
x_1	y_1		$\Delta^2 y_0$	η_1		$\Delta^2 \eta_0$		
		Δy_1			$\Delta \eta_1$		$\Delta^3 \eta_0$	
x_2	y_2		$\Delta^2 y_1$	η_2		$\Delta^2 \eta_1$		$\Delta^4 \eta_0$
		Δy_2			$\Delta \eta_2$		$\Delta^3 \eta_1$	
x_3	y_3		$\Delta^2 y_2$	η_3		$\Delta^2 \eta_2$		$\Delta^4 \eta_1$
		Δy_3			$\Delta \eta_3$		$\Delta^3 \eta_2$	
x_4	y_4			η_4		$\Delta^2 \eta_3$		
					$\Delta \eta_4$			
x_5				η_5				

In the table all the values are known which are located above the heavy stepped line. We need to calculate y_5. We set $k = 3$ and $n = 4$ in formula (10.8):

$$\Delta^2 y_3 = \eta_4 + \tfrac{1}{12}\Delta^2 \eta_3 - \tfrac{1}{240}\Delta^4 \eta_1. \tag{10.9}$$

The differences $\Delta^2 \eta_3$ and $\Delta^4 \eta_1$ are unknown, and are on the diagonal of the table below the heavy stepped line. We indicate initial approximations to the difference $\Delta^4 \eta_1$, supposing, for example, that $\Delta^4 \eta_1^{(0)} = \Delta^4 \eta_0$. This allows one to find the initial approximations

$$\Delta^4 \eta_1^{(0)}, \Delta^3 \eta_2^{(0)}, \Delta^2 \eta_3^{(0)}, \Delta \eta_4^{(0)}, \eta_5^{(0)}.$$

By formula (10.9) we find

$$\Delta^2 y_3^{(1)} = \eta_4 + \tfrac{1}{12}\Delta^2\eta_3^{(0)} - \tfrac{1}{240}\Delta^4\eta_1^{(0)}$$

and determine $y_5^{(1)}$.

We calculate $\eta_5^{(1)} = h^2 f(x_5, y_5^{(1)})$. If $\eta_5^{(1)} = \eta_5^{(0)}$, then we take $y_5 = y_5^{(1)}$, and proceed to the calculation of y_6. If $\eta_5^{(1)} \neq \eta_5^{(0)}$, then we change $\eta_5^{(0)}$ to $\eta_5^{(1)}$ and recalculate all the differences on this diagonal. By formula (10.9), we find $\Delta^2 y_3^{(2)}$. If $\Delta^2 y_3^{(2)} = \Delta^2 y_3^{(1)}$, then we suppose that $y_5 = y_5^{(1)}$. If $\Delta^2 y_3^{(2)} \neq \Delta^2 y_3^{(1)}$, then we find $\eta_5^{(2)}$, we compare $\eta_5^{(2)}$ and $\eta_5^{(1)}$, etc. The step must be such that $y_{n+1}^{(1)}$ or $y_{n+1}^{(2)}$ are the final values of y_{n+1}.

Example. We shall find a numerical solution for the problem (9.18)

$$y'' = y\cos x, \qquad y(0) = 1, \qquad y'(0) = 0$$

by Störmer's interpolation method. We take $h = 0.2$. The beginning of the table was constructed in § 9. The calculation of the values of y_j, $j = 4, 5, \ldots, 10$, we carry out according to formula (10.8) for $k = 5$:

$$\Delta^2 y_{n-1} = \eta_n + \tfrac{1}{12}\Delta^2\eta_{n-1} - \tfrac{1}{240}\Delta^4\eta_{n-3} -$$
$$- \tfrac{1}{240}\Delta^5\eta_{n-4} - \tfrac{221}{60\,480}\Delta^6\eta_{n-5}. \tag{10.10}$$

The results of the calculations are shown in table 40.

We explain the table. For calculating y_4 we must find an initial approximation to the sixth order difference $\Delta^6\eta_{-2}^{(0)}$. Since we have only the one difference $\Delta^6\eta_{-3} = 34$ (in units to the sixth decimal place), we suppose that $\Delta^6\eta_{-2}^{(0)} = 34$. Further, we calculate the differences on the diagonal containing $\Delta^6\eta_{-2}^{(0)}$, and we find $\Delta^2 y_2^{(1)} = 0.03883$ by formula (10.10) for $n = 3$. Now we can calculate $y_4^{(1)}$ and $f(x_4, y_4^{(1)})$. All the quantities to which we just referred and, in particular, the initial approximations to the differences $\Delta^2 y_2^{(1)}$, $\Delta y_3^{(1)}$, $y_4^{(1)}$, $y_4^{(1)}\cos x_4$ are written in the table in ordinary type. In calculating the quantity $\eta_4^{(1)} = 0.036737$, we see that $\eta_4^{(1)} \neq \eta_4^{(0)} = 0.036748$. We find new approximations for the differences: $\Delta^i\eta_{4-i}^{(1)}(i = 0, 1, \ldots, 6)$. The differences $\Delta^i\eta_{4-i}^{(1)}$ in the table are written in bold face above the corresponding differences $\Delta^i\eta_{4-i}^{(0)}$. According to formula (10.10), we find $\Delta^2 y_2^{(2)}$ by using the differences of $\Delta^i\eta_{4-i}^{(1)}$, and it turns out that $\Delta^2 y_2^{(2)} = \Delta^2 y_2^{(1)} = 0.03883$, so that we get $y_4^{(1)}$ as the final value of y_4.

In calculating y_8 (and also y_{10}) the first and second approximations do not coincide: $y_8^{(2)} \neq y_8^{(1)}$, but $\eta_8^{(2)} = \eta_8^{(1)}$ and therefore we take $y_8^{(2)}$ for the

Table 40

x	y	Δy	$\Delta^2 y$	$\cos x$	$y \cos x$	η	$\Delta\eta$	$\Delta^2\eta$	$\Delta^3\eta$	$\Delta^4\eta$	$\Delta^5\eta$	$\Delta^6\eta$
-0.6	1.17968			0.82534	0.97364	0.038946						
		-0.09971					843					
-0.4	1.07997		0.03974	0.92106	0.99472	0.039789		-645				
		-0.05997					198		460			
-0.2	1.02000		0.03997	0.98007	0.99967	0.039987		-185		-301		
		-0.02000					13		159		-17	
0	1		0.04000	1	1	0.040000		-26		-318		34
		0.02000					-13		-159		17	
0.2	1.02000		0.03997	0.98007	0.99967	0.039987		-185		-301		23
		0.05997					-198		-460		40	
0.4	1.07997		0.03974	0.92106	0.99472	0.039789		-645		-261		54
		0.09971					-843		-721	-250	94	
								-1366	-710	-167	63	50
0.6	1.17968		0.03883	0.82534	0.97364	0.038946		-1355		-198	144	54
		0.13854					-2209		-888	-23	148	
						0.036737	-2198	-2254	-919	-19	215	71
0.8	1.31822		0.03655	0.69671	0.91842	0.036748		-2285	-911	192	194	50
		0.17509					-4463		-907			
						0.032274	-4494	-3165				32

Table 40 (continued)

x	y	Δy	Δ²y	cos x	y cos x	η	Δη	Δ²η	Δ³η	Δ⁴η	Δ⁵η	Δ⁶η
1	1.49331		0.03201	0.54030	0.80684	0.032243		−3161		171		71
		0.20710					−7628		−719		247	
							−7624		−740		286	
1.2	1.70041		0.02432	0.36236	0.61616	**0.024646**		**−3884**		**439**		**7**
						0.024650		−3905		478		32
		0.23142					−11512		−280		254	
							−11533		−241		279	
1.4	1.93183		**0.01278**	0.16997	0.32835	**0.013134**		**−4164**		**693**		**−101**
			0.01279			0.013113		−4125		718		7
		0.24420					**−15676**		413		153	
		0.24421					−15637		438		261	
1.6	**2.17603**		−0.00286	−0.02920	−0.06354	**−0.002542**		**−3751**		846		
	2.17604					−0.002503		−3726		954		
		0.24134					−19427		**1259**			
							−19402		1367			
1.8	2.41737		**−0.02218**	−0.22720	−0.54923	**−0.021969**		**−2492**				
			−0.02217			−0.021944		−2384				
		0.21916					−21919					
		0.21917					−21811					
2	**2.63653**			−0.41615	**−1.09719**	−0.043888						
	2.63654				−1.09720	−0.043780						

final value of y_8. The quantities $y_8^{(2)}$, $\Delta y_7^{(2)}$, $\Delta^2 y_6^{(2)}$ are written in the table in bold face.

11. Cowell's method

In order to arrive at Cowell's method, we must replace the function $y''(x)$ in formula (9.6)

$$\Delta^2 y(x_{n-1}) = \int_{x_n-h}^{x_n+h} y''(x)(h-|x-x_n|)dx \tag{11.1}$$

by a representation in terms of the interpolating polynomial constructed from the values $y''(x)$ at the knots (9.7) for $p = k$:

$$x_{n-k}, \ldots, x_{n-1}, x_n, x_{n+1}, \ldots, x_{n+k} \tag{11.2}$$

(see § 9). The choice of the knots (11.2), placed symmetrically with respect to the point x_n, is the best with regard to the minimization of the remainder of interpolation on the points of the interval $[x_n-h, x_n+h]$.

The interpolation polynomial is naturally written according to Stirling's formula [see (8.8) chap. II]:

$$y''(x_n+ht) = y''(x_n) + \frac{t}{1!} \frac{\Delta y''(x_n)+\Delta y''(x_{n-1})}{2} +$$

$$+ \frac{t^2}{2!} \Delta^2 y''(x_{n-1}) + \frac{t(t^2-1^2)}{3!} \frac{\Delta^3 y''(x_{n-1})+\Delta^3 y''(x_{n-2})}{2} + \ldots$$

$$\ldots + \frac{t(t^2-1^2)(t^2-2^2)\ldots(t^2-(k-1)^2)}{(2k-1)!} \times$$

$$\times \frac{\Delta^{2k-1} y''(x_{n-k+1})+\Delta^{2k-1} y''(x_{n-k})}{2} +$$

$$+ \frac{t^2(t^2-1^2)(t^2-2^2)\ldots(t^2-(k-1)^2)}{(2k)!} \Delta^{2k} y''(x_{n-k}) + r(x_n+ht),$$

$$\tag{11.3}$$

where $r(x_n+ht)$ is the remainder term of interpolation.

In the integral (11.1) we will make the change $x = x_n+ht$ of the variable of integration:

$$\Delta^2 y(x_{n-1}) = h^2 \int_{-1}^{1} y''(x_n+ht)(1-|t|)dt. \tag{11.4}$$

We set

$$\tilde{\alpha}_{2j} = \frac{1}{(2j)!} \int_{-1}^{1} t^2(t^2-1^2)\dots(t^2-(j-1)^2)(1-|t|)dt,$$

$$\tilde{\alpha}_{2j+1} = \frac{1}{(2j+1)!} \int_{-1}^{1} t(t^2-1^2)\dots(t^2-j^2)(1-|t|)dt, \tag{11.5}$$

$$j = 0, 1, 2, \dots$$

Obviously, $\tilde{\alpha}_{2j+1} = 0$. We write out several of the first numbers $\tilde{\alpha}_{2j}$:

$$\tilde{\alpha}_0 = 1, \quad \tilde{\alpha}_2 = \tfrac{1}{12}, \quad \tilde{\alpha}_4 = -\tfrac{1}{240}, \quad \tilde{\alpha}_6 = \tfrac{31}{60\,480},$$

$$\tilde{\alpha}_8 = -\tfrac{289}{3\,628\,800}.$$

In (11.4), we substitute the right-hand side of equation (11.3) for $y''(x_n+ht)$ and use the notation (11.5). We get

$$\Delta^2 y(x_{n-1}) = h^2[y''(x_n)+\tfrac{1}{12}\Delta^2 y''(x_{n-1})-\tfrac{1}{240}\Delta^4 y''(x_{n-2})+\dots$$

$$\dots +\tilde{\alpha}_{2k}\Delta^{2k}y''(x_{n-k})]+\tilde{R}_{n,k}, \tag{11.6}$$

where

$$\tilde{R}_{n,k} = h^2 \int_{-1}^{1} r(x_n+ht)(1-|t|)dt.$$

Supposing that $y(x)$ has a derivative of the order $2k+3$, we can write $\tilde{R}_{n,k}$ in the form

$$R_{n,k} = \frac{h^{2k+3}}{(2k+1)!} \int_{-1}^{1} t(t^2-1^2)\dots$$

$$\dots (t^2-k^2)(1-|t|)y^{(2k+3)}(\xi)dt, \tag{11.7}$$

$$x_{n-k} < \xi < x_{n+k}.$$

In formula (11.6) we throw away the remainder term, replace $y''(x_i)$ approximately by $y_i'' = f(x_i, y_i)$, and use the notation (9.16)

$$\eta_i = h^2 y_i'' = h^2 f(x_i, y_i), \quad i = 0, 1, 2, \dots$$

We get the formula for Cowell's method:

$$\Delta^2 y_{n-1} = \eta_n+\tfrac{1}{12}\Delta^2\eta_{n-1}-\tfrac{1}{240}\Delta^4\eta_{n-2}+$$

$$+\tfrac{31}{60\,480}\Delta^6\eta_{n-3}-\dots+\tilde{\alpha}_{2k}\Delta^{2k}\eta_{n-k}. \tag{11.8}$$

All terms on the right-hand side of (11.8), except η_n, are unknown, as they depend on the unknown quantities y_{n+1}, y_{n+2}, ..., y_{n+k}. Formula (11.8) is an equation in k unknowns.

The calculations by formula (11.8) are analogous to the calculations by formula (7.7) for the method of Cowell type. We assume that the values

$$y_0, y_1, \ldots, y_{2k}$$

are known. We construct the table ($k = 2$):

x	y	Δy	$\Delta^2 y$	$\eta = h^2 f(x, y)$	$\Delta\eta$	$\Delta^2\eta$	$\Delta^3\eta$	$\Delta^4\eta$
x_0	y_0			η_0				
		Δy_0			$\Delta\eta_0$			
x_1	y_1		$\Delta^2 y_0$	η_1		$\Delta^2\eta_0$		
		Δy_1			$\Delta\eta_1$		$\Delta^3\eta_0$	
x_2	y_2		$\Delta^2 y_1$	η_2		$\Delta^2\eta_1$		$\Delta^4\eta_0$
		Δy_2			$\Delta\eta_2$		$\Delta^3\eta_1$	
x_3	y_3		$\Delta^2 y_2$	η_3		$\Delta^2\eta_2$		$\Delta^4\eta_1$
		Δy_3			$\Delta\eta_3$		$\Delta^3\eta_2$	
x_4	y_4			η_4		$\Delta^2\eta_3$		$\Delta^4\eta_2$
					$\Delta\eta_4$			
x_5				η_5				

It is necessary to find $\Delta^2 y_3$. We suppose that $k = 2$ and $n = 4$ in (11.8):

$$\Delta^2 y_3 = \eta_4 + \tfrac{1}{12}\Delta^2\eta_3 - \tfrac{1}{240}\Delta^4\eta_2. \tag{11.9}$$

The differences entering this problem are located in the same horizontal row of the table in which one must write $\Delta^2 y_3$. The differences $\Delta^2\eta_3$ and $\Delta^4\eta_2$ are unknown, and are found in the two lower diagonals located below the heavy stepped line.

We indicate the initial values of $\Delta^4\eta_1^{(0)}$ and $\Delta^4\eta_2^{(0)}$ for the higher differences in the unknown diagonals. This allows one to find $\Delta^2\eta_3^{(0)}$ and to use formula (11.9).

We will get $\Delta^2 y_3^{(1)}$ and find $y_5^{(1)}$. From the differential equation, we find

$$\eta_5^{(1)} = h^2 f(x_5, y_5^{(1)}).$$

If $\eta_5^{(1)} = \eta_5^{(0)}$, then we consider that $y_5 = y_5^{(1)}$. If $\eta_5^{(1)} \neq \eta_5^{(0)}$, then we recalculate the differences in the diagonal containing $\eta_5^{(1)}$, and from the new value of $\Delta^4\eta_1^{(1)}$, we find the new value for $\Delta^4\eta_2^{(1)}$. With the help of

Table 41

x	y	Δy	$\Delta^2 y$	$\cos x$	$y \cos x$	η	$\Delta\eta$	$\Delta^2\eta$	$\Delta^3\eta$	$\Delta^4\eta$	$\Delta^5\eta$	$\Delta^6\eta$
−0.6	1.17968			0.82534	0.97364	0.038946						
		−0.09971					843					
−0.4	1.07997		0.03974	0.92106	0.99472	0.039789		−645				
		−0.05997					198		460			
−0.2	1.02000		0.03997	0.98007	0.99967	0.039987		−185		−301		
		−0.02000					13		159		−17	
0	1		0.04000	1	1	0.040000		−26		−318		34
		0.02000					−13		−159		17	
0.2	1.02000		0.03997	0.98007	0.99967	0.039987		−185		−301		23
		0.05997					−198		−460		40	−34
0.4	1.07997		0.03974	0.92106	0.99472	0.039789		−645		−261	51	54
		0.09971					−843		−721	−250	94	34
								−1366	−710	−167	85	50
												51

Table 41 (continued)

x	y	Δy	$\Delta^2 y$	$\cos x$	$y \cos x$	η	$\Delta\eta$	$\Delta^2\eta$	$\Delta^3\eta$	$\Delta^4\eta$	$\Delta^5\eta$	$\Delta^6\eta$
0.6	1.17968		0.03883	0.82534	0.97364	0.038946		−1355		−165		34
		0.13854					−2209		−888		144	
							−2198	−2254	−875	−23	145	34
0.8	1.31822		0.03655	0.69671	0.91842	0.036737		−2230	−911	−22	119	51
						0.036748	−4463	−3165	−910	−46	196	34
		0.17509					−4428	−3164	−921	174	153	
1	1.49331		0.03201	0.54030	0.80684	0.032274		−3151		107		51
			0.03205			0.032309	−7628					34
		0.20710				0.024646	−7627					
		0.20714				0.024647	−7579					
1.2	1.70041				0.61616	0.024730						
	1.70045			0.36236	0.61618							

(11.9), we find $\Delta^2 y_3^{(2)}$. If $\Delta^2 y_3^{(2)} = \Delta^2 y_3^{(1)}$, then we consider that $y_5 = y_5^{(2)} = y_5^{(1)}$. In the contrary case, we find

$$\eta_5^{(2)} = h^2 f(x_5, y_5^{(2)}),$$

we recalculate the differences in the diagonal containing $\eta_5^{(2)}$, getting a new value of $\Delta^4 \eta_2^{(2)}$, we find $\Delta^2 y_3^{(3)}$ etc. The step must be such that $y_j^{(1)}$ or $y_j^{(2)}$ no longer require correction.

We remark that the coefficients of formula (11.8) diminish quickly, therefore the influence of differences on the unknown diagonals not controlling the differential equation is insignificant.

Example. We examine the Cauchy problem (9.18)

$$y'' = y \cos x, \qquad y(0) = 1, \qquad y'(0) = 0.$$

By using the beginning of table 39 constructed in § 9 ($h = 0.2$), we will calculate y_4, y_5 and y_6 by Cowell's method. The results of the calculations are shown in table 41.

The calculations were carried out according to formula (11.8) for $k = 3$:

$$\Delta^2 y_{n-1} = \eta_n + \tfrac{1}{12}\Delta^2 \eta_{n-1} - \tfrac{1}{240}\Delta^4 \eta_{n-2} + \tfrac{31}{60\,480}\Delta^6 \eta_{n-3}. \qquad (11.10)$$

In order to calculate y_4 (or $\Delta^2 y_2$), we must know the differences in the same horizontal row in which $\Delta^2 y_2$ is located. For this it is necessary to find initial approximations to the differences

$$\Delta^6 \eta_{-2}^{(0)}, \Delta^6 \eta_{-1}^{(0)} \quad \text{and} \quad \Delta^6 \eta_0^{(0)}.$$

Intending to calculate all three values y_4, y_5, y_6, we indicate $\Delta^6 \eta_1^{(0)}$ and $\Delta^6 \eta_2^{(0)}$ as well, namely, we suppose that

$$\Delta^6 \eta_{-2}^{(0)} = \Delta^6 \eta_{-1}^{(0)} = \ldots = \Delta^6 \eta_2^{(0)} = 34.$$

This allows one to find the initial values of the differences situated on the lower diagonals containing $\Delta^6 \eta_{-2}^{(0)}, \Delta^6 \eta_{-1}^{(0)}, \ldots, \Delta^6 \eta_2^{(0)}$. Further by formula (11.10), we calculate $\Delta^2 y_j^{(1)}$, $j = 2, 3, 4$. We find $y_j^{(1)}$ and

$$f(x_j, y_j^{(1)}) = y_j^{(1)} \cos x_j, \qquad j = 4, 5, 6.$$

All the quantities we just referred to are written in the table in ordinary type. Having calculated $\eta_j^{(1)}$, we see that $\eta_j^{(1)} \neq \eta_j^{(0)}$, $j = 4, 5, 6$. From the quantities $\eta_j^{(1)}$, $j = 4, 5, 6$, we recalculate the differences in the corresponding three lower diagonals, we suppose $\Delta^6 \eta_1^{(1)} = \Delta^6 \eta_2^{(1)} = \Delta^6 \eta_0^{(1)} = 51$ and calculate $\Delta^5 \eta_2^{(1)}$, $\Delta^4 \eta_3^{(1)}$. The new differences obtained with the help of the first approximations are written in bold face above the corresponding differences of the initial approximations.

According to formula (11.10), we calculate $\Delta^2 y_j^{(2)}$, $j = 2, 3, 4$. Here it turns out that $\Delta^2 y_j^{(2)} = \Delta^2 y_j^{(1)}$ for $j = 2, 3$, therefore we consider the values of $y_j^{(1)}$, $j = 4, 5$, to be final. Since $\Delta^2 y_4^{(2)} \neq \Delta^2 y_4^{(1)}$, we find $y_6^{(2)}$ and calculate $\eta_6^{(2)}$ and the differences $\Delta^i \eta_{6-i}^{(2)}$, $i = 0, 1, 2, \ldots, 6$ (in the table these differences are written in italics). In calculating $\Delta^2 y_4^{(3)}$, we are convinced that $\Delta^2 y_4^{(3)} = \Delta^2 y_4^{(2)}$ and, consequently, one can consider that $y_6 = y_6^{(2)}$.

12. On the estimate of error of the Adams method

We will examine the question of the estimate of error arising from the numerical solution of the Cauchy problem

$$y' = f(x, y), \qquad y(x_0) = y_0 \tag{12.1}$$

by the Adams extrapolation and interpolation methods.

We devote attention to the numerical solution of problem (12.1) on the finite interval $[x_0, X]$, $X > x_0$. We suppose that the solution of $y(x)$ of problem (12.1) exists on $[x_0, X]$. We assume that in a certain closed area D, convex with respect to y and containing the curvilinear strip D_ρ, defined by the inequalities

$$x_0 \leq x \leq X, \qquad -\rho \leq y(x) - y \leq \rho \qquad (\rho > 0), \tag{12.2}$$

$f(x, y)$ is continuous and has continuous partial derivatives up to order $r + 1$ inclusive, $r \geq 0$.

As usual we set

$$x_j = x_0 + jh, \qquad j = 0, 1, 2, \ldots, N,$$

where

$$x_0 + Nh \leq X < x_0 + (N+1)h. \tag{12.3}$$

Formula (3.26) for the Adams extrapolation method and formula (3.32) for the Adams interpolation method will be written by means of the single formula

$$y_{n+1} = y_n + h \sum_{j=-1}^{k} \beta_j f(x_{n-j}, y_{n-j}). \tag{12.4}$$

In the case of the Adams extrapolation method,

$$\beta_{-1} = 0, \qquad \beta_j = b_{kj}, \qquad j = 0, 1, 2, \ldots, k,$$

in the case of the interpolation method,

$$\beta_j = b_{kj}^*, \qquad j = -1, 0, 1, 2, \ldots, k.$$

For simplicity of notation, we do not note the dependence of the coefficients β_j on k.

By virtue of (3.23), (3.30), and (3.31), we have

$$\beta_{-1} \geqq 0, \qquad \beta_0 > 0. \tag{12.5}$$

The Adams method can be considered to be a method of constructing the solution y_n of the difference equation (12.4) obtained by giving for $n = 0, 1, 2, \ldots, k$ the initial values

$$y_0, y_1, \ldots, y_k. \tag{12.6}$$

Namely, supposing $n = k$ in (12.4) and substituting the known values (12.6), we find y_{k+1}. Here, if $\beta_{-1} = 0$, then y_{k+1} is computed directly, but if $\beta_{-1} > 0$, then y_{k+1} must be calculated as a solution of the equation of the form

$$y_{k+1} = h\beta_{-1}f(x_{k+1}, y_{k+1}) + F(y_0, y_1, \ldots, y_k).$$

Similarly, supposing that $n = k+1$ in (12.4) and substituting the known values $y_1, y_2, \ldots, y_k$ and the y_{k+1} already found, we find y_{k+2}, etc. The exact solution of the difference equation (12.4) is denoted by y_n, $n = k+1$, $k+2, \ldots, N$.

In actual calculation by this method, computation is carried out with rounding off, therefore, instead of the exact values

$$y_{k+1}, y_{k+2}, \ldots, y_N,$$

we get the approximate values

$$\tilde{y}_{k+1}, \tilde{y}_{k+2}, \ldots, \tilde{y}_N.$$

The initial values (12.6) will also be denoted by $\tilde{y}_n$:

$$\tilde{y}_n = y_n, \qquad n = 0, 1, 2, \ldots, k. \tag{12.7}$$

We suppose that

$$(x_n, \tilde{y}_n) \in D, \qquad n = 0, 1, 2, \ldots, N. \tag{12.8}$$

Thus, $\tilde{y}_n$ does not satisfy the difference equation (12.4) exactly, and we have

$$\tilde{y}_{n+1} = \tilde{y}_n + h \sum_{j=-1}^{k} \beta_j f(x_{n-j}, \tilde{y}_{n-j}) + \Gamma_n. \tag{12.9}$$

The quantity Γ_n will be called the round-off error.

On the basis of (3.20) and (3.27), the following equality is correct:

$$y(x_{n+1}) = y(x_n) + h \sum_{j=-1}^{k} \beta_j f(x_{n-j}, y(x_{n-j})) + R_n, \tag{12.10}$$

where R_n is the remainder term or error of the Adams method (12.4). For R_n in the case of the Adams extrapolation method, the representation (3.22) was indicated, in the case of the interpolation method, the representation is (3.29). We write these representations in one formula.

$$R_n = ah^{r+2}y^{(r+2)}(\xi), \qquad x_{n-k} \leq \xi \leq x_{n+1}, \tag{12.11}$$

where

$$r = k + \text{sign } \beta_{-1} \geq 0.$$

The coefficient a in the case of the Adams extrapolation method is equal to a_{k+1} where a_{k+1} is defined by formula (4.5), and in the case of the Adams interpolation method, $a = a_{k+2}^{*}$ where a_{k+2}^{*} is defined by formula (6.6).

Equations (12.10) and (12.11) show in general that the $y(x_n)$ do not satisfy equation (12.4); $y(x_n)$ satisfies equation (12.4), if $y(x)$ is a polynomial of a degree not greater than $r+1$. Since $r \geq 0$, then equation (12.4) is always satisfied by $y(x_n)$, where $y(x) = x - x_0$. In this case, $f(x, y(x)) = y'(x) = 1$, $y(x_n) = nh$ and equation (12.4) is written thusly:

$$(n+1)h = nh + h \sum_{j=-1}^{k} \beta_j$$

or

$$\sum_{j=-1}^{k} \beta_j = 1. \tag{12.12}$$

We denote the error of $\tilde{y}_n$ as an approximation to solution $y(x)$ of problem (12.1) by ε_n:

$$\varepsilon_n = y(x_n) - \tilde{y}_n, \qquad n = 0, 1, 2, \ldots, N. \tag{12.13}$$

We write out equation (12.9) term by term from equation (12.10) and take into consideration the notation of (12.13). We get the difference equation

$$\varepsilon_{n+1} = \varepsilon_n + h \sum_{j=-1}^{k} \beta_j [f(x_{n-j}, y(x_{n-j})) - f(x_{n-j}, \tilde{y}_{n-j})] + R_n - \Gamma_n,$$
(12.14)

which is satisfied by the error ε_n.

We note that $\varepsilon_0, \varepsilon_1, \ldots, \varepsilon_k$, in view of (12.7), represent the errors of the initial values of (12.6). An upper bound for the absolute values of these errors is usually known. We denote it by ε:

$$|\varepsilon_j| \leq \varepsilon, \quad j = 0, 1, 2, \ldots, k.$$
(12.15)

We also suppose that upper bounds are known for the absolute value of error of the method

$$|R_n| \leq R$$
(12.16)

and for the absolute value of the round-off error

$$|\Gamma_n| \leq \Gamma$$
(12.17)

for $n = k, k+1, k+2, \ldots, N-1$.

An upper bound for R_n may be indicated if one uses the representation (12.11):

$$|R_n| \leq |a| h^{r+2} \max_{x_0 \leq x \leq X} |y^{(r+2)}(x)|.$$
(12.18)

We note that the determination of an upper bound for $\max |y^{(r+2)}(x)|$, which does not greatly exceed this maximum, presents considerable difficulty.

The satisfaction of inequality (12.17) can be obtained if one carries out the calculations with sufficient exactness.

Our problem consists of the fact that we must indicate an estimate for $|\varepsilon_n|$ for $n = k+1, k+2, \ldots, N$. By virtue of the convexity of D and condition (12.8) we can use the formula for finite increments,

$$f(x_i, y(x_i)) - f(x_i, \tilde{y}_i) = \frac{\partial f(x_i, \xi_i)}{\partial y} [y(x_i) - \tilde{y}_i] = \varepsilon_i \frac{\partial f(x_i, \xi_i)}{\partial y}.$$

The point (x_i, ξ_i) belongs to the area D.

Using this equation, we rewrite equation (12.14) in the form

$$\left[1 - h\beta_{-1}\frac{\partial f(x_{n+1}, \xi_{n+1})}{\partial y}\right]\varepsilon_{n+1} =$$

$$= \varepsilon_n + h\sum_{j=0}^{k}\beta_j\frac{\partial f(x_{n-j}, \xi_{n-j})}{\partial y}\varepsilon_{n-j} + R_n - \Gamma_n. \qquad (12.19)$$

We denote by K a constant satisfying the inequality

$$\left|\frac{\partial f(x, y)}{\partial y}\right| \leq K \qquad (12.20)$$

with (x, y) in the area D. If $\beta_{-1} > 0$, we consider h to be so small that the following inequality is satisfied

$$h\beta_{-1}K < 1. \qquad (12.21)$$

With the help of (12.20) and (12.21) we get from (12.19),

$$(1 - h\beta_{-1}K)|\varepsilon_{n+1}| \leq |\varepsilon_n| + hK\sum_{j=0}^{k}|\beta_j||\varepsilon_{n-j}| + |R_n| + |\Gamma_n|. \qquad (12.22)$$

We consider the finite difference equation

$$(1 - h\beta_{-1}K)E_{n+1} = E_n + hK\sum_{j=0}^{k}|\beta_j|E_{n-j} + R + \Gamma. \qquad (12.23)$$

If

$$|\varepsilon_n| \leq E_n, |\varepsilon_{n-1}| \leq E_{n-1}, \ldots, |\varepsilon_{n-k}| \leq E_{n-k},$$

then from inequality (12.22), we get

$$(1 - h\beta_{-1}K)|\varepsilon_{n+1}| \leq E_n + hK\sum_{j=0}^{k}|\beta_j|E_{n-j} + |R_n| + |\Gamma_n|.$$

Since, by virtue of (12.16) and (12.17),

$$|R_n| + |\Gamma_n| \leq R + \Gamma,$$

then the right-hand side of the last inequality may be exchanged for the left-hand side of equality (12.23), and we get

$$(1 - h\beta_{-1}K)|\varepsilon_{n+1}| \leq (1 - h\beta_{-1}K)E_{n+1}$$

or, according to (12.21),

$$|\varepsilon_{n+1}| \leqq E_{n+1}.$$

Thus, if E_n is a solution of the difference equation (12.23) which satisfies the inequalities

$$|\varepsilon_n| \leqq E_n, \qquad n = 0, 1, 2, \ldots, k, \tag{12.24}$$

then inequalities (12.24) will be valid also for $n = k+1, k+2, \ldots, N$.
We proceed to the construction of a solution of the difference equation (12.23) which satisfies inequalities (12.24). Since, according to (12.15),

$$|\varepsilon_j| \leqq \varepsilon, \qquad j = 0, 1, 2, \ldots, k,$$

then it is sufficient to construct a solution E_n (12.23) satisfying the inequalities

$$\varepsilon \leqq E_n, \qquad n = 0, 1, 2, \ldots, k. \tag{12.25}$$

Equation (12.23) is non-homogeneous, with the right-hand side equal to the constant $R + \Gamma$. We seek a particular solution of this equation equal to the constant A. Substituting $E_n = A$ in (12.23), we get

$$A = -\frac{R+\Gamma}{hK\sigma}, \tag{12.26}$$

where

$$\sigma = \sum_{j=-1}^{k} |\beta_j|. \tag{12.27}$$

We remark that $\sigma \geqq 1$, on the basis of (12.12).
Since $A < 0$, then the particular solution found does not satisfy conditions (12.25). We shall find the solution of the homogeneous equation

$$(1 - h\beta_{-1}K)E_{n+1} = E_n + hK \sum_{j=0}^{k} |\beta_j| E_{n-j}. \tag{12.28}$$

We write down the characteristic equation for (12.28),

$$\chi(z) \equiv (1 - h\beta_{-1}K)z^{k+1} - (1 + h\beta_0 K)z^k - hK \sum_{j=1}^{k} |\beta_j| z^{k-j} = 0. \tag{12.29}$$

We have

$$\chi(1) = -hK \sum_{j=-1}^{k} |\beta_j| = -hK\sigma < 0$$

and $\chi(+\infty) = +\infty$, therefore, equation (12.29) has a real root z_1, and $z_1 > 1$. We note that equation (12.29) lacks other positive solutions which satisfy (12.21). $E_n = z_1^n$ is the solution of homogeneous equation (12.28). Since the constant A, determined by equality (12.26), satisfies the non-homogeneous equation (12.23), then

$$E_n = Bz_1^n - \frac{R+\Gamma}{hK\sigma} \tag{12.30}$$

satisfies equation (12.23) for any constant B. We choose the constant B so that the solution (12.30) satisfies inequalities (12.25). It is sufficient to require that $E_0 = \varepsilon$. This leads to the following value for B:

$$B = \varepsilon + \frac{R+\Gamma}{hK\sigma}, \tag{12.31}$$

and the solution (12.30) is written in the form

$$E_n = \varepsilon z_1^n + \frac{R+\Gamma}{hK\sigma}(z_1^n - 1). \tag{12.32}$$

Since $z_1 > 1$, the solution (12.32) satisfies conditions (12.25) and, consequently, the inequality

$$|\varepsilon_n| \leqq \varepsilon z_1^n + \frac{R+\Gamma}{hK\sigma}(z_1^n - 1) \tag{12.33}$$

is satisfied for all $n = k+1, k+2, \ldots, N$. Inequality (12.33) also represents the estimate which we wanted to obtain.

The solution z_1 of the characteristic equation (12.29) is a function of h: $z_1 = z_1(h)$. Setting $h = 0$ in (12.29), we get $z_1(0) = 1$. We shall calculate the derivative with respect to h of the left-hand side of (12.29):

$$-\beta_{-1} K z^{k+1} + (1-h\beta_{-1}K)(k+1)z^k z' - \beta_0 K z^k - (1+h\beta_0 K)kz^{k-1}z' -$$
$$- K \sum_{j=1}^{k} |\beta_j| z^{k-j} - hK \sum_{j=1}^{k} |\beta_j|(k-j)z^{k-j-1}z' = 0.$$

Here we set $h = 0$ and take into account that $z_1(0) = 1$. We get

$$-\beta_{-1}K + (k+1)z_1'(0) - \beta_0 K - kz_1'(0) - K\sum_{j=1}^{k}|\beta_j| = 0$$

or

$$z_1'(0) = K\sigma.$$

Thus, the expansion of z_1 in powers of h has the form

$$z_1 = 1 + K\sigma h + \ldots \tag{12.34}$$

It follows from (12.34) that the right-hand side of inequality (12.33) increases with n exponentially. In fact,

$$z_1^n = (1 + K\sigma h + \ldots)^{(x_n - x_0)/h} \approx e^{K\sigma(x_n - x_0)}.$$

The estimate (12.33) turns out, as a rule, to be greatly excessive. This explains the fact that it is applicable to a wide class of problems (12.1). We note the estimate (12.33) is *a posteriori* since we assumed that $(x_i, \tilde{y}_i) \in D$. We shall show that for rather small h, $0 < h \leq h_0$, and for all $n = 0, 1, 2, \ldots, N$, the following inequality is satisfied,

$$z_1^n \leq C, \tag{12.35}$$

where the constant C depends on the problem (12.1) and the method (12.4) and not on h. On the basis of (12.3) and (12.34), we have

$$z_1^n \leq z_1^N \leq z_1^{(X-x_0)/h} = (1 + K\sigma h + \ldots)^{(X-x_0)/h}.$$

Since the last term of this chain of inequalities tends to

$$e^{K\sigma(X-x_0)},$$

as $h \to 0$, then inequality (12.35) is established. From inequality (12.33) with the help of (12.35), we get

$$\max_{0 \leq n \leq N} |\varepsilon_n| \leq \varepsilon C + \frac{R+\Gamma}{hK\sigma}(C-1) \tag{12.36}$$

for all $0 < h \leq h_0$.
We assume that

$$\varepsilon = \alpha h^p, \qquad \Gamma = \alpha h^{p+1},$$

where α and p are certain positive constants. It follows from inequality

(12.18) that for R one can take

$$R = \gamma h^{r+2}, \qquad \gamma > 0.$$

By substituting these values ε, Γ and R in (12.36), we get

$$\max_{0 \leqq n \leqq N} |\varepsilon_n| \leqq C\alpha h^p + \frac{C-1}{K\sigma}(\gamma h^{r+1} + \alpha h^p).$$

Since $p > 0$ and $r \geqq 0$, then the right-hand side of this inequality tends to zero as $h \to 0$. From this follows the uniform convergence of the numerical solution $\tilde{y}_n$ to $y(x)$ on the interval $[x_0, X]$ as $h \to 0$:

$$\lim_{h \to 0} \max_{0 \leqq n \leqq N} |\varepsilon_n| = 0 \tag{12.37}$$

for any errors $\varepsilon_0, \varepsilon_1, \ldots, \varepsilon_k$ of the initial values and any round-off errors of Γ_n, satisfying the inequalities

$$|\varepsilon_j| \leqq \alpha h^p, \qquad j = 0, 1, 2, \ldots, k,$$

$$|\Gamma_n| \leqq \alpha h^{p+1}, \qquad n = k, k+1, \ldots, N-1.$$

The area D contains the band D_ρ defined by inequalities (12.2), therefore, for all sufficiently small h, the points $(x_n, \tilde{y}_n)$ belong to the area D.

We suppose now that the right-hand side of the differential equation (12.1) has a negative derivative with respect to y and the following inequalities are fulfilled in area D

$$q \leqq -\frac{\partial f(x, y)}{\partial y} \leqq K, \tag{12.38}$$

where q and K are positive contants. In this case, one can indicate a more exact estimate for $|\varepsilon_n|$ than the one in (12.33).

We write equation (12.19) in the form

$$\left[1 - h\beta_{-1}\frac{\partial f(x_{n+1}, \xi_{n+1})}{\partial y}\right]\varepsilon_{n+1} = \left[1 + h\beta_0\frac{\partial f(x_n, \xi_n)}{\partial y}\right]\varepsilon_n +$$

$$+ h\sum_{j=1}^{k}\beta_j\frac{\partial f(x_{n-j}, \xi_{n-j})}{\partial y}\varepsilon_{n-j} + R_n - \Gamma_n. \tag{12.39}$$

We suppose that

$$h\beta_0 q < 1. \tag{12.40}$$

By virtue of (12.38) and (12.40), from (12.39) we get the inequality

$$(1+h\beta_{-1}q)|\varepsilon_{n+1}| \leqq (1-h\beta_0 q)|\varepsilon_n|+$$

$$+hK\sum_{j=1}^{k}|\beta_j||\varepsilon_{n-j}|+|R_n|+|\Gamma_n|. \tag{12.41}$$

We assume, as above, that $(x_n, \tilde{y}_n) \in D$, where $n = 0, 1, 2, \ldots, N$. We consider the difference equation

$$(1+h\beta_{-1}q)E_{n+1} = (1-h\beta_0 q)E_n + hK\sum_{j=1}^{k}|\beta_j|E_{n-j}+R+\Gamma. \tag{12.42}$$

It is not hard to verify, by using inequality (12.41), that if E_n is the solution of the difference equation (12.42) which satisfies inequalities (12.24),

$$|\varepsilon_n| \leqq E_n, \qquad n = 0, 1, 2, \ldots, k,$$

then inequality (12.24) will also be satisfied for $n = k+1, k+2, \ldots, N$. By virtue of (12.15), it is sufficient to construct a solution E_n of equation (12.42), which satisfies the inequalities

$$\varepsilon \leqq E_n \quad \text{where} \quad n = 0, 1, 2, \ldots, k. \tag{12.43}$$

The particular solution of equation (12.42), equal to the constant A, is

$$A = \frac{R+\Gamma}{h\delta}, \tag{12.44}$$

where

$$\delta = q(\beta_{-1}+\beta_0)-K\sum_{j=1}^{k}|\beta_j|. \tag{12.45}$$

We suppose that

$$\delta > 0. \tag{12.46}$$

We consider the characteristic equation for the difference equation (12.42),

$$\chi(z) = (1+h\beta_{-1}q)z^{k+1}-(1-h\beta_0 q)z^k-hK\sum_{j=1}^{k}|\beta_j|z^{k-j} = 0. \tag{12.47}$$

Under conditions (12.40), equation (12.47) has a single positive solution. We denote it by z_1. Since

$$\chi(0) = -hK|\beta_k| < 0,$$
$$\chi(1) = h\delta > 0,$$

then

$$0 < z_1 < 1. \tag{12.48}$$

It is not hard to get that the first two terms of the expansion of z_1 in powers of h:

$$z_1 = 1 - \delta h + \ldots$$

Thus

$$E_n = A + Bz_1^n, \tag{12.49}$$

where A is defined by the solution of the difference equation (12.42). We choose the constant B so that condition (12.43) will be satisfied. Here, one must distinguish two cases: $\varepsilon < A$ and $\varepsilon > A$.

In the case $\varepsilon < A$, the following inequality will be satisfied

$$|\varepsilon_n| \leqq A \tag{12.50}$$

for all $n = 0, 1, 2, \ldots, N$. But the evaluation (12.50) may be made more precise by selecting for B a negative number such that

$$\varepsilon \leqq A + Bz_1^n, \qquad n = 0, 1, 2, \ldots, k. \tag{12.51}$$

It is obvious that if one takes $B = \varepsilon - A$, then on the basis of (12.48), inequalities (12.51) will be satisfied, and, consequently,

$$|\varepsilon_n| \leqq A + (\varepsilon - A)z_1^n$$

or, by virtue of (12.44),

$$|\varepsilon_n| \leqq \varepsilon z_1^n + \frac{R + \Gamma}{h\delta}(1 - z_1^n), \qquad n = 0, 1, 2, \ldots, N. \tag{12.52}$$

This is the required estimate for the case $\varepsilon < A$.

Now, let $\varepsilon > A$. In order that inequalities (12.51) be fulfilled, it is sufficient to determine B from the inequality

$$\varepsilon = A + Bz_1^k.$$

We get

$$B = (\varepsilon - A)z_1^{-k}.$$

Thus,

$$|\varepsilon_n| \leqq A + (\varepsilon - A)z_1^{n-k},$$

and the estimate for the case $\varepsilon > A$ has the form

$$|\varepsilon_n| \leqq \varepsilon z_1^{n-k} + \frac{R+\Gamma}{h\delta}(1 - z_1^{n-k}), \qquad n = 0, 1, 2, \ldots, N. \tag{12.53}$$

For $\varepsilon = A$, inequalities (12.52) and (12.53) become inequality (12.50). We note the special case: $k = 0$, for which formula (12.4) has the form

$$y_{n+1} = y_n + h\beta_{-1}f(x_{n+1}, y_{n+1}) + h\beta_0 f(x_n, y_n).$$

In this case, the quantity δ defined by equation (12.45) is simply:

$$\delta = q(\beta_{-1} + \beta_0),$$

and the condition (12.46), $\delta > 0$, is satisfied by virtue of (12.5). It is easy to see that for $k = 0$,

$$z_1 = \frac{1 - h\beta_0 q}{1 + h\beta_{-1}q}.$$

If we take into account that $\varepsilon_0 = y(x_0) - y_0 = 0$ and, consequently, one can set $\varepsilon = 0$, then the estimate (12.52) can be rewritten thusly:

$$|\varepsilon_n| \leqq \frac{R+\Gamma}{hq(\beta_{-1} + \beta_0)}\left[1 - \left(\frac{1 - h\beta_0 q}{1 + h\beta_{-1}q}\right)^n\right]. \tag{12.54}$$

In particular, for $\beta_{-1} = 0$, $\beta_0 = 1$, (12.54) represents the estimate for Euler's method.

The right-hand sides of inequalities (12.52) and (12.53) do not exceed

$$\varepsilon + \frac{R+\Gamma}{h\delta}.$$

If the step h is chosen so that

$$\varepsilon + \frac{R+\Gamma}{h\delta} \leqq \rho,$$

then $(x_n, \tilde{y}_n) \in D_\rho \subset D$ and the estimates (12.52) and (12.53) can be used *a priori*.

We shall consider the method for the numerical solution of the Cauchy problem (12.1), defined by the formula

$$y_{n+1} = \sum_{j=0}^{k} \alpha_{j+1} y_{n-j} + h \sum_{j=-1}^{k} \beta_j f(x_{n-j}, y_{n-j}). \qquad (12.55)$$

The coefficients α_j and β_j are assumed to be real. Formula (12.55) contains in itself, as a particular case, formula (12.4) for the Adams method ($\alpha_1 = 1, \alpha_2 = \alpha_3 = \ldots = \alpha_{k+1} = 0$).

There are also other methods of numerical integration of the problem (12.1), defined by formulas of the form (12.55). For example,

$$y_{n+1} = y_{n-1} + \frac{h}{3} [f(x_{n+1}, y_{n+1}) + 4f(x_n, y_n) + f(x_{n-1}, y_{n-1})]. \qquad (12.56)$$

This formula can be obtained if in the equation

$$y(x_{n+1}) = y(x_{n-1}) + \int_{x_{n-1}}^{x_{n+1}} y'(x) dx$$

the integral is replaced by the quadrature sum of Simpson's rule.

We note another method defined by the formula

$$y_{n+1} = -4y_n + 5y_{n-1} + h[4f(x_n, y_n) + 2f(x_{n-1}, y_{n-1})]. \qquad (12.57)$$

This formula can be obtained in the following way. We represent the function $y(x)$ by its Hermite interpolation polynomial $P_3(x)$, constructed from the values $y(x)$ and $y'(x)$ at the knots x_{n-1} and x_n:

$$y(x) = P_3(x) + \frac{1}{4!} (x - x_{n-1})^2 (x - x_n)^2 y^{(IV)}(\xi). \qquad (12.58)$$

$P_3(x)$ is written according to formula (10.23) and the remainder term of integration by (10.24) from chapter II. If one puts $x = x_{n+1}$ in (12.58), one obtains

$$y(x_{n+1}) = -4y(x_n) + 5y(x_{n-1}) + h[4y'(x_n) + 2y'(x_{n-1})] +$$
$$+ \frac{h^4}{6} y^{(IV)}(\xi), \qquad x_{n-1} \leq \xi \leq x_{n+1}. \qquad (12.59)$$

Substituting here $y(x_j)$ for y_j, $y'(x_j)$ for $f(x_j, y_j)$ and throwing away the remainder term, one gets formula (12.57).

Let $y(x)$ be the solution of problem (12.1). The quantity R_n, defined by the equation

$$y(x_{n+1}) = \sum_{j=0}^{k} \alpha_{j+1} y(x_{n-j}) + h \sum_{j=-1}^{k} \beta_j f(x_{n-j}, y(x_{n-j})) + R_n, \quad (12.60)$$

will be called the error of method (12.55).

We suppose that the following representation is valid for R_n,

$$R_n = ah^{r+2} y^{(r+2)}(\xi), \qquad x_{n-k} \leq \xi \leq x_{n+1}, \tag{12.61}$$

where r is a non-negative integer and a is a constant. We assume that $f(x, y)$ has continuous partial derivatives up to order $r+1$.

It is easy to verify, by using the representation of the remainder term of Simpson's rule (2.25) from chap. III, that in the case of method (12.56), $r = 3$ and $a = -\frac{1}{90}$. From formula (12.59), it can be seen that for method (12.57), $r = 2$ and $a = \frac{1}{6}$.

From formulas (12.60) and (12.61) it follows that $y(x_n)$ satisfies the difference equation (12.55), if $y(x)$ is any polynomial of degree $r+1$. Since $r \geq 0$, then $y(x_n) = 1$ and $y(x_n) = nh$ satisfy the difference equation (12.55). This leads to the following constraints on the coefficients α_j and β_j:

$$\sum_{j=0}^{k} \alpha_{j+1} = 1, \qquad n+1 = \sum_{j=0}^{k} \alpha_{j+1}(n-j) + \sum_{j=-1}^{k} \beta_j.$$

We introduce the polynomial

$$\Lambda(z) = z^{k+1} - \alpha_1 z^k - \alpha_2 z^{k-1} - \ldots - \alpha_k z - \alpha_{k+1}, \tag{12.62}$$

the first equation means that one is a zero of the polynomial $\Lambda(z)$, and the second equation may be written in the form

$$\Lambda'(1) = \sum_{j=-1}^{k} \beta_j.$$

We will denote by $\tilde{y}_n$, $n = 0, 1, 2, \ldots, N$, the approximate solution of the difference equation (12.55). The initial values $\tilde{y}_n = y_n$, $n = 0, 1, 2, \ldots, k$ are assumed to be given. Further values $\tilde{y}_{k+1}, \tilde{y}_{k+2}, \ldots, \tilde{y}_N$ are calculated sequentially by means of (12.55). In view of the fact that the calculations are carried out with rounding off, $\tilde{y}_n$ does not satisfy the difference equation (12.55) exactly, and we have

$$\tilde{y}_{n+1} = \sum_{j=0}^{k} \alpha_{j+1} \tilde{y}_{n-j} + h \sum_{j=-1}^{k} \beta_j f(x_{n-j}, \tilde{y}_{n-j}) + \Gamma_n, \tag{12.63}$$

where Γ_n is the round-off error.

The error $\varepsilon_n = y(x_n) - \tilde{y}_n$ satisfies the difference equation

$$\varepsilon_{n+1} = \sum_{j=0}^{k} \alpha_{j+1} \varepsilon_{n-j} + \tag{12.64}$$

$$+ h \sum_{j=-1}^{k} \beta_j [f(x_{n-j}, y(x_{n-j})) - f(x_{n-j}, \tilde{y}_{n-j})] + R_n - \Gamma_n = 0.$$

Suppose that the following inequalities are satisfied,

$$|\varepsilon_n| \leq \alpha h^p, \qquad n = 0, 1, 2, \ldots, k,$$

$$|\Gamma_n| \leq \alpha h^{p+1}, \qquad n = k, k+1, \ldots, N-1, \tag{12.65}$$

where $\alpha > 0$ and $p > 0$. From (12.61) we get

$$|R_n| \leq \gamma h^{r+2}, \qquad \gamma > 0. \tag{12.66}$$

In the case of the Adams method (12.4), inequalities (12.65) and (12.66) provide the uniform convergence of the numerical solution $\tilde{y}_n$ to $y(x)$ on $[x_0, X]$ as $h \to 0$. For the general method (12.55), this assertion does not hold without additional assumptions.

It is not difficult to point out the conditions necessary for the limiting relation

$$\lim_{h \to 0} \max_{0 \leq n \leq N} |\varepsilon_n| = 0 \tag{12.67}$$

to hold by examining inequalities (12.65) and (12.66). The limit relation (12.67) must hold, in particular, for the Cauchy problem

$$y' = 0, \qquad y(x_0) = y_0.$$

By virtue of (12.61), $R_n = 0$. The calculation of $f(x, y) = 0$ is carried out without round-off, so that $\Gamma_n = 0$. Equation (12.64) is written thusly:

$$\varepsilon_{n+1} = \sum_{j=0}^{k} \alpha_{j+1} \varepsilon_{n-j}.$$

The solutions of this difference equation must be bounded for any initial values $\varepsilon_0, \varepsilon_1, \ldots, \varepsilon_n$, as $n \to \infty$. As is known from the theory of finite difference equations, it is necessary and sufficient for this that all solutions of the characteristic equation $\Lambda(z) = 0$, where $\Lambda(z)$ is defined by formula (12.62), be located in the unit circular disk $|z| \leq 1$ of the complex plane, moreover, the solutions on the unit circle $|z| = 1$ must be simple. If the solutions of the equation $\Lambda(z) = 0$ possess the indicated properties, then the method of numerical integration (12.55) is called *stable*.

Adam's method (12.4) is stable since the polynomial (12.62) $\Lambda(z) = z^{k+1} - z^k$ has the simple zero 1 and the k-multiple zero 0. The method (12.56) is also stable since, in this case, $\Lambda(z) = z^2 - 1$ has two simple zeros: 1 and -1. The method (12.57) is not stable, since the characteristic trinomial $\Lambda(z) = z^2 + 4z - 5$ has zeros 1 and 5.

It turns out that a sufficient condition for the stability of method (12.55) is for relationship (12.67) to hold given the satisfaction of inequalities (12.65) and (12.66). We shall not carry out the proof of this assertion. It can be found in [4] (see also [2], vol. II).

The problem of the estimation of error for the methods of Adams type for numerical solution of the Cauchy problem for a system of ordinary differential equations is considered in the paper [22]. The convergence and stability of numerical methods for solving the Cauchy problem for differential equations of second order of the particular form

$$y'' = f(x, y), \qquad y(x_0) = y_0, \qquad y'(x_0) = y'_0$$

is considered in work [19].

EXERCISES FOR CHAPTER IV

1.

1. Show that the Cauchy problem

$$y'' = f(x, y, y'),$$
$$y(x_0) = y_0, \qquad y'(x_0) = y_0',$$

can be written as a Cauchy problem for a system of two first order differential equations.

Hint: Set $z(x) = y'(x)$.

2. Show that the Cauchy problem $(1.1) - (1.2)$ can be written as a Cauchy problem for a system of n first order differential equations.

2.

1. Use Picard's theorem to show that the Cauchy problem

$$y' = 1 - y^2, y(0) = 0,$$

has a solution which lies in the rectangle $|x| \leq 1, |y| \leq 1$. Use the Runge-Kutta method to calculate $y(1.0)$ for $h = 1.0, 0.5, 0.25$, carrying five significant digits. Compare the results with the exact value

$$y(1.0) = \tanh (1.0) = 0.76159 \ldots.$$

2. Write the first six terms in the Taylor series expansion

$$y(h) = y(0) + y'(0)h + \tfrac{1}{2}y''(0)h^2 + \ldots$$

of the function $y(x)$ defined by the Cauchy problem

$$y' = 1 - y^2, y(0) = 0.$$

Estimate the error in retention of the terms containing powers of h up to h^4, inclusive, for $h = 1.0, 0.5, 0.25$. Find a value of h for which these terms apparently give five decimal place accuracy.

3. The Cauchy problem

$$y''+y = 0, y(0) = 0, y'(0) = 1$$

for a second order equation is equivalent to the Cauchy problem

$$y' = z, \quad y(0) = 0,$$
$$z' = -y, z(0) = 1,$$

for the indicated first order system. Use the Runge-Kutta method with $h = 0.1$ to calculate $y(1.0)$, $z(1.0)$. Compare the results with the exact values

$$y(1.0) = \sin(1.0) = 0.84147\ldots,$$
$$z(1.0) = \cos(1.0) = 0.54030\ldots.$$

Also, estimate the error by the Taylor series method, and compare with the actual error.

3.

1. Calculate the numbers $b_{4j}, j = 0, 1, 2, 3, 4$, defined by (3.21).

2. Calculate the numbers $b_{3j}, j = -1, 0, 1, 2, 3$, defined by (3.28).

3. Write formulas (3.26) and (3.32) explicitly for $k = 2$ for the differential equation

$$y' = 2x - (x^2+1)y + y^2.$$

Discuss the calculation of y_{n+1} by each formula.

4.

1. Calculate the numbers a_7, a_8 defined by (4.5).

2. Use the Euler formula (4.9) to obtain numerical values for $y(1.0)$, where $y(x)$ is defined by the Cauchy problem

$$y' = 1 - y^2, y(0) = 0,$$

for $h = 0.5, 0.25, 0.125$. Carry five decimal placed in the calculations. Compare the results obtained with the exact value

$$y(1.0) = \tanh(1.0) = 0.76159\ldots.$$

3. For the Cauchy problem stated in Problem 2, given $y(0.2) = 0.19738$, use (4.8) with $k = 1$ to find an approximate value of $y(1.0)$, carrying five decimal places in the computation. Given also that $y(0.1) = 0.09967$,

use (4.8) with $k = 2$ to find another approximation to $y(1.0)$, once again carrying five decimal places. Compare both results with the exact value given in Problem 2.

5.

1. Use (5.6) and Cauchy's theorem to determine the radius of convergence of the series solution for

$$y' = 1+y^2$$

with $x_0 = 0$, $y(0) = 0$. Find a value of h for which the start of a table of the numerical solution of this problem can be obtained to five decimal places for $x = 0, h, 2h, 3h, 4h$ by the series.

2. With $h = 0.05$, use Euler's method and Krylov's method to construct the start of the table for the numerical solution of the Cauchy problem

$$y' = 1+y^2, y(0) = 0,$$

for $x = 0.00(0.05)2.0$ to five decimal places.

3. Perform the construction indicated in Problem 2, using the Runge-Kutta method to obtain the initial approximations in place of Euler's method.

6.

1. Calculate the numbers a_7^*, a_8^* defined by formula (6.6).

2. Carry the calculations in table 35 forward to $x = 1.5$, using the initial approximations (6.11) obtained by taking $\Delta^4\eta_1^{(0)} = \Delta^4\eta_0$, and using the formulas

$$\Delta^3\eta_2^{(0)} = \Delta^3\eta_1+\Delta^4\eta_1^{(0)},$$
$$\Delta^2\eta_3^{(0)} = \Delta^2\eta_2+\Delta^3\eta_2^{(0)},$$
$$\Delta\eta_4^{(0)} = \Delta\eta_3+\Delta^2\eta_3^{(0)},$$
$$\eta_5^{(0)} = \eta_4+\Delta\eta_4^{(0)}.$$

3. Carry the calculations in table 35 forward to $x = 1.5$, using the Adams

extrapolation formula (4.8) with $k = 4$ to obtain the initial approximation $y_5^{(0)}$, and thus the initial values

$$\eta_5^{(0)}, \Delta\eta_4^{(0)}, \Delta^2\eta_3^{(0)}, \Delta^3\eta_2^{(0)}, \Delta^4\eta_1^{(0)}.$$

7.

1. Calculate the numbers $\tilde{a}_8$, $\tilde{a}_{10}$ defined by formula (7.5).

2. Carry the calculations in table 38 forward to $x = 2.0$. Compare the result with the exact answer $y(2.0) = \sqrt{3e^2 - 6} = 4.02084 \ldots$.

3. Write the Cowell formula (7.7) for $k = 3$. Using the values in table 36 to start the calculation, carry the solution forward to $x = 0.9$ by this formula. Compare the results obtained with the ones given in table 37.

8.

1. Write the formulas corresponding to the Adams interpolation formula (6.9) for the system (8.1). Indicate on the corresponding schematic difference tables the location of the differences needed in the computation.

2. Write the formulas corresponding to the Cowell formula (7.8) for the system (8.1). Indicate on the corresponding schematic difference tables the locations of the differences needed in the computation.

3. Apply the Adams extrapolation formula given in this section to the system

$$y' = z,$$
$$z' = -y,$$
$$y(0) = 0, z(0) = 1,$$

given the starting values:

x	y	z
0.00	0.00000	1.00000
0.01	0.01000	0.99995
0.02	0.02000	0.99980
0.03	0.03000	0.99955
0.04	0.03999	0.99920

Carry the computations forward to $x = 0.1$, and compare the results with the exact values

$$y(0.1) = \sin 0.1 = 0.09983 \ldots, z(0.1) = \cos 0.1 = 0.99500 \ldots.$$

9.

1. Calculate the numbers α_7, α_8 defined by formula (9.10).

2. Carry the calculations in table 39 forward to $x = 2.5$, using the Störmer extrapolation formula (9.17) with $k = 6$.

3. Using the values for y and η from table 39 corresponding to $x = -0.6$, -0.4, -0.2, 0.0 as starting values, carry the computation forward to $x = 1.0$ by the Störmer extrapolation formula (9.17) with $k = 3$. Compare the results obtained with those in table 39.

10.

1. Calculate the numbers α_7^*, α_8^* defined by formula (10.4).

2. Carry the calculations in table 40 forward to $x = 2.0$ by Störmer's interpolation formula (10.10).

3. Using the values given in table 40 for $x = -0.6$, -0.4, -0.2, 0.0, 0.2 as starting values, carry the computation forward to $x = 1.0$ by the Störmer interpolation formula (10.9). Compare the results obtained with those given in table 40.

11.

1. Calculate the numbers $\tilde{\alpha}_{10}$, $\tilde{\alpha}_{12}$ defined by formula (11.5).

2. Carry the calculations in table 41 forward to $x = 2.0$ by the Cowell formula (11.10).

3. Using the value in table 41 for $x = -0.6$, -0.4, -0.2, 0.0, 0.2 as starting values, carry the calculations forward to $x = 1.2$ by the Cowell formula (11.9). Compare the results obtained with those given in table 41.

12.

1. Show the uniform convergence on $[x_0, X]$ of the numerical solution $\tilde{y}_n$ obtained by the Adams method (12.4) to the solution $y(x)$ of the Cauchy problem (12.1), using inequalities (12.65) and (12.66).

2. Estimate the error in the calculation of $y(1.0)$ in table 34.

3. Estimate the error in the calculation of $y(1.0)$ in table 35.

BIBLIOGRAPHY

1. Aitken, A. On the factorization of polynomials by iterative methods. Uspehi Matem. Nauk **8** (1953), 71–86.
2. Berezin, I. S. and Židkov, N. P. Computing methods. Russian edition: Moscow, 1959. English translation by O. M. Blunn, Pergamon Press, 1965.
3. Collatz, L. Numerical treatment of differential equations. Russian edition: Moscow, 1953. German (2nd) edition: Springer-Verlag, 1955. English (3rd) edition: Springer-Verlag, 1960.
4. Dahlquist, G. Convergence and stability in the numerical integration of ordinary differential equations. Mathematica Scandinavica **4** (1956), 33–53.
5. Erugin, N. P. and Sobolev, S. L. Approximate integration of some oscillatory functions. Prikladnaja matematika i mehanika **14** (1950), 193–196. (Russian).
6. Faddeeva, V. N. Computational methods of linear algebra. Russian edition: Moscow, 1950. English translation by C. D. Benster, Dover Publications, 1959.
7. Gel'fond, A. O. The calculus of finite differences. Moscow, 1952. (Russian).
8. Hildebrand, F. B. Introduction to numerical analysis. McGraw-Hill Book Company, 1956.
9. Householder, A. S. Principles of numerical analysis. McGraw-Hill Book Company, 1953. Russian edition: Moscow, 1956.
10. Kantorovič, L. V. On Newton's method. Trudy Mat. Inst. Im. V. A. Steklova **28** (1949), 104–144. (Russian).
11. Kantorovič, L. V. The principle of majorants and Newton's method. Dokl. Akad. Nauk SSSR **76** (1951), 17–20. (Russian).
12. Kantorovič, L. V. Some further applications of Newton's method to functional equations. Vestnik Leningrad. Gos. Univ. No. 7. ser. matem., meh. i astr., V. **2** (1957), 68–103. (Russian).
13. Kantorovič, L. V. and Akilov, G. P. Functional analysis in normed spaces. Russian edition: Moscow, 1959. English translation: Pergamon Press, 1964.
14. Kantorovič, L. V. and Krylov, V. I. Approximate methods of higher analysis. Russian edition: Moscow, 1949. English translation: Interscience Publishers, 1958.

15. Karmazina, L. N. and Kuročkina, L. V. Tables of interpolation coefficients. Press of the Academy of Sciences of the USSR, Moscow, 1956.
16. Krylov, A. N. Lectures on approximate computation. Moscow, 1950. (Russian).
17. Krylov, V. I. Numerical calculation of integrals of functions which are products of rapidly oscillating factors. Dokl. Akad. Nauk SSSR **108** (1956), 1014–1017. (Russian)
18. Krylov, V. I. Approximate calculation of integrals. Russian edition: Moscow, 1959. English translation by A. H. Stroud: Macmillan, 1962.
19. Krylov, V. I. Convergence and stability of numerical solutions of second-order differential equations. Dokl. Akad. Nauk BSSR **4** (1960), 187–189. (Russian).
20. Lin, Shih-nge. A method of successive approximations for evaluating the real and complex roots of cubic and higher-order equations. Journal of Mathematics and Physics **20** (1941), 231–242.
21. Lin. Shih-nge. A method for finding roots of algebraic equations. Journal of Mathematics and Physics **22** (1943), 60–77.
22. Lozinskiĭ, S. M. Error estimates for the numerical integration of differential equations I. Izv. Vuzov, matematika **5** (6) (1958), 52–90. (Russian).
23. Markov, A. A. Differenzenrechnung. German edition: Leipzig, 1898. Russian edition: Odessa, 1910.
24. Segal. B. I. and Semendjaev, K. A. Five-place mathematical tables. Press of the Academy of Sciences of the USSR, Moscow, 1950. Translator's note: Any standard set of five-place tables, such as those published by the Chemical Rubber Publishing Company, Cleveland, Ohio, may be used in place of these.
25. Milne, W. E. Numerical calculus. Princeton University Press, 1949. Russian edition: Moscow, 1951.
26. Milne, W. E. Numerical solution of differential equations. John Wiley and Sons, 1953. Russian edition: Moscow, 1955.
27. Natanson, I. P. The constructive theory of functions. Moscow, 1949. (Russian).
28. Nikolaeva, M. V. On the approximate calculation of oscillatory integrals. Trudy Mat. Inst. im. V. A. Steklova **28** (1949), 26–32. (Russian).

29. Ostrowski, A. M. Über die Konvergenz und die Abrundungsfestigkeit des Newtonschen Verfahrens. Matem. Sbornik **2** (1937), 1073–1095.
30. Salzer, H. E. Tables for facilitating the use of Chebyshev's quadrature formula. Journal of Mathematics and Physics **26** (1947), 191–194.
31. Salzer, H. E., Zucker, R., and Capuano, R. Table of the zeros and weight factors of the first twenty Hermite polynomials. Journal of Research of the National Bureau of Standards **48** (1952), 111–116.
32. Subbotin, M. F. A course in celestial mechanics, vol. 2. Moscow, 1937. (Russian).
33. Zaguskin, V. L. Handbook of numerical methods for the solution of equations. Russian edition: Moscow, 1960. English translation by G. O. Harding, Pergamon Press, 1961.